K. Linga Murty and
Indrajit Charit

An Introduction to Nuclear Materials

Related Titles

Prussin, S. G.

Nuclear Physics for Applications

A Model Approach

2007

ISBN: 978-3-527-40700-2

Bokhan, P. A., Buchanov, V. V., Fateev, N. V., Kalugin, M. M., Kazaryan, M. A., Prokhorov, A. M., Zakrevskii, D. E.

Laser Isotope Separation in Atomic Vapor

2006

ISBN: 978-3-527-40621-0

Kutz, M. (ed.)

Mechanical Engineers' Handbook

Four Volume Set

2005

ISBN: 978-0-471-44990-4

K. Linga Murty and Indrajit Charit

An Introduction to Nuclear Materials

Fundamentals and Applications

WILEY-VCH Verlag GmbH & Co. KGaA

The Authors

Dr.K.L. Murty
Mat.Science & Eng.
North Carolina State Univ.
1110 Burlington Eng. Labs
Raleigh NC27695-7909
USA

Indrajit Charit
University of Idaho
Dept. of Materials Science
McClure Hall, Room 405D
Moscow, ID 83844-3024
USA

Cover Image:
Nuclear reactor core of the Advanced Test Reactor (ATR) at the Idaho National Engineering and Environmental Lab (INEEL) at Idaho Falls, USA. This large reactor has been in use since 1967. It operates with a thermal power of 250,000 kilowatts. The reactor is water-cooled and the blue glow results from Cerenkov radiation, emitted when energetic charged particles travel faster through the water than light. This reactor is used for research into the effects of radiation on materials, and for the production of rare and valuable medical and industrial isotopes.
© US DEPARTMENT OF ENERGY/SPL/ Agentur Focus

All books published by **Wiley-VCH** are carefully produced. Nevertheless, authors, editors, and publisher do not warrant the information contained in these books, including this book, to be free of errors. Readers are advised to keep in mind that statements, data, illustrations, procedural details or other items may inadvertently be inaccurate.

Library of Congress Card No.: applied for

British Library Cataloguing-in-Publication Data
A catalogue record for this book is available from the British Library.

Bibliographic information published by the Deutsche Nationalbibliothek
The Deutsche Nationalbibliothek lists this publication in the Deutsche Nationalbibliografie; detailed bibliographic data are available on the Internet at http://dnb.d-nb.de.

© 2013 Wiley-VCH Verlag & Co. KGaA, Boschstr. 12, 69469 Weinheim, Germany

All rights reserved (including those of translation into other languages). No part of this book may be reproduced in any form – by photoprinting, microfilm, or any other means – nor transmitted or translated into a machine language without written permission from the publishers. Registered names, trademarks, etc. used in this book, even when not specifically marked as such, are not to be considered unprotected by law.

Hardcover ISBN: 978-3-527-41201-3
Softcover ISBN: 978-3-527-40767-5

Cover Design Adam-Design, Weinheim

Typesetting Thomson Digital, Noida, India

Printing and Binding Markono Print Media Pte Ltd, Singapore

Printed in Singapore

Printed on acid-free paper

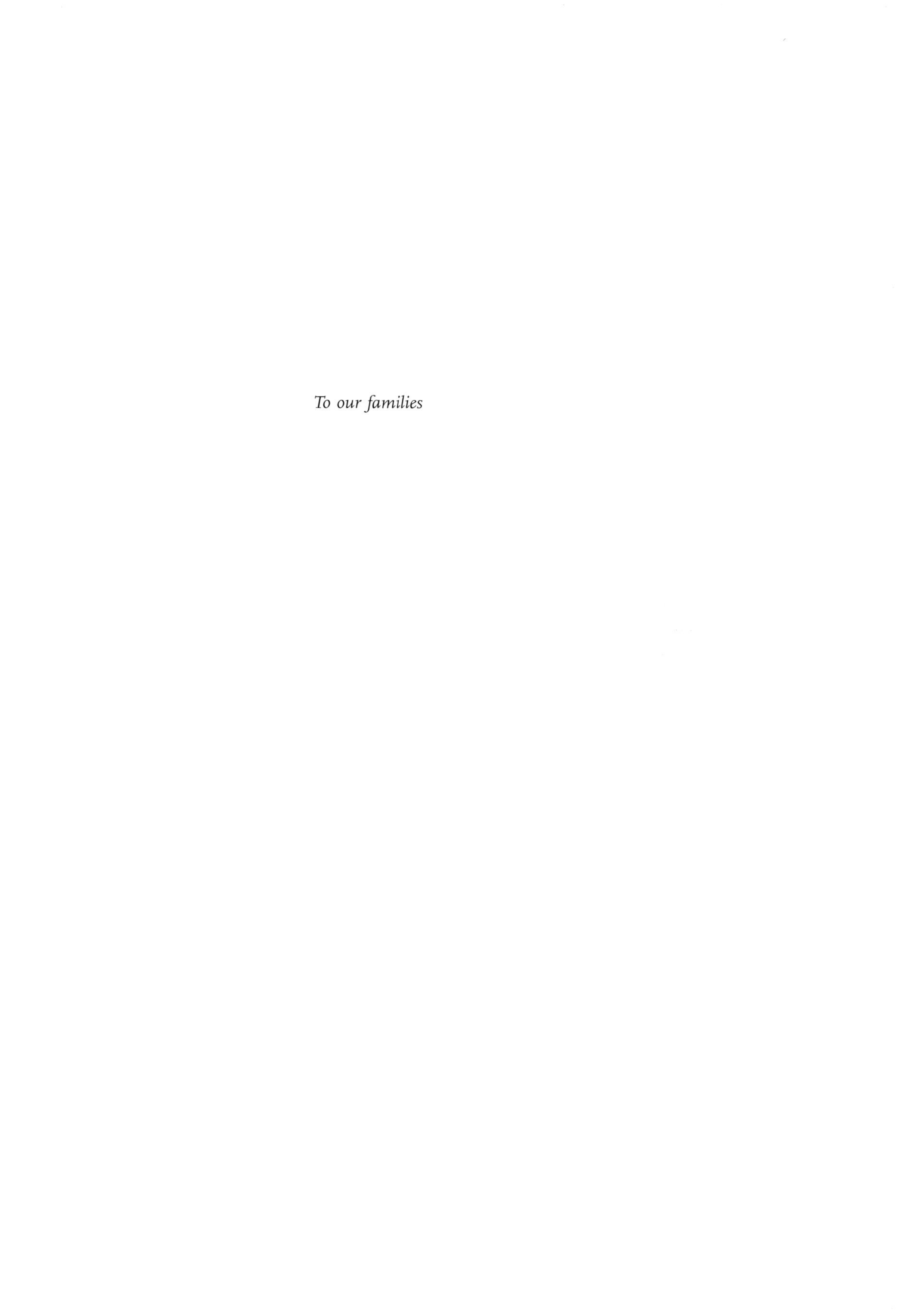

To our families

Contents

Preface

This textbook was first conceived from our observation that there had been no suitable textbook that caters towards the needs of undergraduate students in nuclear engineering for learning about nuclear reactor materials from materials science perspective. We have come across some books which are heavy on the 'nitty gritty' details that assume significant prior knowledge of basics of materials science, whereas others miss to highlight important materials science principles for students' perspective to grow into the field of nuclear materials. Moreover, most of them are out-of-print. Thus, the book has primarily been written for undergraduate students; however it can also be used by beginning graduate students and professionals who are interested in learning about nuclear materials from a rudimentary level but have very little background in materials science. We also hope that materials science and engineering (MSE) students will take delight into learning the topics if they wish to enter the nuclear materials field even though they may be quite conversed with the contents of some of the chapters involving materials science fundamentals.

Instructors who teach nuclear materials always have a difficult task of teaching the fundamentals of nuclear materials. Origin of the contents of the book has roots in the course on Nuclear Materials taught during more than three decades by the senior author (KLM) at NC State and recent teaching by the co-author (IC) at the University of Idaho. Copies taken of various chapters from different books have been used as course notes with various references that the students found not very convenient especially since only a few copies are available in the library. In addition, they were found to be disjointed especially in the varied nomenclatures used by different authors.

It has been our experience that many of the nuclear engineering students find that it is required to repeat basic materials science aspects including crystal structures especially because hcp and fluorite structures are not well covered in the first course on Elements of Materials Science. Along with all basic phenomena of materials science, the book deals with dislocation theory in more detail because of the significance of dislocations in the understanding of radiation damage and radiation effects. During the class teaching, we emphasize from the very beginning the significance of these fundamental materials science principles in our ability to be able to predict/estimate effects of radiation using detailed quantitative microstructural features.

Following an introductory chapter on Nuclear Reactor Systems and Fundamentals, Chapter 2 starts with Crystal Structure followed by Crystal Imperfections (2.2) and Diffusion in Solids (2.3). Radiation Damage fundamentals are covered in Chapter 3 with the major aim of dpa (displacements per atom) calculation. Dislocation Theory comprises the Chapter 4 while Mechanical Properties, Fracture, Fatigue, Creep; some fundamentals of thermophysical properties; and Corrosion and SCC are all included in Chapter 5. Chapter 6 covers both Radiation Effects and Reactor Materials while the book is concluded with the Chapter 7 on Reactor Fuels. For first year graduate course, these sections were followed up with a term-long project on topics related to materials issues in reactor systems with in-class presentations by the students. Generally, we did not restrict on the students' selection of topics; a list given in the semester-beginning as well as the students' choice but on materials issues related to reactor applications.

We are not saying that this book is going to be a 'panacea' for curing all the problems faced by these instructors. Also, we would like to accept that many issues of importance to nuclear field remained untouched because we did not wish the book to be so voluminous making it difficult to cover the subject matter in a single semester. Moreover, many of the aspects such as radiation creep, embrittlement of pressure vessel steels, etc., were dealt in a rather simplistic fashion albeit they have extensive pedagogic advantages. Our experience indicated that even the contents of chapters included here are quite large for a one-semester course and often the last chapters on reactor materials and fuels are covered rather briefly. For making the course manageable for one semester, we could not include/cover phase diagrams and phase transformations in this course and the text.

We would like to gratefully acknowledge several past and present colleagues and students whose work has been incorporated in this book in one way or the other. Special thanks go to Mr. Brian Marple for giving inputs on the Gen-III+ reactors for Chapter 1. We would also like to acknowledge the support of Drs. Louis Mansur, Donald Olander, and Sheikh T. Mahmood for supporting our book proposal at the early stage. The authors are indebted to many authors and publishers who gave consent to the reproduction of appropriate figures etc. We would like to also acknowledge incredible efforts by Anja Tschörtner of Wiley-VCH for driving us to complete the book and we commend her sustained efforts in this regard. Acknowledgements are due to our current students and colleagues who went through the proofs for corrections and many comments. Finally, the book would not have been written without the much needed emotional support of our families, in particular to our spouses, Ratnaveni Murty and Mohar B. Charit.

K. Linga Murty
Indrajit Charit
May 5, 2012

1
Overview of Nuclear Reactor Systems and Fundamentals

"Someday man will harness the rise and fall of the tides, imprison the power of the sun, and *release atomic power.*"

—*Thomas Alva Edison*

1.1
Introduction

There is no doubt that *energy* has been driving and will drive the technological progress of the human civilization. It is a very vital component for the economic development and growth, and thus our modern way of life. Energy has also been tied to the national security concerns. It has been projected that the world energy demand will almost double by the year 2040 (based on 2010 energy usage), which must be met by utilizing the energy sources other than the fossil fuels such as coal and oil. Fossil fuel power generation contributes to significant greenhouse gas emissions into the atmosphere and influences the climate change trend. Although several research and development programs (e.g., carbon sequestration and ultrasupercritical steam turbine programs) have been initiated to make the fossil power generation much cleaner, they alone will not be enough to fend off the bigger problem. Therefore, many countries worldwide have recognized the importance of clean (i.e., emission-free) nuclear energy, and there are proven technologies that are more than ready for deployment. The use of nuclear energy for the power generation varies widely in different parts of the world. The United States produces about 19% (2005 estimate) of its total energy from nuclear sources, whereas France produces ~79% and Brazil and India rely on the nuclear energy for only about 2.5% and 2.8% of their energy needs, respectively. Japan, South Korea, Switzerland, and Ukraine produce 30%, 35%, 48%, and 40%, respectively, of their energy requirements from the nuclear sources. It is important to note that the fast growing economies like China, India, and Brazil produce relatively less electricity from the nuclear sources. Hence, there are tremendous opportunities for nuclear energy growth in these emerging economies as well as many other countries. Nuclear reactors have been built for the primary purpose of electricity production, although they are used for desalination and radioisotope production.

An Introduction to Nuclear Materials: Fundamentals and Applications, First Edition.
K. Linga Murty and Indrajit Charit.
© 2013 Wiley-VCH Verlag GmbH & Co. KGaA. Published 2013 by Wiley-VCH Verlag GmbH & Co. KGaA.

There are now about 440 nuclear power reactors worldwide generating almost 16% of the world electricity needs; among them, 104 nuclear reactors are in the United States. Since the first radioactive chain reaction that was successfully initiated at the University of Chicago research reactor in the 1940s, the field has seen an impressive growth until Three Mile Island and Chernobyl accidents happened. Following these incidents, the public confidence in the nuclear power dwindled, and the nuclear power industry saw a long stagnation. However, the US government's decision to increase energy security and diversity by encouraging nuclear energy generation (as laid out in the US government's Advanced Energy Initiative in 2005) has rekindled much hope for the revival of the nuclear power industry in the United States, and as a matter of fact, the *Nuclear Renaissance* has already begun – the US Nuclear Regulatory Commission (NRC) approved an early site permit application for the Clinton Power Station in Illinois (Exelon Power Corporation) in March 2007. As the scope of the nuclear energy is expanded, the role of materials is at the front and center. Recent (2011) accidents in Japan due to earthquakes and tsunami are now pointing toward further safeguards and development of more resistant materials. Thus, this book is devoted to addressing various important fundamental and application aspects of materials that are used in nuclear reactors.

1.2 Types of Nuclear Energy

Nuclear energy can be derived from many forms such as nuclear fission energy, fusion energy, and radioisotopic energy.

1.2.1 Nuclear Fission Energy

The essence of nuclear fission energy is that the heat produced by the splitting of heavy radioactive atoms (nuclear fission) during the chain reaction is used to generate steam (or other process fluid) that helps rotate the steam turbine generator, thus producing electricity. Nuclear fission energy is the most common mode of producing the bulk of the nuclear energy.

1.2.2 Nuclear Fusion Energy

A huge amount of energy (much higher than fission) can be produced using the nuclear fusion reaction (deuterium–tritium reaction). There is currently no commercial fusion reactors and is not envisioned to be set up for many years. A prototype fusion reactor known as ITER (International Thermonuclear Experimental Reactor) is being built in France and scheduled to produce the first plasma by 2018.

1.2.3 Radioisotopic Energy

Either radioactive isotopes (e.g., ^{238}Pu, ^{210}Po) or radioactive fission products (e.g., ^{85}Kr, ^{90}Sr) can produce decay heat that can be utilized to produce electric power. These types of power sources are mainly used in remote space applications.

1.3 Neutron Classification

Chadwick discovered neutron in 1932. Generally, neutrons are generated during radioactive chain reactions in a power reactor. Neutron is subatomic particle present in almost all nuclides (except normal hydrogen isotope or protium) with a mass of 1.67×10^{-27} kg and has no electrical charge.

Neutrons are classified based on their kinetic energies. Although there is no clear boundary between the categories, the following limits can be used as a useful guideline:

Cold neutrons (<0.003 eV), slow (thermal) neutrons (0.003–0.4 eV), slow (epithermal) neutrons (0.4–100 eV), intermediate neutrons (100 eV–200 keV), fast neutrons (200 keV–10 MeV), high-energy (relativistic) neutrons (>10 MeV). *Note*: $1\,\text{eV} = 1.6 \times 10^{-19}\,\text{J}$.

Generally, thermal neutrons are associated with a kinetic energy of 0.025 eV that translates into a neutron speed of $2200\,\text{m s}^{-1}$!

1.4 Neutron Sources

Various radiation types are produced in a nuclear environment. It could be alpha particles, beta particles, gamma rays, or neutrons. In this book, we are primarily concerned with the radiation damage and effects caused by neutrons. There could be several sources of neutrons, including alpha particle-induced fission, spontaneous fission, neutron-induced fission, accelerator-based sources, spallation neutron source, photoneutron source, and nuclear fusion.

1.5 Interactions of Neutrons with Matter

Collision of neutrons with atom nuclei may lead to different scenarios – scattering of the neutrons and recoil of nuclei with conservation of momentum (elastic scattering) or loss of kinetic energy of the neutron resulting in gamma radiation (inelastic scattering). The capture of neutrons may result in the formation of new nuclei (transmutation), or may lead to the fragmentation of the nucleus (fission) or the emission of other nuclear particles from the nucleus. We shall discuss some of the effects in more detail in Chapter 3.

a) **Elastic Scattering**
Elastic scattering refers to a neutron–nucleus event in which the kinetic energy and momentum are conserved.
b) **Inelastic Scattering**
This interaction refers to neutron–nuclide interaction event when the kinetic energy is not conserved, while momentum is conserved.
c) **Transmutation**
When a nuclide captures neutrons, one result could be the start of a sequence of events that could lead to the formation of new nuclide. The true examples of this type of reaction are (n, α), (n, p), (n, β^+), (n, β^-), and (n, f). Reactions like (n, γ) and $(n, 2n)$ do not result in new elements, but only produce isotopes of the original nuclide.
d) **Fission**
Fission is a case of (n, f) reaction, a special case of transmutation reaction. Uranium is the most important nuclear fuel. The natural uranium contains about 0.7% U^{235}, 99.3% U^{238}, and a trace amount of U^{234}. Here, we discuss the neutron-induced nuclear fission, which is perhaps the most significant nuclear reaction. When a slow (thermal) neutron gets absorbed by a U^{235} atom, it leads to the formation of an unstable radionuclide U^{236}, which acts like an unstable oscillating droplet, immediately followed by the creation of two smaller atoms known as fission fragments (not necessarily of equal mass). About 2.5 neutrons on average are also released per fission reaction of U^{235}. An average energy of 193.5 MeV is liberated. A bulk of the energy (~160 MeV or ~83%) is carried out by the fission fragments, while the rest by the emitted neutrons, gamma rays, and eventual radioactive decay of fission products. Fission fragments rarely move more than 0.0127 mm from the fission point and most of the kinetic energy is transformed to heat in the process. As all of these newly formed particles (mostly fission fragments) collide with the atoms in the surroundings, the kinetic energy is converted to heat. The fission reaction of U^{235} can occur in 30 different ways leading to the possibility of 60 different kinds of fission fragments. A generally accepted equation for a fission reaction is given below:

$$U_{92}^{235} + n_0^1 \rightarrow Kr_{36}^{92} + Ba_{56}^{142} + 2n_0^1 + \text{Energy}, \tag{1.1}$$

which represents the fission of one U^{235} atom by a thermal neutron resulting into the fission products (Kr and Ba) with an average release of two neutrons and an average amount of energy (see above). It is clear from the atomic masses of the reactant and products, that a small amount of mass is converted into an equivalent energy following Einstein's famous equation $E = mc^2$.

U^{235} is the one and only naturally occurring radioisotope (fissile atom) in which fission can be induced by thermal neutrons. There are two other fissile atoms (Pu^{239} and U^{233}) that are not naturally occurring. They are created during the neutron absorption reactions of U^{238} and Th^{232}, respectively. Each event consists of (n, γ) reactions followed by beta decays. Examples are shown below:

$$U_{92}^{238} + n_0^1 \rightarrow U_{92}^{239} + \gamma \tag{1.2a}$$

$$U_{92}^{239} \rightarrow Np_{93}^{239} + \beta^-, \quad t_{1/2} = 23.5\ \text{min} \tag{1.2b}$$

$$Np_{93}^{239} \rightarrow Pu_{94}^{239} + \beta^-, \quad t_{1/2} = 23.5\ \text{days} \tag{1.2c}$$

$$Th_{90}^{232} + n_0^1 \rightarrow Th_{90}^{233} + \gamma \tag{1.3a}$$

$$Th_{90}^{233} \rightarrow Pa_{91}^{233} + \beta^-, \quad t_{1/2} = 22.4\ \text{min} \tag{1.3b}$$

$$Pa_{91}^{233} \rightarrow U_{92}^{233} + \beta^-, \quad t_{1/2} = 27.0\ \text{days} \tag{1.3c}$$

The concept of the "breeder" reactors is based on the preceding nuclear reactions, and U^{238} and Th^{232} are known as "fertile" atoms. Heavy radioisotopes such as Th^{232}, U^{238}, and Np^{237} can also undergo neutron-induced fission, however, only by fast neutrons with energy in excess of 1 MeV. That is why these radionuclides are sometimes referred to as "fissionable."

1.5.1 Fission Chain Reaction

As the preceding section on fission emphasized, each fission reaction of an U^{235} atom leads to the release of an average of 2.5 neutrons. Hence, to sustain a continuous fission reaction (i.e., a *chain reaction*), these neutrons should be able to initiate the fission of at least another fissile atom. Note the schematic of a chain reaction involving U^{235} atoms in Figure 1.1. A majority of the neutrons (~99.25%) produced due to the fission reaction of U^{235}, known as *prompt neutrons*, are released instantaneously (within 10^{-14} s). But there are about 0.75% neutrons that are released over

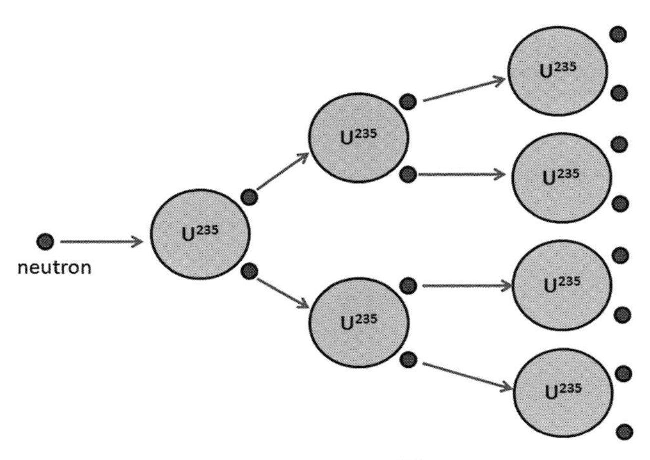

Figure 1.1 A schematic chain reaction of U^{235} fissile atoms in progress (for the sake of simplicity, it is assumed that two neutrons are released due to the fission of one U^{235} atom and fission fragments created in each fission are also not shown).

a longer period (over ~20 s) and these neutrons are called *delayed neutrons*. These delayed neutrons play a very important role in controlling the fission chain reaction.

There is always a competition for neutrons between various processes, namely, (i) fission reaction of fissile atom nuclei, (ii) nonfission capture of neutrons by uranium and other reactions, (iii) nonfission capture of neutrons by other components in the reactor core, and (iv) leakage of neutrons from the core. The reaction can be termed as a chain reaction when the number of neutrons consumed in the processes (ii)–(iv) is at least equal to or less than that consumed in the process (i). Thus, *neutron economy* plays a very important role in the design of a nuclear reactor. The need for a favorable neutron economy necessitates certain conditions to be met by a chain reacting system. For a given geometry, there is a certain minimum size of a chain reacting system, called the *critical size* (in terms of volume), for which the production of neutrons by fission just balances the loss of neutrons by leakage and so on, and the chain reaction can be sustained independently. The mass corresponding to the critical size is called *critical mass*. Dependent on the relative generation of fission neutrons and their loss, the reactor is said to be in different stages: subcritical (neutron loss more than the production), critical (balance between the neutron production and loss, $\kappa = 1$), or supercritical (the neutron production is more than the loss, $\kappa > 1$). The multiplication factor κ is often used to express the criticality condition of a reactor. This factor is basically the net number of neutrons per initial neutron.

1.6 Definition of Neutron Flux and Fluence

The concept of neutron flux is very similar to heat flux or electromagnetic flux. Neutron flux (ϕ) is simply defined as the density of neutrons n (i.e., number of neutrons per unit volume) multiplied by the velocity of neutrons v. Hence, nv represents the *neutron flux*, which is the number of neutrons passing through a unit cross-sectional area per second perpendicular to the neutron beam direction. However, sometimes this is called *current* if we consider that neutrons moving in one direction only. Neutron flux, in general terms, should take into account all the neutrons moving in all directions and be defined as the number of neutrons crossing a sphere of unit projected area per second. Total neutron flux (ϕ) is expressed by the following integral:

$$\phi = \int_0^\infty \phi(E_i)\mathrm{d}E_i, \tag{1.4}$$

where $\phi(E_i)\mathrm{d}E_i$ is the flux of neutrons with energies between E_i and $E_i + \mathrm{d}E_i$.

The term "*nvt*" represents the neutron fluence, that is, neutrons per unit cross-sectional area over a specified period of time (here t). Thus, the units of neutron flux and fluence are n cm^{-2} s^{-1} and n cm^{-2}, respectively.

1.7 Neutron Cross Section

We have already discussed different ways in which a neutron can interact with nuclides (or specifically nuclei). This is indeed a probabilistic event that depends on the energy of the incident neutrons and the type of nuclei involved in the interaction. Therefore, one can define this probability of interactions in terms of *cross section* that is a measure of the degree to which a particular material will interact with neutrons of a particular energy. But remember that the neutron cross section for a particular element has nothing to do with the actual physical size of the atoms. The range of neutrons (the distance traveled by the neutron before being stopped) is a function of the neutron energy (recall the classification based on neutron energy) as well as the capture cross section of the medium/material through which the neutrons traverse.

To understand it easily, one may consider a simple case shown in Figure 1.2 with a beam of neutrons impinging on a material of unit area (in cm^2) and thickness x (in cm). Thus, the intensity of neutrons traveling beyond the material will be diminished depending on the number of nuclei per unit volume of the material (n') and the "effective area of obstruction" (in cm^2) presented by a single nucleus. This area of obstruction is generally called "microscopic cross section" (σ) of the material. Like any other absorption equation, one can write

$$I = I_0 \exp\left(-n'\sigma x\right), \tag{1.5}$$

where the quantity $n'\sigma$ is called the macroscopic cross section or obstruction coefficient $\sum$ (unit in cm^{-1}). This represents the overall effect of nuclei (n') in the neuron beam path and the power of the nuclei to take part in the interaction. Equation (1.5) is specifically for only one type of reaction when the absorber material contains only one type of pure nuclide. But actually the material could consist of several types of nuclei, and in that case, we should add all the neutron cross sections for all possible reactions to obtain the total neutron cross section.

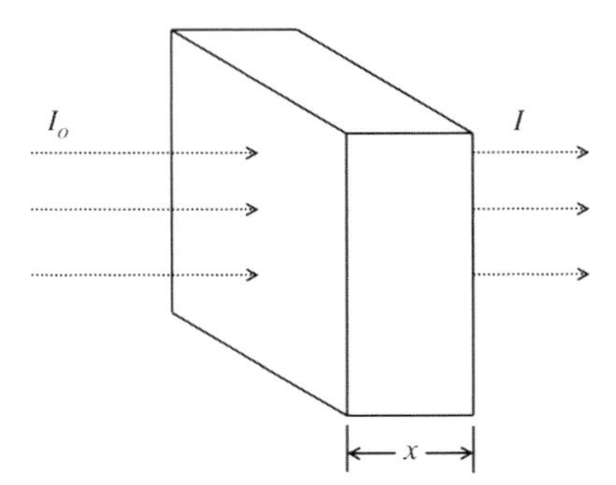

Figure 1.2 Attenuation of an incident neutron beam of intensity I_0 by an absorber material.

From Eq. (1.5), a half-value thickness ($x_{1/2}$) for neutron beam attenuation can be derived using the following relation:

$$x_{1/2} = 0.693/(n' \cdot \sigma). \tag{1.6}$$

The values of neutron microscopic cross section (σ) are typically between 10^{-22} and $10^{-26}\,\text{cm}^2$, leading to the development of a convenient unit called *barn* (1 barn $= 10^{-24}\,\text{cm}^2$).

Note: Macroscopic neutron cross section ($\sum$) can be calculated with the knowledge of n' that can be calculated from the following relation:

$$n' = (\varrho/M) \times (6.023 \times 10^{23}), \tag{1.7}$$

where ϱ is the density (g cm^{-3}) and M is the atomic weight of the element.

$$\text{Thus, } \sum = (\varrho/M) \times (6.023 \times 10^{23})\sigma. \tag{1.8}$$

In order of increasing cross section for absorption of thermal neutrons, various metals can be classified as follows (normalized to Be):

Be 1, Mg 7, Zr 20, Al 24, Nb 122, Mo 278, Fe 281, Cr 322, Cu 410, and Ni 512

We will see in later chapters that neutron capture cross section has a significant role to play in the selection of reactor materials coupled with other considerations. See Table 1.1 for some representative values of neutron capture cross sections for several important nuclides.

Most nuclides exhibit both the $1/v$ (v = velocity of neutron) dependence of neutron cross section and the resonance effects over the entire possible neutron energy spectrum. We should not forget that the neutron cross sections heavily depend on the type of reactions they take part in, such as alpha particle producing reaction (σ_α), fission reactions (σ_f), neutron capture cross sections (σ_c), and so on. As discussed above, the total cross section (σ_t) is a linear summation of all neutronic reactions possible at the specific neutron energy level.

Table 1.1 Neutron cross sections (in barn) for capture of thermal neutrons (i.e., of average kinetic energy 0.025 eV) of a few nuclides.

Nuclide	Neutron cross section (b)
$^{1}_{1}H$	0.332
$^{2}_{1}H$	0.00052
$^{12}_{6}C$	0.0035
$^{238}_{92}U$	2.7
$^{235}_{92}U$	586

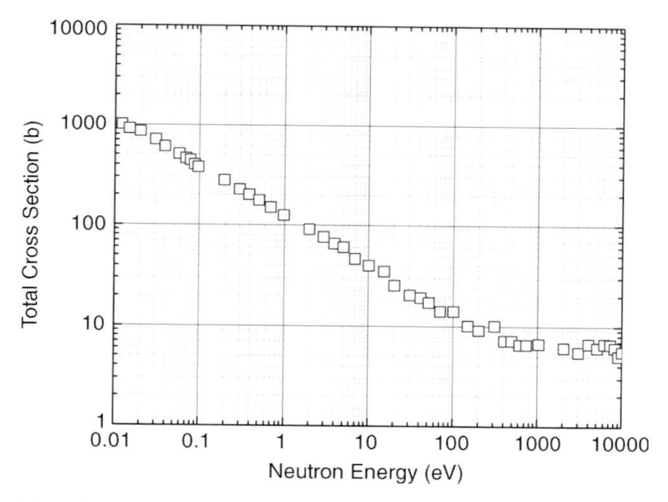

Figure 1.3 Variation of total cross section of elemental boron as a function of neutron energy.

The inverse proportionality of total neutron cross section in elemental boron as a function of neutron energy is shown in Figure 1.3. However, this is not always the case. Many nuclides show abrupt increases in neutron cross section at certain narrow energy ranges due to resonance effects, which happens when the energy of the incident neutron corresponds to the quantum state of the excited compound nucleus. An example of such a situation for Mn-55 is shown in Figure 1.4.

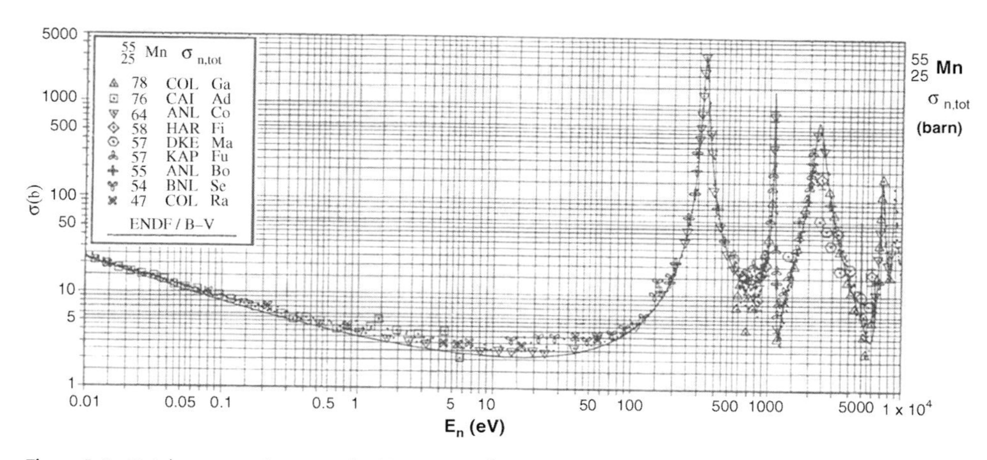

Figure 1.4 Total cross section curve for Mn-55 over the neutron energy range of 0.01 eV–10 keV. (From M.F.L' Annunziata, Handbook of Radioactivity Analysis, Academic Press, New York, 1998; with permission.)

1.7.1
Reactor Flux Spectrum

The neutron energy spectrum is affected by various factors, including reactor type, position in the reactor, and immediate surroundings, such as adjacent fuel, control rods, and empty surroundings. The overall shape of the neutron spectrum is influenced by the specific type of reactor. For reactors using moderators, such as heavy water, light water, or graphite, Figure 1.5a depicts idealized curves for a normalized flux of neutrons as a function of neutron energy.

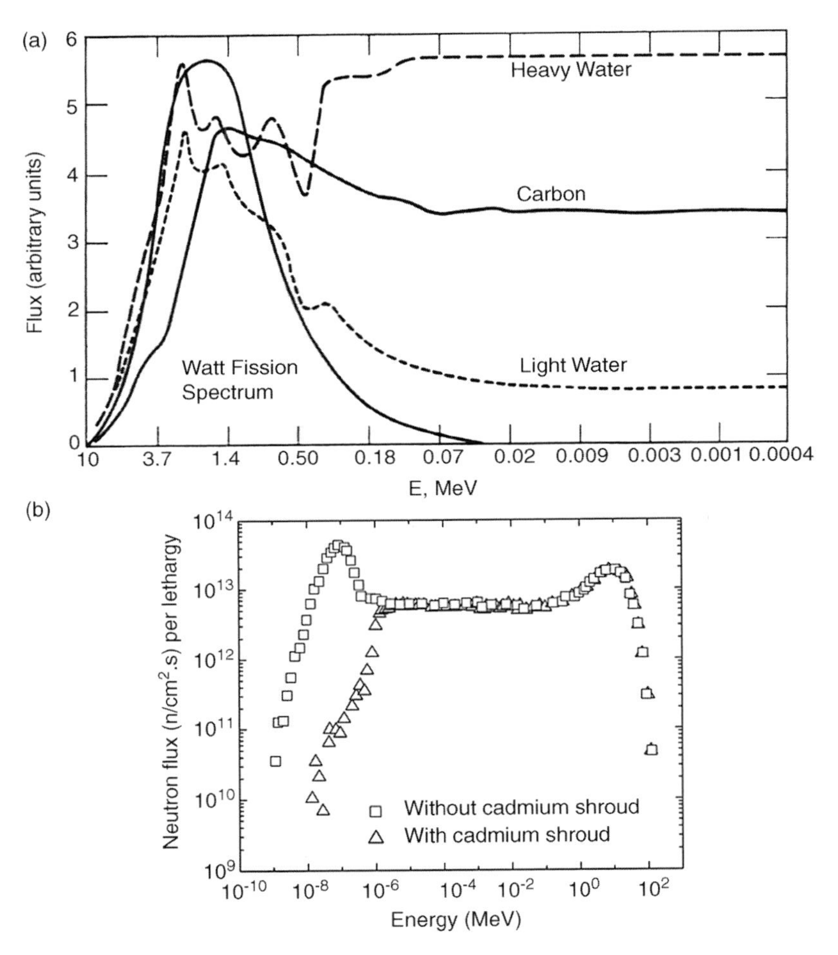

Figure 1.5 (a) Representative flux (energy) spectra for the slowing down of neutrons in infinite carbon, light, and heavy water media compared to a Watt fission spectrum. (From S. H. Bush, Irradiation Effects in Cladding and Structural Materials, Rowman & Littlefield Inc., New York, 1965; with permission.) (b) The reactor flux spectra in the ATR with and without cadmium shroud.

The neutron fission spectrum calculated by Watt is also superimposed on the graph for comparison. Convenient techniques such as assuming monoenergetic neutron flux and the arbitrary selection of neutron flux cutoff level (>1 MeV) are mostly general approximations. Remember that most neutron fluxes cited at irradiation damage studies are expressed in terms of >1 MeV. Figure 1.5b depicts the two flux spectra obtained from the Advanced Test Reactor (ATR). One spectrum is without the use of cadmium shroud and another one is with the cadmium shroud (of ~1.14 mm thickness). It is clear that fast (hard) spectrum is achieved with the use of cadmium shroud (i.e., irradiation jig wrapped into cadmium foil) due to its absorption of thermal neutrons, but not fast ones. Dosimetric experiments followed by calculations can generate the flux spectrum for a specific position in the reactor.

1.8 Types of Reactors

Even though a nuclear reactor can be defined in different ways, almost all reactors except fusion reactor (commercially nonexistent) can be defined as follows: "A nuclear reactor is a device, designed to produce and sustain a long term, controlled fission chain reaction, and made with carefully selected and strategically placed collection of various materials." The classification of reactors vary and are generally based on the following: type of fission reaction (thermal, epithermal, and fast reactors), purpose of the reactor (power reactors, research reactors, and test reactors), type of the coolant present (such as light/heavy water reactors, gas-cooled reactors, and liquid metal-cooled reactors), type of core construction (cubical, cylindrical, octagonal, and spherical reactors), and so forth.

1.8.1 A Simple Reactor Design

Almost all the reactors in the United States and a majority in the world are thermal reactors wherein *thermal neutrons* cause the bulk of the fission reactions. If one starts to think about designing a prototype reactor, the several design elements need to be flawlessly integrated. Figure 1.6 shows such a schematic for a primitive thermal reactor. The tubes containing fuels are generally made of metallic alloys (also known as fuel cladding). The radioactive fuels (such as uranium) could be in metallic, alloy, or ceramic forms. The fuel cladding serves many purposes: it provides mechanical support to the fuel, keeps the fission products from leaving the fuel element, and protects the fuels from corrosion from the coolant. The fuel elements are arranged in a distinct regular pattern (square, hexagon, etc., dictated by neutronics and other factors) with the moderator. Moderator slows down the neutron to sustain the fission reaction with thermal neutrons. The fuel–moderator assembly is surrounded by a

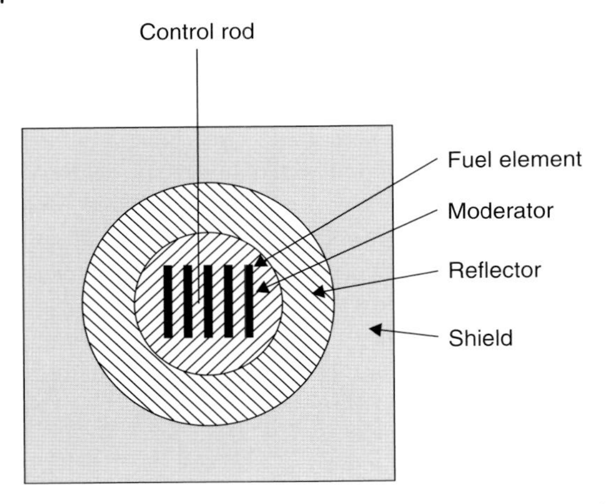

Figure 1.6 A schematic of a simple reactor design. (adapted from C.O. Smith, Nuclear Reactor Materials, Addison-Wesley, Reading, MA, 1967)

reflector. The purpose of a *reflector* is to direct all neutrons generated toward the core so that neutron leakage can be controlled, thus improving the neutron economy. On the outside, the reactor is lined by *shielding materials* that absorb neutrons and gamma rays that escape the core and reduce the radiation intensity to a tolerable level so that people near the reactor are not exposed to these radiations. The control rod (usually an assembly) helps control the chain reaction by absorbing neutrons, maintaining the steady state of operation. Hence, the control materials are neutron-absorbing materials (boron, hafnium, and so forth), and are generally fabricated in the form of rods (in some cases, plates). A reactor is typically equipped with two types of control rods – regulating rods for routine control reasons and safety rods (to permit shutdown in the case of emergency). Even though coolant is not shown in Figure 1.6, it is an important component of a reactor. As a huge amount of heat is generated in the fuel elements, the heat needs to be removed continuously in an efficient manner in order to maintain a safe, steady-state reactor operation. This means an efficient coolant is needed. The coolant can be a gas or liquid (such as light or heavy water, carbon dioxide, liquid metals, and molten salts). However, it is important to remember that the presence of any coolant tends to adversely affect the neutron economy. Hence, the balance between the reduction in the neutron economy and the efficiency of heat removal needs to be carefully considered.

1.8.2
Examples of Nuclear Reactors

A good point to start the discussion on various examples of nuclear reactors is to understand the evolution of nuclear power over the past six to seven

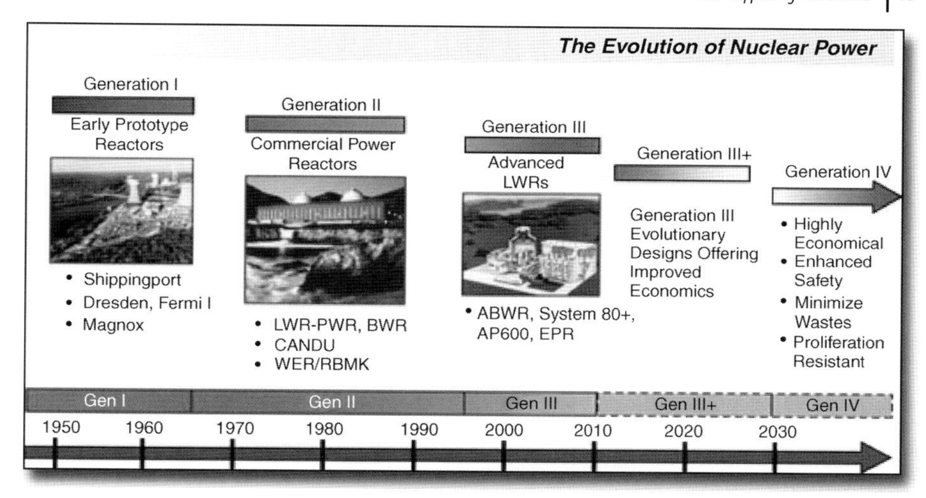

Figure 1.7 Nuclear power evolution in the world. *Courtesy:* The US Department of Energy Gen-IV Initiative.

decades. As shown in Figure 1.7, there are different generations of reactors. Thus, a chronological approach has been taken to start an overview of various nuclear power reactors. However, it is, by no means, exhaustive account. The first nuclear reactor known as Chicago Pile 1 (CP-1) was built during the heydays of the World War II (criticality was achieved on December 2, 1942) at the University of Chicago, Chicago, IL. It was designed and built by a team led by Enrico Fermi. This reactor was a thermal reactor with graphite moderators and natural uranium dioxide fuel. No coolant or shielding was used. The reactor could produce only ~200 W of heat. However, the primary aim of the reactor was to demonstrate the occurrence of the fission chain reaction. It was dismantled in February 1943, and CP-2 reactor was installed at the Argonne National Laboratory based on the experience gained with CP-1.

1.8.2.1 Generation-I Reactors

Generation-I reactors were built in the initial period of nuclear power expansion and generally had primitive design features. Most of these reactors have either been shut down or will be soon done so. Examples of such reactors are Magnox reactor (Calder Hall reactor in the United Kingdom) and first commercial power reactor at Shippingport in 1957 (in the state of Pennsylvania in the United States).

Magnox Reactor

This is a notable Generation-I gas-cooled reactor. Early breed of this reactor was used for the purpose of plutonium production (for atomic weapons) as well as electricity generation. Figure 1.8a shows a cross section of a typical Magnox reactor. The

Calder Hall station in the United Kingdom was a Magnox type of reactor starting successful operation in 1956. Following that, several of these reactors were built and operated in the United Kingdom and a few elsewhere (e.g., Italy, France, and Japan). Generally, Magnox reactors were graphite moderated, and used the natural uranium as fuel clad in thin cylindrical tubes of a magnesium alloy (Magnox comes from the name of the magnesium-based alloy with a small amount of aluminum and other minor elements, magnesium nonoxidizing, for example, Mg–0.8Al–0.005Be) and carbon dioxide (CO_2) as coolant (heat transfer medium). Magnesium-based alloy was chosen since Mg has a very low thermal neutron capture cross section (0.059 b; lower than Zr or Al). The fuel elements were impact extruded with the integral cooling fins or machined from finned extrusions (Figure 1.8b). Also,

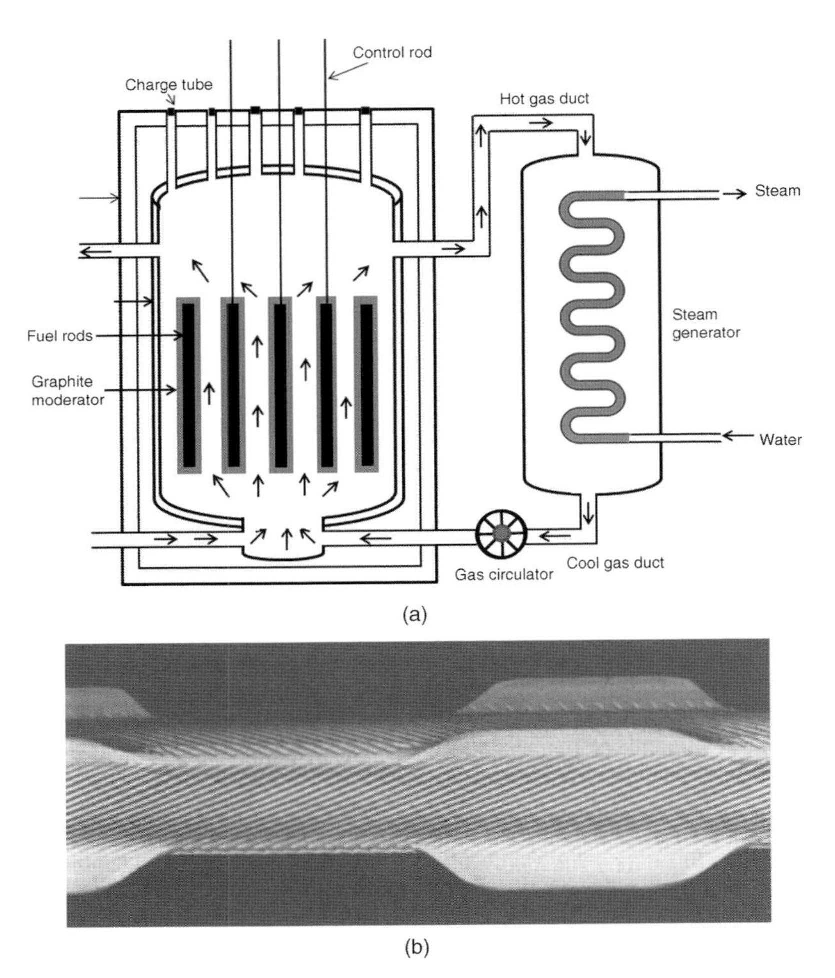

Figure 1.8 (a) A schematic of a Magnox reactor. (b) A part of the magnesium alloy fuel can of a British Magnox reactor. *Courtesy:* Light Alloys by Ian Polmear.

the alloy was resistant to creep and corrosion from CO_2 atmosphere in the operating temperature range, and contrary to Al, the alloy did not react with the uranium fuel. The addition of Al in the alloy provided solid solution strengthening, while the presence of minor amounts of Be helped improve the oxidation resistance. CO_2 was circulated under pressure through the reactor core and sent to the steam generator to produce steam that is then passed through a turbo generator system generating electricity. These reactors could sustain lower temperatures (maximum coolant temperature of 345 °C) and, thus, has a limited plant efficiency and power capacity. This was mainly out of the concern of the possible reaction of CO_2 with graphite at higher temperatures and the lower melting point of uranium fuel (1132 °C). Another problem was that the spent fuel from these reactors could not be safely stored under water because of its chemical reactivity in the presence of water. Thus, the spent fuels needed to be reprocessed immediately after taking out of the reactor and expensive handling of equipment was required. Only two Magnox reactors still operating in the United Kingdom are scheduled to be decommissioned soon. Magnox reactors were followed by an improved version of gas-cooled reactor known as Advanced Gas-Cooled Reactor (AGR) operating at higher temperatures and thus improving the plant efficiency. The magnesium fuel element was replaced by stainless steels.

1.8.2.2 **Generation-II Reactors**

Most of the commercial nuclear power plants operating today are of Generation-II type. Also, the reactors employed in naval vessels (such as aircraft carriers and submarines) and many research/test reactors are of this type. The Generation-II reactors incorporated improved design and safety features and productivity over Generation-I reactors. In the Western Hemisphere, a majority of commercial nuclear power plants have light water reactor (LWR), both pressurized water reactor (PWR) and boiling water reactor (BWR). It is important to remember that LWRs were also built as Generation-I reactors (such as Shippingport facility with 60 MWe power capacity), however most of them are no longer in operation. Another variety is the CANDU (Canadian Deuterium Uranium) reactor, which is basically a pressurized heavy water reactor (PHWR). There are a few different versions of pressurized water reactors (e.g., RBMK type) in Russia and former Soviet-block countries, but discussion on those reactors is outside the scope of this book.

Light Water Reactors

As the name implies, LWRs use light water as the coolant and the moderator, and in many cases as the reflector material. These are typical thermal reactors as they utilize thermalized neutrons to cause nuclear fission reaction of the U^{235} atoms. The thermal efficiency of these reactors hover around 30%. Two main types of LWR are PWR and BWR. These two types are created mainly because of the difference in approaches of the steam generating process (good quality steam should not contain more than 0.2% of condensed water). LWRs have routinely been designed with 1000 MWe capacity.

Pressurized Water Reactor PWRs were designed and implemented commercially much sooner than the BWRs due to the earlier notion that the pressurized liquid water would somehow be much safer to handle than the steam in the reactor core and would add to the stability of the core during the operation. That is why the first commercial reactor in Shippingport was a PWR. PWRs are designed and installed by companies such as Westinghouse and Areva.

A schematic design of a typical PWR plant is shown in Figure 1.9a. A PWR plant consists of two separate light water (coolant) loops, primary and secondary. The PWR core is located inside a reactor pressure vessel (RPV) made of a low-alloy ferritic steel (SA533 Gr. B) shell (typical dimensions: outside diameter ~5 m, height ~12 m, and wall thickness 30 cm), which is internally lined by a reactor cladding of 308-type stainless steel or Inconel 617 to provide adequate corrosion resistance against coolant in contact with the RPV. The PWR primary loop works at an average pressure of 15–16 MPa with the help of a set of pressurizers so that the water does not boil even at temperatures of 320–350 °C. The PWR core contains an array of fuel elements with stacks of a slightly enriched (2.5–4%) UO_2 fuel pellets clad in Zircaloy-4 alloy (new alloy is Zirlo or M5). Individual cladding tubes are generally about ~10 mm in outer diameter and ~0.7 mm in thickness. The fuel cladding tube stacked with fuel pellets inside and sealed from outside is called a *fuel rod* or *fuel pin*. About 200 of such fuel rods are bundled together to form a *fuel element*. Then, about 180 of such fuel elements are grouped together to form an array to create the *reactor core* of various shapes – square, cylindrical, hexagonal, and so on (Figure 1.9b). The reactor core is mounted on a core-support structure inside the RPV. Depending on the specific design, the above-mentioned dimensions of the various reactors may vary.

The control rod used is typically an Ag–In–Cd alloy or a B_4C compound, which is used for rapid control (start-up or shutdown). Boric acid is also added to the primary loop water to control both the water chemistry acting as "poison" and the long-term reactivity changes. This primary loop water is transported to the steam generator where the heat is transferred to the secondary loop system forming steam. The steam generator is basically a heat exchanger containing thousands of tubes made from a nickel-bearing alloy (Incoloy 800) or nickel-based superalloy (e.g., Inconel 600) supported by carbon steel plates (SA515 Gr.60).

Boiling Water Reactor BWR design embodies a direct cycle system of cooling, that is, only one water loop, and hence no steam generator (Figure 1.10a). Early boiling water experiments (BORAX I, II, III, etc.) and development of experimental boiling water reactor (EBWR) at the Argonne National Laboratory were the basis of the future commercial BWR power plants. Dresden Power Station (200 MWe), located at the south of Chicago, IL, was a BWR power plant that started operating in 1960. It is of note that this was a Generation-I BWR reactor. However, most BWRs operating today are of Generation-II type and most significant features are discussed below. The reactors (Fukushima Daiichi) that underwent core melting following the unprecedented earthquake and tsunami in Japan during 2011 were all of BWR type.

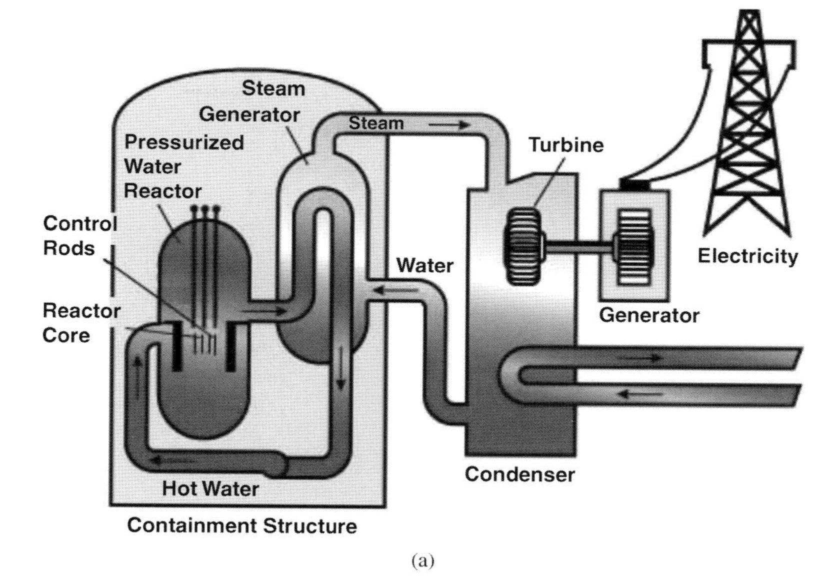

(a)

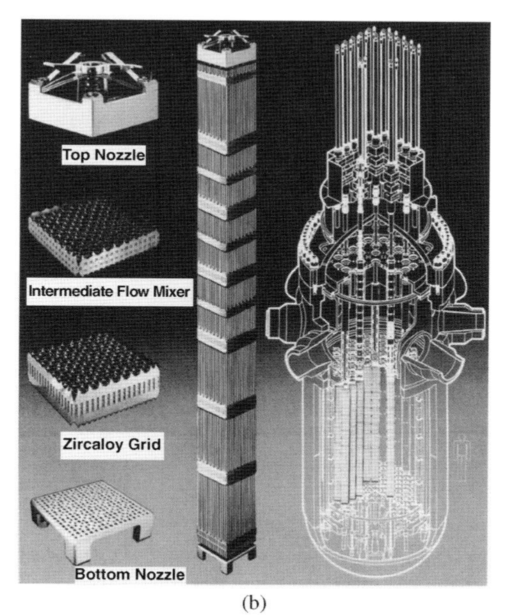

(b)

Figure 1.9 (a) A view of a typical PWR plant. Courtesy: US Nuclear Regulatory Commission (b) A view of various PWR components. Courtesy: Westinghouse Electric Corporation.

The BWR reactor core is located near the bottom end of the reactor pressure vessel. Details of various components in a typical BWR are shown in Figure 1.10b. The BWR RPV is more or less similar to the PWR one. The BWR core is made of fuel assembly consisting of slightly enriched UO_2 fuels clad with recrystallized Zircaloy-

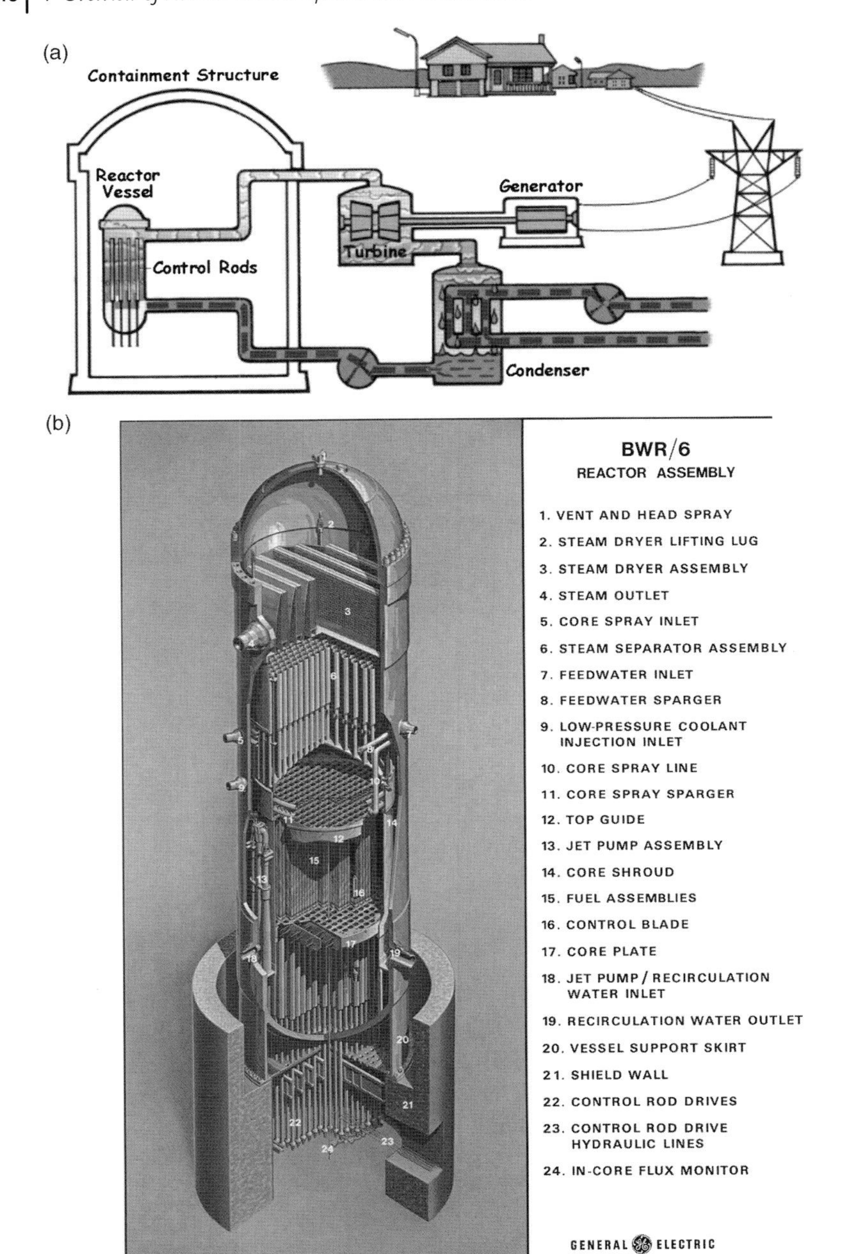

Figure 1.10 (a) A view of a BWR plant. (b) A view of the cut section of a typical BWR. *Courtesy:* The US Nuclear Regulatory Commission and GE.

Table 1.2 PWR versus BWR.

PWR	BWR
Principle of steam generation	
RPV pressure ~15 MPa	RPV pressure ~7 MPa
RPV temperature ~326 °C	RPV temperature ~290 °C
Steam generated in steam generator (via secondary loop)	Steam generated in RPV (with separator and dryer)
No bulk boiling in RPV	Bulk boiling allowed in RPV
Major components	
RPV	RPV with separator and dryer
Two–four steam generators	No steam generator
One pressurizer	No pressurizer
Top entry control rod clusters	Bottom entry control rod drives
Zircaloy-4 fuel cladding tubes	Zircaloy-2 fuel cladding tubes

2 cladding tubes (about 12.5 mm in outer diameter). For a BWR core of 8 × 8 type, each fuel assembly contains about 62 fuel rods and 2 water rods, which are sealed in a Zircaloy-2 *channel box*. The control material is in the shape of blades arranged through the fuel assembly in the form of a cruciform and is generally made of B_4C dispersed in 304-type stainless steel matrix or hafnium, or a combination of both. Water is passed through the reactor core producing high-quality steam and dried at the top of the reactor vessel. The BWR operates at a pressure of about 7 MPa and the normal steam temperature is 290–330 °C.

Note

Tables 1.2 and 1.3 contain relevant information on PWRs and various types of BWRs.

Table 1.3 Operating parameters and design features of BWRs.

Parameter/feature	BWR (Browns Ferry 3)	ABWR	ESBWR[a]
Power (MWt/MWe)	3293/1098	3926/1350	4500/1590
Vessel height/diameter (m)	21.9/6.4	21.7/7.1	27.6/7.1
Fuel bundles (number)	764	872	1132
Active fuel height (m)	3.7	3.7	3.0
Recirculation pumps (number)	2 (Large)	10	Zero
Number of control drive rods	185	205	269
Safety diesel generator	2	3	Zero
Safety system pumps	9	18	Zero
Safety building volume	120	180	135

Courtesy: GE Global Research.

a) ESBWR – economic simplified boiling water reactor – of Generation-III+ category, developed by the GE.

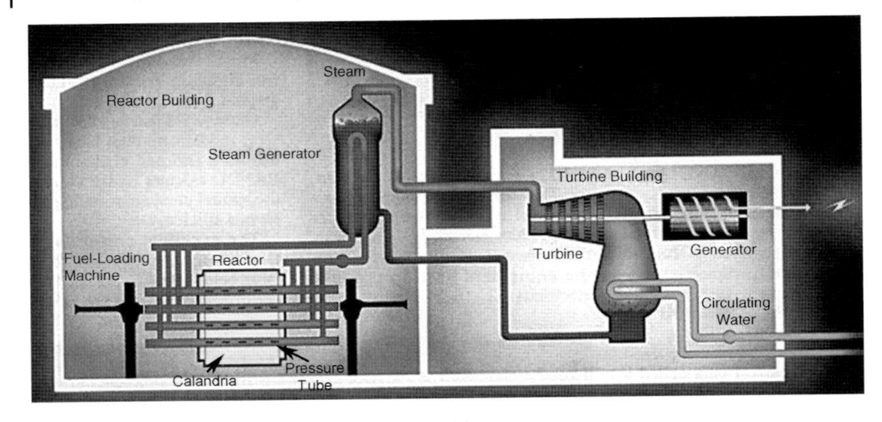

(a)

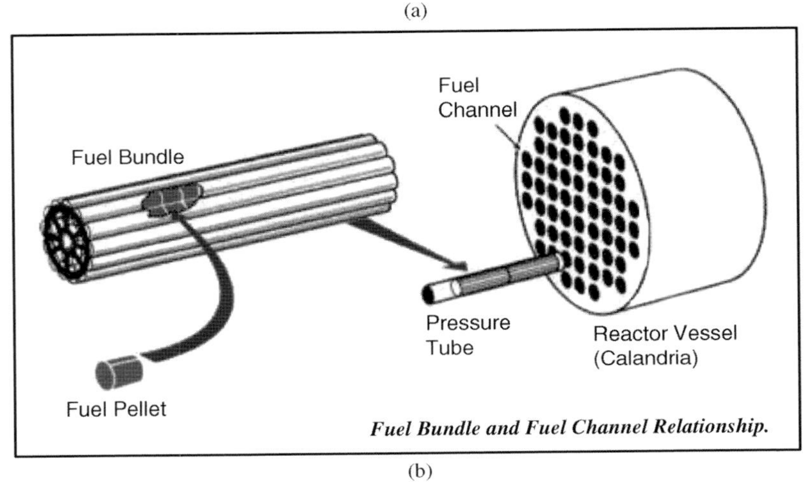

(b)

Figure 1.11 (a) A simplified schematic view of a CANDU reactor. *Courtesy:* Canadian Nuclear Association. (b) The configuration of fuel bundles in the fuel channel. *Courtesy:* www.cameco.com.

Pressurized Heavy Water Reactor The PHWR reactors were mainly developed by a partnership between the Atomic Energy of Canada Limited (AECL) and Hydro-Electric Power Commission of Ontario in 1960s. The reactors were of Generation-II type. Notably, these reactors are also called CANDU reactors (Figure 1.11a). They are so named because they use heavy water (deuterium oxide) as the moderator and natural uranium as the fuel. These reactors are located mainly in Canada, India, China, and few other countries. The CANDU reactor design does not require a reactor pressure vessel as in LWRs, and hence not a single CANDU reactor operates in the United States since the nuclear safety regulations of the US Nuclear Regulatory Commission specifically call for an RPV in a compliant reactor design.

Unlike LWRs, the natural uranium (0.7% U^{235}) oxide fuel clad in zirconium alloy tubes (known as pressure tubes, made of Zr–2.5Nb) is used in this reactor. These hundreds of pressure tubes are kept inside a calandria shell made of an austenitic stainless steel and reinforced by outer stiffening rings. The shell also keeps channels for the pressurized coolant (hot heavy water or light water) and moderator (heavy water). If light water is used to moderate the neutrons, it would adversely affect the neutron economy due to the absorption of neutrons. That is why cold heavy water is used as the moderator. The pressure tubes along with moderator and cooling tubes are arranged in a horizontal fashion (not vertical as in LWRs) and the fuels can be replaced and reloaded without shutting down the whole reactor. Note the fuel and associated structural configuration in Figure 1.11b. The pressurized coolant stays typically at about 290 °C. This reactor system requires a steam generator to produce steam as does a conventional PWR. The control rods are dropped vertically into the fuel zones in case total shutdown or controlling of reactivity is necessary. Gadolinium nitrate is added in the moderator system that acts as a secondary shutdown system.

Liquid Metal Fast Breeder Reactor Liquid metal (generally sodium) is used in liquid metal fast breeder reactors (LMFBRs) to transport the heat generated in the core. These reactors are called "breeder" because more new fuels are produced than consumed during its operation. The reactor can convert fertile material (containing U^{238} and Th^{232}) into fissile materials (Pu^{239} and U^{233}), respectively. The concept of this reactor type is very practical from the fuel utilization viewpoint. The natural uranium contains only about 0.7% of fissile U^{235}. The majority of the natural uranium contains U^{238} isotope. These reactors are characterized by high power density (i.e., power per unit volume) due to the lack of a moderator (i.e., much more improved neutron economy). The reactor cores are typically small because of the high power density requirements. The temperature attained in this type of reactors is higher and thus leads to higher efficiency of electric power generation ($\sim$42% in LMFBRs versus $\sim$30% in LWRs). The use of Th^{232} in LMFBRs is particularly advantageous for countries like India that do not have a large deposit of uranium, but has plenty of thorium. It should be noted that sodium used in this type of reactor transfers heat to the steam generators. The system containing sodium should be leak-proof since sodium reacts with oxygen and water vapor very fast. Furthermore, it becomes radioactive as the coolant passes through the reactor core. The whole primary coolant system should be put in the shielded protection to keep the operating personnel safe.

The first prototype LMFBR reactor named EBR-I (Experimental Breeder Reactor) was built at the present-day site of the Idaho National Laboratory near Idaho Falls, ID. This was the first reactor to demonstrate that the electricity can be generated using the nuclear energy. Also, it was the first "fast breeder" reactor. It used sodium–potassium (NaK) as coolant. It started its operation in 1951, and was decommissioned by 1964. By this definition, it was a Generation-I fast reactor

Figure 1.12 A view of EBR-II reactor.

design. Following the decommissioning of EBR-I, another fast breeder reactor (EBR-II) was installed and started operation in 1963. EBR-II (Figure 1.12) was operated very successfully before it was shut down in 1994. The fuel used was a mixture of 48–51% of U^{235}, 45–48% U^{238}, and the rest a mixture of fissium metals (Mo, Zr, Ru, etc.) and plutonium.

1.8.2.3 Generation-III and III+ Reactors

The new reactors that are being built or will be built within a few years are of Generation-III category. These are mainly advanced LWRs. Examples include advanced boiling water reactor (ABWR) and evolutionary or European power reactor (EPR). In the same line, Generation III+ category aims to provide reactor systems that have much improved designs and safety features, and much greater capacities. Notably, all these reactors are thermal in nature. No fast reactor is in the pipeline in these categories. An improved version of the US EPR® is being designed and developed as a Generation III+ reactor by AREVA NP. It is a four-loop plant with a rated thermal power of 4590 MW_{th}. Figure 1.13a and b shows the details of a EPR power plant.

The primary system design and main components are similar to those of PWRs currently operating in the United States. However, the US EPR incorporates several new advancements in materials technology as well as new uses for existing materials. Some notable examples include M5® fuel cladding and the stainless steel heavy neutron reflector. Some details of these are discussed below:

a) Zircaloy-4 (Zr-4) has been used extensively for many years in PWR fuel cladding applications. Zr-4 is a cold worked stress-relieved (CWSR) alloy with a zirconium base containing tin, iron, chromium, and oxygen. This offers good corrosion resistance and mechanical properties. AREVA NP has developed an advanced zirconium alloy for PWR fuel rod cladding and fuel assembly structural components, known as M5® (Zr-0.8–0.12Nb–0.09–0.13O). M5® makes high-burnup fuel cycles possible in the increasingly higher duty operating environments of PWRs. Introduced commercially in the 1990s after a rigorous development program, M5® was a breakthrough in zirconium alloy development. This fully recrystallized, ternary Zr–Nb–O alloy produces improved in-reactor corrosion,

(a)

(b)

Figure 1.13 (a) A layout of a US EPR® power plant. (b) Main coolant line components in the US EPR®. *Courtesy:* Areva NP.

hydrogen, growth, and creep behaviors. The remarkably stable microstructure responsible for these performance improvements is the result of carefully controlled ingot chemistry and product manufacturing parameters. M5® helps utilities achieve significant fuel cycle cost savings and enhanced operating margins

Figure 1.14 A 14 × 14 matrix fuel assembly made of M5® alloy. *Courtesy:* Areva NP.

by allowing higher burnups and higher duty cycles. Excellent corrosion resistance and low irradiation growth allow higher burnups, extended fuel cycle operation, and fuel assembly design modifications that enhance operating flexibility. The corrosion resistance of M5® allows operation in high pH environments, eliminating the risk of oxide spalling and helping to minimize dose rates. M5® has been proven to withstand severe operating conditions: high neutron flux, heat flux, and high operating temperatures. As of January 2007, over 1 576 000 fuel rods in designs from 14 × 14 (Figure 1.14) to 18 × 18 matrices have been irradiated in commercial PWRs to burnups exceeding 80 000 MWd per mtU. In fuel rod cladding with hydrogen concentrations above the solubility limit, excess hydrogen precipitates as brittle hydrides, reducing the ability of the cladding to cope with pellet-to-clad mechanical interactions during reactivity insertion accidents. During a LOCA, (Loss Of Coolant Accident) high hydrogen levels increase the transport and solubility of oxygen at high temperatures, leading to substantial embrittlement of the cladding as it cools and the oxygen precipitates. M5® cladding absorbs approximately 85% less hydrogen than other Zr-4-based alloys. As a result, M5® will not reach hydrogen levels sufficient enough to precipitate hydrides and will not lead to excess oxygen absorption during a LOCA.

b) The space between the multicornered radial periphery of the reactor core and the cylindrical core barrel is filled with an austenitic stainless steel structure, called the *heavy reflector*. The primary purpose of the heavy reflector is to reduce fast neutron leakage and flatten the power distribution of the core, thus improving the neutron utilization. It also serves to reduce the reactor vessel fluence ($\sim 1 \times 10^{19}\,\mathrm{n\,cm^{-2}}$, $E > 1\,\mathrm{MeV}$ after 60 years). The reflector is inside the core barrel above the lower core support plate. The reflector consists of stacked forged slabs (rings) positioned one above the other with keys, and axially restrained by tie rods bolted to the lower core.

1.8.2.4 Generation-IV Reactors

Gen-IV reactors are the futuristic reactors for which research and development efforts are currently in progress. These reactors will be more efficient, safer, longer lasting (60 years and beyond), proliferation-resistant, and economically viable compared to the present nuclear reactors. Six reactor designs were selected at the outset. They are summarized in Table 1.4 including information on the type (fast or thermal spectrum) of the reactor, coolants, and approximate core outlet temperatures. Two reactor concepts, *sodium fast reactor* (SFR) – along with the *advanced burner reactor* (ABR) concept under the erstwhile Advanced Fuel Cycle Initiative Program) – and *very high-temperature reactor* (VHTR) under the Next Generation Nuclear Plant (NGNP) are of the highest priority in the United States (Figure 1.15a and b, respectively). VHTR is the reactor concept employed for the Next Generation Nuclear Plant program where the heat generated will cogenerate electricity and hydrogen.

The demanding service conditions (higher neutron doses, exposure to higher temperatures, and corrosive environments) that the structural components will experience in these reactors would pose a significant challenge for the structural materials selection and qualification efforts. Because of the stringent requirements noted above, the materials employed in today's commercial reactors are not suitable for use in Gen-IV reactors. For example, zirconium alloys (Zircaloy-2, Zircaloy-4, Zr-2.5Nb, M5) have been used routinely as fuel cladding and other reactor internals in both light and heavy water reactors because of their low neutron capture cross section and acceptable mechanical and corrosion resistance in high temperature

Table 1.4 A summary of Gen-IV reactor system designs.

Reactor system	Priority/timeline	Coolant	Neutron spectrum	Core outlet temperature (°C)
Gas-cooled fast reactor (GFR)	Medium/long term	Gas (e.g., He)	Fast	~850
Lead-cooled reactor (LFR)	Medium/long term	Liquid metal (e.g., Pb, Pb–Bi)	Fast	550–800
Molten salt reactor (MSR)	Low/long term	Molten salt (e.g., fluoride salts)	Thermal	700–800
Sodium-cooled fast reactor (SFR)	High/mid term	Liquid metal (Na)	Fast	~550
Very high-temperature reactor (VHTR)	High/mid term	Gas (e.g., He), CO_2	Thermal	> 900
Supercritical water-cooled reactor (SCWR)	Medium/long term	Water	Thermal/fast	280–620

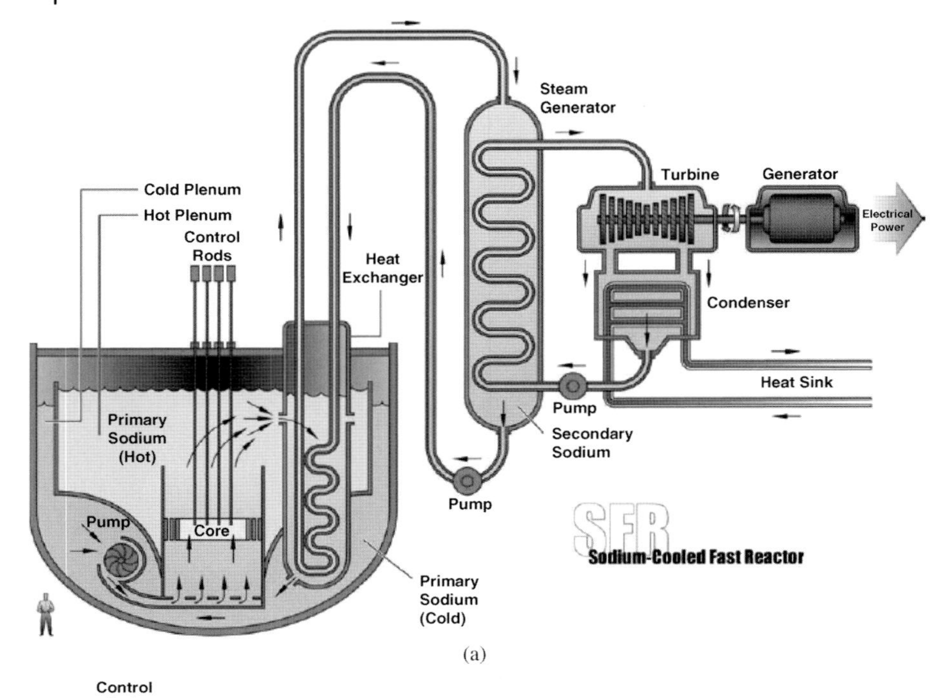

(a)

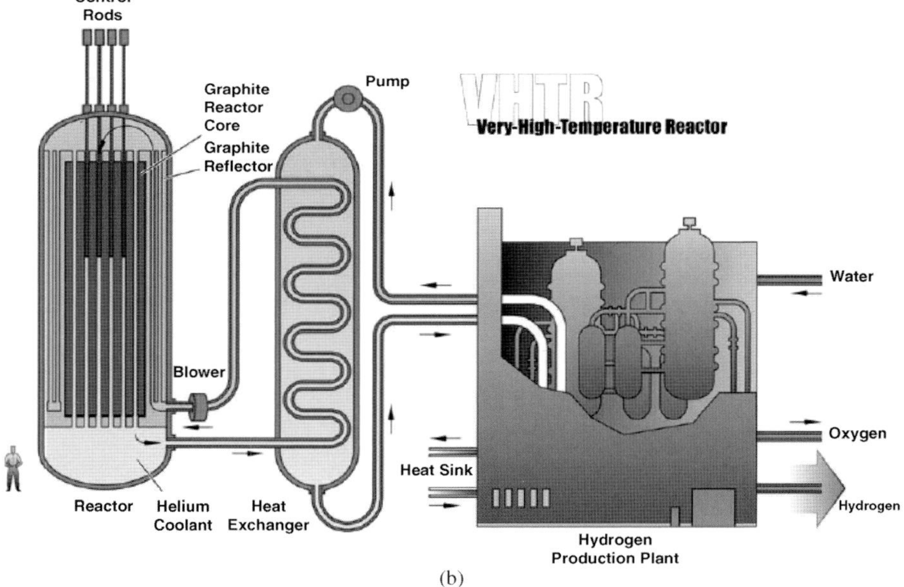

(b)

Figure 1.15 Schematics of (a) a sodium fast reactor and (b) a very high-temperature reactor. (Courtesy: US Department of Energy Gen-IV Initiative) (probably never exceeding 350–380 °C under normal service conditions) aqueous environment. However, higher temperatures envisioned in Gen-IV reactors would limit the use of zirconium alloys because of increased susceptibility to hydrogen embrittlement due to severe hydride formation, allotropic phase changes at higher temperatures (α–β phase), poor creep properties, and oxidation. It is instructive to note that some high-performance zirconium alloys may be of possible use in relatively low-temperature Gen-IV reactor design (such as SCWR). Furthermore, the out-of-core components (pressure vessel, piping, etc.) may need to be made from materials other than the low-alloy ferritic steels (e.g., A533B) currently employed primarily because similar components in Gen-IV reactors are expected to withstand much higher temperatures and neutron doses. Some of the fabrication difficulty involved in the VHTR construction demands mention here. For example, the pressure vessel for VHTR reactor as shown in Figure 1.16 is about double the size of the currently operating PWRs. Heavy component forgings will be needed in the construction of these huge pressure vessels.

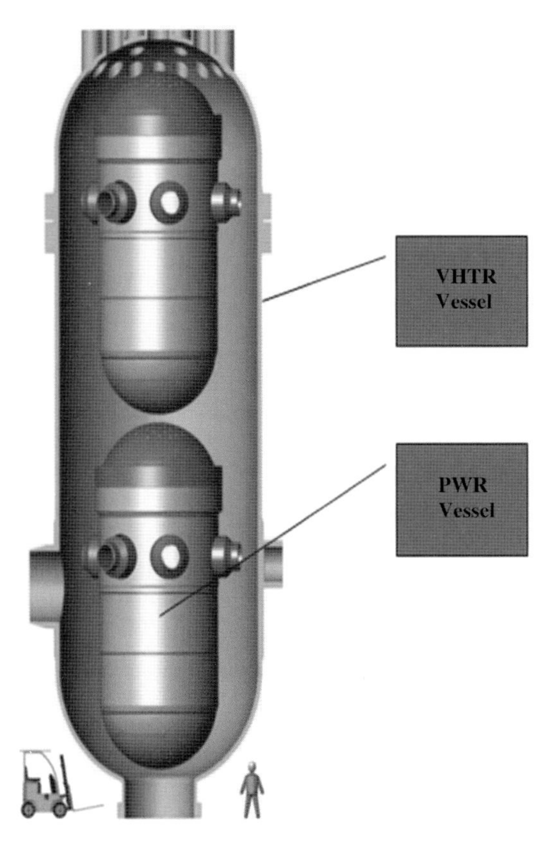

Figure 1.16 A schematic size comparison between the reactor pressure vessels in a typical PWR and a conceptual VHTR. *Courtesy:* Nuclear News.

Test Reactors

So far we have discussed the past, current, and future nuclear reactors in detail. Now let us discuss about a test reactor. One of the well-known test reactors in the United States is the Advanced Test Reactor located at the Idaho National Laboratory near Idaho Falls, ID. ATR is a flagship reactor that serves the US Department of Energy, Naval Nuclear Propulsion Program, and different other governmental and commercial entities. It also acts as a user facility that the university-led teams can use to irradiate materials and perform postirradiation examination (PIE) upon going through a competitive proposal process. ATR started its operation in 1967 and is being operated continuously since then with 250+ days of operation in an average year. A view section of the ATR is shown in Figure 1.17.

ATR is a pressurized, light water cooled and moderated, 250 MW_{th} reactor with beryllium reflectors and hafnium control drums. The metallic fuel (U or U–Mo) is in the plate morphology clad in an aluminum alloy. There are 40 fuel assemblies in the reactor core; each core contains 19 fuel plates. At 250 MW, maximum thermal neutron flux is $\sim 10^{15}\,\mathrm{n\,cm^{-2}\,s^{-1}}$, and the maximum fast neutron flux could reach $5 \times 10^{14}\,\mathrm{n\,cm^{-2}\,s^{-1}}$. Thus, ATR can be used to study the radiation damage under the fast neutron spectrum. The ATR has 77 irradiation positions (4 flux traps, 5 in-pile tubes, and 68 in-reflector) (for details, see Figure 1.18). The reactor pressure vessel is made of stainless steel, and is 3.65 m in diameter and $\sim$10.67 m in height. Table 1.5 lists some of the differences between ATR and a typical PWR.

ATR even though used for neutron irradiation experiments is not a fast reactor facility. Note that at present the United States does not have any operating/underconstruction fast reactor test facility (EBR-II and FFTF facilities were shut down during 1990s) as opposed to countries like France (Phenix), India (prototype fast breeder reactor (PFBR)), Japan (Joyo and Monju), Russia (BOR-60), and China (China experimental fast reactor (CEFR). The lack of a fast reactor facility is a challenge for the US nuclear R&D community. The proposed Advanced Burner Test Reactor (ABTR), a sodium-cooled fast reactor, is still under the planning stage, and there is no further confirmation of its installation yet.

1.9 Materials Selection Criteria

"We physicists can dream up and work out all the details of power reactors based on dozens of combinations of the essentials, but it's only a paper reactor until the metallurgist tells us whether it can be built and from what.

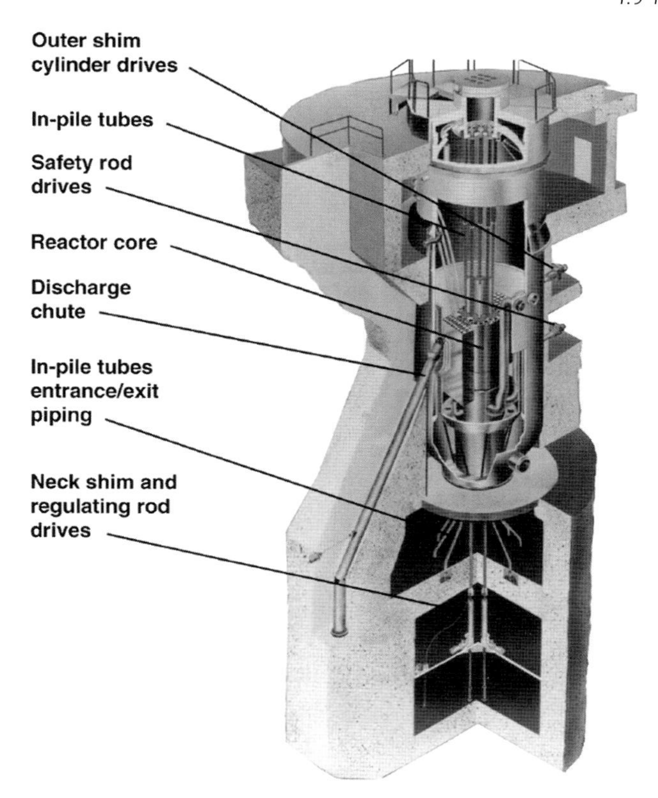

Figure 1.17 A view section of the Advanced Test Reactor. *Courtesy:* Idaho National Laboratory.

Table 1.5 Comparison between ATR and a typical PWR.

Reactor Features	ATR	PWR
Power (MW_{th})	250 (maximum design)	~3800
Operating pressure (MPa per psig)	~2.5/~355	~15.5/~2235
Inlet temperature (°C)	~52	~288
Outlet temperature (°C)	~71	~327
Power density (kW per ft^3)	~28 300	~2800
Fuel element shape	Plate	Tubular
Fuel	Enriched U^{235}	3–4% Enriched U^{235}
Fuel temperature (°C)	~462	>538

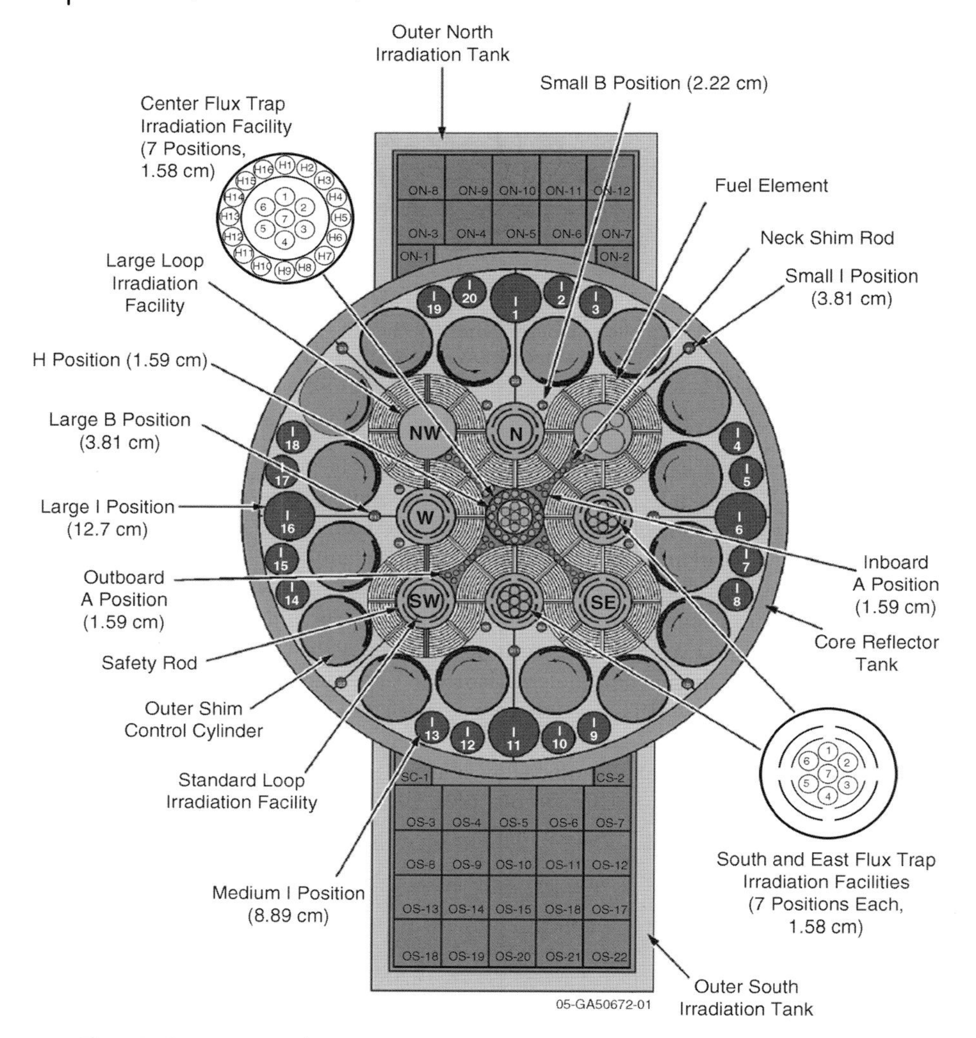

Figure 1.18 Various irradiation positions. *Courtesy:* Idaho National Laboratory.

> Then only can we figure whether there is any hope that they can produce power at a price."
>
> —*Dr. Norman Hillberry*
> *(former director of Argonne National Laboratory, 1957–1961)*
> *(excerpt from the book by C.O. Smith)*

The above statement by Dr. Hillberry (a physicist by training) says it all! Erstwhile metallurgical engineering field has now largely morphed into the field of materials science and engineering (MSE). MSE as a field of study is based on a common theme of finding out the interlinkage between processing, structure, and properties in various types of materials. If the interlinkage is clearly understood and established, the performance of these materials under service conditions could be ensured. Hence, the materials selection process in any structure design is a very important step. It is very common to encounter various *tradeoffs* during the materials selection process for a given application, and most times compromise is called for. Moreover, a nuclear reactor design entails complex procedures in itself given the multiple challenges. Different components of a reactor may require different types of materials. This is mostly done with the help of engineering expertise (experience and judgment).

There could be two broader types of materials selection considerations – *general* and *special*. General considerations involve factors such as mechanical strength, ductility, toughness, dimensional stability, fabricability, cost and availability, heat transfer properties, and so on. General properties come from the general engineering considerations as they would be applicable in most engineering designs. On the other hand, special properties are considered solely because the materials are to be used in a nuclear reactor. These include properties like the neutron absorbing characteristics, susceptibility to induced radioactivity, radiation damage resistance, and ease of reprocessing of materials. Each of the material characteristics is evaluated following standard (sometimes nonconventional) testing procedures. The knowledge of service conditions and broader goals of the future reactor is a must for a successful materials selection process. This information may come from utilizing predictive capabilities (modeling and simulation tools) and/or known data/experiences from previous reactor design and operations, if available. Brief discussions on these properties have been made in the following sections. Some of these properties will be again elaborated in the subsequent chapters.

1.9.1 General Considerations

1.9.1.1 General Mechanical Properties

Important general mechanical properties include tensile strength, ductility, and toughness. The material should be strong enough to bear the loads of the structure and also sustain any internal or external stresses generated during service. Also, the material should have enough ductility (a measure of percentage elongation or reduction in area in standard tensile specimens) to avoid any catastrophic failure. Usually, as a rule of thumb, a percentage elongation of 5% is considered a minimum requirement for a load-bearing engineering structure. But one must admit that this often changes with the type of application at hand. In some cases, the materials should have sufficient ductility in order to be formed into different components. Toughness is defined as the ability of a material to absorb energy without failure, and that dictates how tough a material is for use. Generally, tensile strength and ductility combined is referred to as toughness. However, generally impact tests and fracture toughness tests are conducted to evaluate toughness properties of materials. All these affect the mechanical integrity of the reactor components.

1.9.1.2 **Fabricability**

Fabricability includes a host of characteristics such as formability, weldability, machinability, and so on. If fabricability issues are not dealt with during the first stage, it may cause problems at the later stages. In many cases, some parts of the nuclear power plant are to be built at the site (also called *field fabrication*) from smaller parts. If the materials do not have the requisite fabricability, it would not be possible to use the material no matter what fantastic properties it may have!

1.9.1.3 **Dimensional Stability**

The material should have adequate stability in properties. For example, many nuclear components would work at higher temperatures for extended period of time. So, the creep deformation (i.e., time-dependent plastic deformation) may cause dimensional stability problems.

One should also recognize that the microstructure of a material changes as a function of temperature, time, and stresses. So, the effects of these factors on microstructure and the consequent effects on the properties need to be taken into account carefully.

1.9.1.4 **Corrosion Resistance**

Corrosion is an electrochemical process that causes the surface of the metals/alloys degrade over time in the presence of a chemical environment. Corrosion resistance of materials used in nuclear components is important in many applications to ensure that they serve as desired. The "cost of corrosion" can result in immediate property and life endangerment and increased downtime, leading to substantial losses. Many nuclear components inside the reactor stay in close contact with reactor fluids (e.g., coolant in the form of liquid or gas). These effects get exacerbated due to the presence of radiation fields.

1.9.1.5 **Design**

Although design does not generally fall under the purview of a materials engineer, he/she is in a unique position to figure out early whether the faulty design would pose a problem. Designs leaving stress concentration sites (sharp recesses, keyholes, and the like) are typically unwarranted in load-bearing applications since it may interfere with the capability of the component to serve properly. For example, fatigue properties are especially prone to the presence of stress concentration sites.

1.9.1.6 **Heat Transfer Properties**

As we know from our fundamental physics classes, heat transfer modes are of three types – conduction, convection, and radiation. The first two processes are of importance in nuclear reactor materials selection. Most important example in nuclear reactor is the choice of fuel and cladding materials. The safety and efficiency of the reactor depends on how efficiently the heat generated inside the fuel can be removed. Hence, thermal conductivity is an important property. Otherwise, the fuel will melt such as in a "loss of coolant accident" (LOCA) scenario. Similarly, heat transfer properties are also important for various balance-of-plant features, such as heat exchangers, condensers, and other ancillary equipment (such as steam generator in a PWR system).

1.9.1.7 **Availability and Cost**

This is the last but not the least general consideration in the materials selection process. Availability and cost are the economic consideration that may trump technical considerations. If a material is not available in the form or at an allowable price, the prudent engineering decision would be to find an alternative material with similar properties or a different form of material or make changes in design to allow different characteristics. Cost-benefit analysis must be at the heart of that process. In this regard, the fair question becomes, "If it were your money, would you refuse to buy the item because it costs too much?"

1.9.2 Special Considerations

1.9.2.1 **Neutronic Properties**

Neutronic properties are of significant consideration in the design and development of nuclear reactors. As discussed earlier, the fission chain reaction requires continued supply of neutrons for it to proceed and that is why neutron economy plays an important role. Fuel cladding materials need to have a lower neutron absorption cross section and that is why zirconium alloys are used in LWRs (see Appendix 1.A). On the other hand, to control the chain reaction, the control materials should have high neutron absorption cross section. The same consideration would also apply to shielding materials.

1.9.2.2 **Susceptibility to Induced Radioactivity**

The materials in the reactor can absorb fast/thermal neutrons and undergo reactions that may lead to the production of different radioactive isotopes of the constituent elements of the materials. These reactions can induce radioactivity as these isotopes would decay by emitting gamma rays, beta rays, and alpha rays of different energy levels. While selecting an alloy, we should be concerned about the following factors: (a) quantity of the impurities/alloying elements, (b) abundance of the isotopes and corresponding cross section, (c) half-life of the product nuclide, and (d) the nature of the radiation produced. If the produced isotope has a short half-life and emit radiation of low energy, it should not be a cause for great concern. However, if the isotope is long-lived and produces radiation of high energy, all precautions must be taken. For example, the main isotope of iron (Fe^{56}) that accounts for almost 92% of the natural iron forms a stable isotope (Fe^{57}) upon absorbing neutrons. The absorption of neutrons in Fe^{54} and Fe^{58} yielding Fe^{55} (half-life: 2.9 years) and Fe^{59} (half-life: 47 days) results in activation. However, the impurities or alloying elements cause more induced radioactivity than iron itself. Generally, the test samples irradiated in a reactor are not examined immediately after taking out from the reactor because they remain literally hot and continue to be hot due to the decay heat produced by various reactions even if the fission chain reaction no longer occurs. The Fukushima Daiichi accident in Japan did show the severity of the heat produced due to these decay reactions even after the emergency shutdown of the reactor, leading to very high temperatures (in the absence of proper coolant) and eventually resulting in the cladding breach and perhaps some form of core melting.

Note

The development of *reduced activation steels* comes from the consideration of the induced radioactivity. In the mid-1980s, the international fusion reactor program initiated the development of these steels first in Europe and Japan and later in the United States. The rationale behind developing these materials stems from the easier hands-on maintenance and improved safety of operation requirements that the materials used to build the fusion reactor would not activate when irradiated by neutrons or even if it gets activated may develop only low level of activation and would decay fast. Thus, the program to produce reduced activation steels that would require only shallow burial as opposed to putting them in deep geologic repository was pursued. Researchers found out that replacing or minimizing the amount of molybdenum, niobium, nickel, copper, and nitrogen in the alloy steels would help in developing reduced activation steels. Tungsten, vanadium, and/or tantalum (low activating) have been added to these steels. Table 1.6 shows the nominal compositions of three reduced activation steels. Although the approach has originated in the fusion reactor program, it can be equally applicable to fission reactor systems.

Table 1.6 Nominal compositions of reduced activation steels (in wt%, balance Fe).

Steel	Region	C	Si	Mn	Cr	W	V	Ta	N	B
JLF-1	Japan	0.1	0.08	0.45	9.0	2.0	0.2	0.07	0.05	—
Eurofer	Europe	0.11	0.05	0.5	8.5	1.0	0.25	0.08	0.03	0.005
9Cr–2WVTa	USA	0.10	0.30	0.40	9.0	2.0	0.25	0.07	—	—

1.9.2.3 Radiation Stability

In the subsequent chapters, we will see more detailed accounts of how energetic radiation plays a significant role in modifying the microstructure of the materials involved. Radiation damage under the fast neutron flux involves atomic displacements (i.e., displacement damage) leading to the creation of a host of defects in the material. The effects of radiation can be diverse, including radiation hardening, radiation embrittlement, void swelling, irradiation creep, and so forth, with all having significant effects on the performance of the reactor components. Another interesting effect of radiation is the radiolytic decomposition of coolant (e.g., water molecule is radiolyzed into more active radicals) that may definitely affect the corrosion behavior of the reactor components. Fission fragments also cause damage, but they are mostly limited to the fuel. So, for selecting materials for a nuclear reactor, we must know the concomitant radiation effects on these materials. That is why millions of dollars are spent to wage materials irradiation campaigns in test reactors followed by careful postirradiation examination to ascertain fitness-for-service quality of the materials to be used in nuclear reactors.

1.9.3 Application of Materials Selection Criteria to Reactor Components

Here, we summarize the criteria for materials selection for different nuclear components. Let us take the example of fuel cladding material for the LWR. As noted before, cladding materials are used to encapsulate the fuel and separate it from the coolant. The requirements for fuel cladding material are as follows: (a) low cross section for absorption of thermal neutrons, (b) higher melting point, adequate strength and ductility, (c) adequate thermal conductivity, (d) compatibility with fuel, and (e) corrosion resistance to water. Following the first factor, we have discussed in Section 1.7 how different metals have different cross sections for absorption of thermal neutrons. Although Be, Mg, and Al all have lower cross sections for absorption of thermal neutrons, other nonnuclear factors become the impediment for their use in commercial power reactors. Even though Be has a high melting point (1278 °C), it is scarce, expensive, difficult to fabricate, and toxic. Mg has a low melting point (650 °C), is not strong at higher temperatures, and has poor resistance to hot water corrosion. Al has a low melting point (660 °C) and poor high-temperature strength. Even though an Al-based alloy has been used as fuel cladding materials in reactors like ATR, and in the past a magnesium-based alloy was used in Magnox reactors, their use remains very limited. This leaves zirconium-based materials as the mainstay of fuel cladding materials for LWRs. Zirconium has various favorable features: (a) relatively abundant, (b) not prohibitively expensive, (c) good corrosion resistance, (d) reasonable high-temperature strength, and (e) good fabricability. Some of the properties could be further improved through appropriate alloying. More detailed discussion on the development of zirconium alloys is included in Appendix 1.A at the end of the chapter.

1.9.3.1 Structural/Fuel Cladding Materials

Major requirements	Possible materials
Low neutron absorption	Al, Be, Mg, and Zr
Stability under heat and radiation	Stainless steels
Mechanical strength	Superalloys (Ni-based)
Corrosion resistance	Refractory metals (Mo, Nb, Ti, W, etc.)
Good heat transfer properties	

1.9.3.2 Moderators and Reflectors

Major requirements	Possible materials
Low neutron absorption	Water (H_2O, D_2O)
Large energy loss by neutron per collision	Beryllium (BeO)
High neutron scattering	Graphite (C)

1.9.3.3 Control Materials

Major requirements	Possible materials
High neutron absorption	Boron
Adequate strength	Cadmium
Low mass (for rapid movement)	Hafnium
Corrosion resistance	Hafnium
Stability under heat and radiation	Rare earths (Gadolinium, Gd; Europium, Eu)

1.9.3.4 Coolants

Major requirements	Possible materials
Low neutron absorption	Gases (air, hydrogen, helium, carbon dioxide, and water)
Good heat transfer properties	Water (H_2O and D_2O)
Low pumping power (i.e., low melting point)	Liquid metals (Na, Na—K, Bi)
Stability under heat and radiation	Molten salts (—Cl, —OH, —F)
Low induced radioactivity	Organic liquids
Noncorrosiveness	

(Example:
The world's first nuclear power plant was EBR-1.
It carried a coolant, an alloy of sodium (Na) and potassium (K), called Na—K ("nack").
The following are the coolant characteristics:

- Stays liquid over a wide range of temperatures without boiling away.
- Transfers heat very efficiently taking heat away from the reactor core and keeping it cool.
- Allows neutrons from the reactor core to collide with U-238 in the breeding blanket and produce more fuels.)

1.9.3.5 Shielding Materials

Major requirements	Possible materials
Capacity to slow down neutrons	Light water (H_2O)
Absorption of gamma radiation	Concrete, most control materials, and metals (Fe, Pb, Bi, Ta, W, and Broal – a B and Al alloy)
Absorb neutrons	

1.10 Summary

In this introductory chapter, we first introduced nuclear energy and discussed its significance in the modern civilization. We also discussed some nuclear physics fundamentals such as half-value thickness for neutron beam attenuation, nuclear cross sections, neutron flux and fluence, and other concepts. A detailed overview of different reactors is presented. The material selection criteria for nuclear components are also discussed.

Appendix 1.A

Zirconium-based alloys are commonly used in water reactors for cladding UO_2, while Zircaloy-2 and Zircaloy-4 are used in BWRs and PWRs, respectively. The following are the reasoning and historical development of these cladding materials:

The fuel (UO_2) is inserted in canning tubes that separate the radioactive fuel from the coolant water. The requirements for cladding materials thus are as follows:

- Low cross section for absorption of *thermal* neutrons
- Adequate strength and ductility
- Compatibility with fuel

- Adequate thermal conductivity
- Corrosion resistance to water

In order of increasing cross section for absorption $\left\{\sum_a^{th}\right\}$ of thermal neutrons, the various metals can be classified [normalized to Be] as follows:

Relative to Be	1	7	20	24	122	278	281	322	410	512
	Be	Mg	Zr	Al	Nb	Mo	Fe	Cr	Cu	Ni
Melting point {°C}	1283	650	1845	660	2415	2625	1539	1890	1083	1455

Be: scarce, expensive, difficult to fabricate, and toxic
Mg: not strong at high temperatures and poor resistance to hot water corrosion
Al: low melting point and poor high temperature strength

Zirconium is relatively abundant, is not prohibitively expensive, has good corrosion resistance, has *reasonable* high-temperature strength, good fabricability, and can be further improved by *proper alloying*. Processing of Zr metal from ore requires removal of hafnium [Hf], which is always associated with Zr. Hf has relatively high absorption of thermal neutrons. This Kroll process was relatively more expensive ⇒ Mg treated. The elements used in alloying for increasing strength are O, Sn, Fe, and [Cr, Ni] and for improving corrosion are Cr, Ni, and Fe.

Thus, the Zircaloys were developed (mainly from the US Navy in 1950s). For a nice description of the history of Zry development, refer to the following:

- Krishnan, R. and Asundi, M.K. (1981) Zirconium alloys in nuclear technology, in *Alloy Design* (eds. S. Ranganathan *et al.*), Indian Academy of Sciences.
- Rickover, H.G. (1975) *History of Development of Zirconium Alloys for Use in Nuclear Power Reactors*, US ERDA, NR&D.

Element [w/o]	Sn	Fe	Cr	Ni	O	Zr
Zircaloy-2	1.5	0.12	0.10	0.05	0.01	Balance – BWRs
Zircaloy-4	1.5	0.21	0.10	–*	0.01	Balance – PWRs

*Ni enhances hydrogen pickup and thus was removed for PWR applications. To compensate for the corrosion and strength improvements realized by adding Ni, it was replaced by Fe.

Element [w/o]	Sn	Fe	Zr	
Zircaloy-1	2.5	–	Balance	Break-away transition *not improved*
Zircaloy-3	0.25	0.25	Balance	Poor mechanical strength

• Recent Developments •

Barrier Cladding _ Zircaloy-2 with Zr liner (ID) for PCI resistance
Zirlo@ alloys _ Zircaloy-4 + 0.5 to 1% Nb _ good long-term corrosion
Duplex alloy _ low Sn on the surface (OD) but still in ASTM spec

• Zr + Fe liner in lieu of Zr • Zry-2/Zr/Zry-2 (Tri-clad) • M5 (Zr-1Nb

Crystal bar (Kroll) Zr – free of Hf ⇒ neutron economy

Sn, Ta, and Nb (in decreasing order of effectiveness) circumvent damaging effect of impurities such as nitrogen and improve corrosion resistance

_ Sn selected since neutron economy is little affected and

2.5%Sn _ improved corrosion, strength, and fabricability

⇓

Zircaloy-1 : Zr 2.5%Sn

⇓

Breakway transition in corrosion (wt gain versus time) was not improved

⇓

Zircaloy-2 was developed as an accidental finding at Bettis labs–corrosion improvement noted by contamination from SS

⇒ new alloy developed

with Fe and Cr similar to impurity levels in Zr + 0.05% Ni

Sn reduced to 1.5%–enough to counteract the nitrogen effect

⇓

Zircaloy-2 : Zr 1.5%Sn 0.15%Fe 0.05%Ni 0.1Cr *(BWRs)*

• has as good a strength as Zry-1 but improved corrosion •

⇓

Sn is known to be bad for long-time corrosion under PWR conditions ∅

reduced Sn to 0.25% and simultaneously increased Fe to 0.25%

lead to the most corrosion resistant alloy in Zircaloy family

⇓

Zircaloy-3 : Zr 0.25%Sn 0.25%Fe

• the mechanical strength was not adequate and thus Zry-3 was abandoned •

⇓

Hydrogen effects on mechanical properties were just emerging and zirconium hydrides resulted in reduced impact energy ∅

again, accidentally the effect of Ni on hydrogen absorption was noted (during work on eutectic diffusion bonded plates) ∅

Ni-free Zircaloy-2 was developed but with poor corrosion resistance

⇓

increased Fe to 0.24% with no Ni → almost as good steam corrosion resistance as Zry-2 but with reduced (by 50%) hydrogen absorption

Zircaloy-4 : Zr 1.5%Sn 0.24%Fe 0.1%Cr *(PWRs)*

Recent Trends

Barrier cladding → Zry-2 with Zr liner for PCI resistance ∅ *Tri-Clad*

Zirlo@ alloys → Zry + 0.5 to 1% Nb → good for long-term corrosion

Duplex alloy → low Sn on the surface (but still in ASTM spec) versus bulk

Problems

1.1 a) What is the percentage of U^{235} in naturally occurring uranium and what is the rest made of?

b) A nuclear fission reaction of an U^{235} atom caused by a neutron produces one barium atom, one Krypton atom, and three more neutrons. Evaluate approximately how much energy is liberated by this reaction.

Approximately how much percentage of energy is carried by the fission fragments (no calculation necessary for the last part of the question)?

1.2 What is the difference between fissile and fertile isotopes? Give two examples of each. What is the role of fertile isotopes in a breeder reactor?

1.3 Define a nuclear reactor? What is the basic difference between an atomic bomb and a power-producing reactor?

1.4 What are the prime differences between LWR and CANDU reactors (comment mostly on materials aspects)?

1.5 Describe the importance of control materials with respect to reactor safety and control. What are the primary requirements for a control material? Give at least four examples of control materials.

1.6 Categorize neutrons based on their kinetic energy. What is the major difference between a thermal reactor and a fast reactor?

1.7 a) Zirconium and hafnium both have crystal structures (HCP) in the general operating regimes of LWRs. Naturally occurring Zr always has some Hf (1–3 wt%) in it. Hf-containing Zr alloys are very common in chemical industries but not in nuclear industries. Why?

b) What is the main application of Zr alloys in LWRs? What are the various functions of this reactor component? What are the reasons that make Zr alloys suitable for such use?

1.8 What are the two main zirconium alloys used in light water reactors? Give their compositions. Name two recently developed zirconium alloys with their compositions.

1.9 What is neutron economy? What significance does it have? How much influence does it exert in the selection of materials used in nuclear reactors?

1.10 a) Define neutron flux and neutron fluence. What are their units?

b) Define neutron cross section? Briefly comment on the importance of neutron cross section from a reactor perspective.

c) Neutrons of 10 keV energy are incident on a light water barrier. The neutron cross section for hydrogen (protium) at 10 keV is about 20 b and that of oxygen is only 3.7 b. Determine the half-value thickness of neutron attenuation for the water barrier (assume that neutron interaction with oxygen in water molecule is negligible). Find out the half-thickness value for 1 MeV neutrons traveling through the water barrier (neutron cross section for protium is 4.1 b for 1 MeV neutrons). Comment on the significance of the results.

Additional Reading Materials

1 L'Annunziata, M.F. (1998) *Handbook of Radioactivity Analysis*, Academic Press, New York.

2 Murray, R.L. (1955) *Introduction to Nuclear Engineering*, Prentice Hall, New York.

3 Smith, C.O. (1967) *Nuclear Reactor Materials*, Addison-Wesley, Massachusetts.

4 Ma, B. (1983) *Nuclear Reactor Materials and Applications*, Van Nostrand Reinhold Company, New York.

5 Klueh, R.L. (2005) Elevated temperature ferritic and martensitic steels and their application to future nuclear reactors. *International Materials Review*, **50** (5), 287–310.
6 Marshall, F.M. (2009) *The Advanced Test Reactor Capabilities Overview*, ATR User Facilities Workshop, Idaho Falls, ID.
7 Hinds, D. and Maslak, C. (January 2006) *Next Generation Nuclear Energy: The ESBWR*, Nuclear News, pp. 35–40.
8 Charit, I. and Murty, K.L. (2008) Structural materials for Gen-IV nuclear reactors: challenges and opportunities. *Journal of Nuclear Materials*, **383**, 189–195.

2
Fundamental Nature of Materials

"FORTUNATELY crystals are seldom, if ever, perfect."

—*Anonymous*

2.1
Crystal Structure

Engineering materials (metals and alloys and ceramics) used in nuclear applications are almost always crystalline. That is why the understanding of crystal structure basics is very important in the context of nuclear materials. As a matter of fact, an overwhelming majority of materials crystallize when they solidify, that is, atoms get arranged in a periodic three-dimensional pattern leading to a long-range order and symmetry, while minimizing the overall free energy of the solid. However, before we start discussing the details of a crystal structure, let us assess broadly what are the different length scales in a material system. It has long been established that only a single length scale cannot adequately describe all the behaviors of a material, and that is why multiscale methodologies (Figure 2.1) are being increasingly implemented in modeling various materials systems, including nuclear materials. For instance, subatomic scale involves the interaction between the subatomic particles (neutrons, protons, and electrons). On the other hand, a single crystal entails an ensemble of several atoms (the smallest unit being called a *unit cell*), while several such single crystals can create a polycrystalline material (microscale), and, finally, the macrostructure (basically the components, machines, etc.) can be seen by our bare eyes at the top of the length scale. Hence, to describe the behavior of a material, one needs to rely on several length scales, not just one.

An underlying theme of the materials science and engineering (MSE) field is to understand the interrelationships of *processing–structure–property*, which gives one greater opportunity for predicting to a reasonable degree the materials performance under real service conditions. This is exemplified by the materials science tetrahedron, as depicted in Figure 2.2.

This theme is equally applicable to the nuclear materials. For example, materials scientists and engineers study microstructural features (grain size, type of second phases and their relative proportions, grain boundary character distribution to name a few) to elucidate the behavior of a material. These are structural features

An Introduction to Nuclear Materials: Fundamentals and Applications, First Edition.
K. Linga Murty and Indrajit Charit.
© 2013 Wiley-VCH Verlag GmbH & Co. KGaA. Published 2013 by Wiley-VCH Verlag GmbH & Co. KGaA.

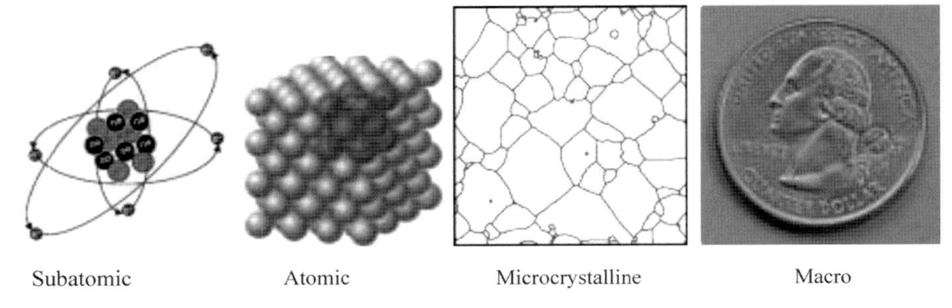

Figure 2.1 Different length scales present in a material.

that are influenced by the nature of the processing techniques (casting, rolling, forging, powder metallurgy, and so forth) employed, leading to changes in properties. This understanding will be very helpful as we wade through the subsequent chapters.

A *lattice* is an array of points in three dimensions such that each point has identical surroundings. When such lattice point is assigned one or more atoms/ions (i.e., basis), a crystal is formed. In this chapter, we present a simple treatment of the crystal structure and relate it to its importance with respect to nuclear materials. There are 7 basic crystal systems and a total of 14 unique crystal structures (Bravais lattices) that can be found in most elemental solids. These are based on the crystal symmetry and the arrangement of atoms as described in the following sections.

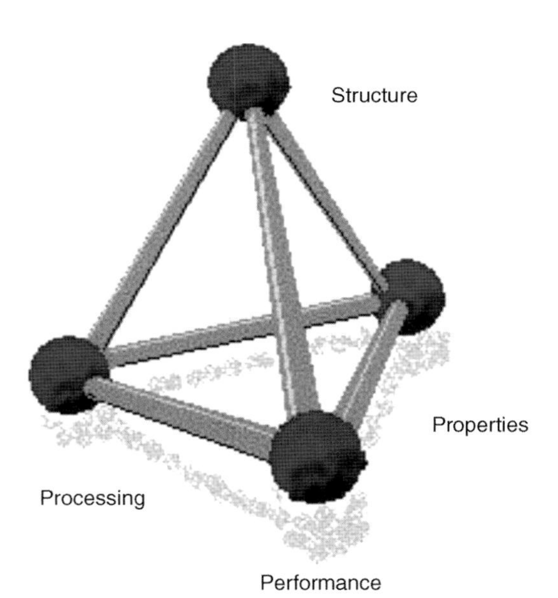

Figure 2.2 Materials science and engineering tetrahedron.

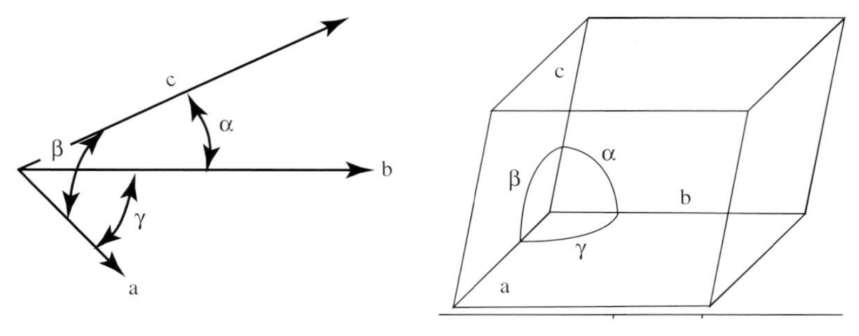

Figure 2.3 Unit cell defining the axes *a*, *b*, *c* and angles α, β, γ.

2.1.1
Unit Cell

A *unit cell* is the smallest building block of a crystal, which when repeated in translation (i.e., with no rotation) in three-dimension can create a single crystal. Therefore, a single crystal or a "grain" in a polycrystalline material would contain many of these unit cells. A general unit cell can be created based on three lattice translation vectors (a, b, and c) on three orthogonal axes and interaxial angles (α, β, and γ), which are also known as lattice parameters or lattice constants. Figure 2.3 illustrates the definitions of the lattice parameters and the angles. There are *seven* basic crystal systems. They are summarized in Table 2.1 with their lattice parameters. The least symmetric crystal structure is the triclinic and the most symmetric one is cubic.

There are a total of 14 Bravais lattices (space lattice) based on the atom positions in the unit cells of the basic crystal systems (Figure 2.4). The scheme consists of three cubic systems (simple, body-centered, and face-centered), four orthorhombic systems (simple, base-centered, body-centered, and face-centered), two tetragonal (simple and body-centered) and two monoclinic systems (simple and base-centered), and one each from triclinic, rhombohedral, and hexagonal systems. In each system, simple system is also known as primitive unit cells. A primitive unit cell contains 1 atom per unit cell. Note that during the discussion of crystal structure, we would not consider any crystal imperfection/defects, and we will introduce the topic in Chapter 2.3.)

Table 2.1 Seven basic crystal systems.

Crystal system	**Axial lengths**	**Axial angles**
Cubic	$a=b=c$	$\alpha=\beta=\gamma=90^\circ$
Rhombohedral	$a=b=c$	$\alpha=\beta=\gamma\neq 90^\circ$
Tetragonal	$a=b\neq c$	$\alpha=\beta=\gamma=90^\circ$
Hexagonal	$a=b\neq c$	$\alpha=\beta=90^\circ,\ \gamma=120^\circ$
Orthorhombic	$a\neq b\neq c$	$\alpha=\beta=\gamma=90^\circ$
Monoclinic	$a\neq b\neq c$	$\alpha=\beta=90^\circ\neq\gamma$
Triclinic	$a\neq b\neq c$	$\alpha\neq\beta\neq\gamma\neq 90^\circ$

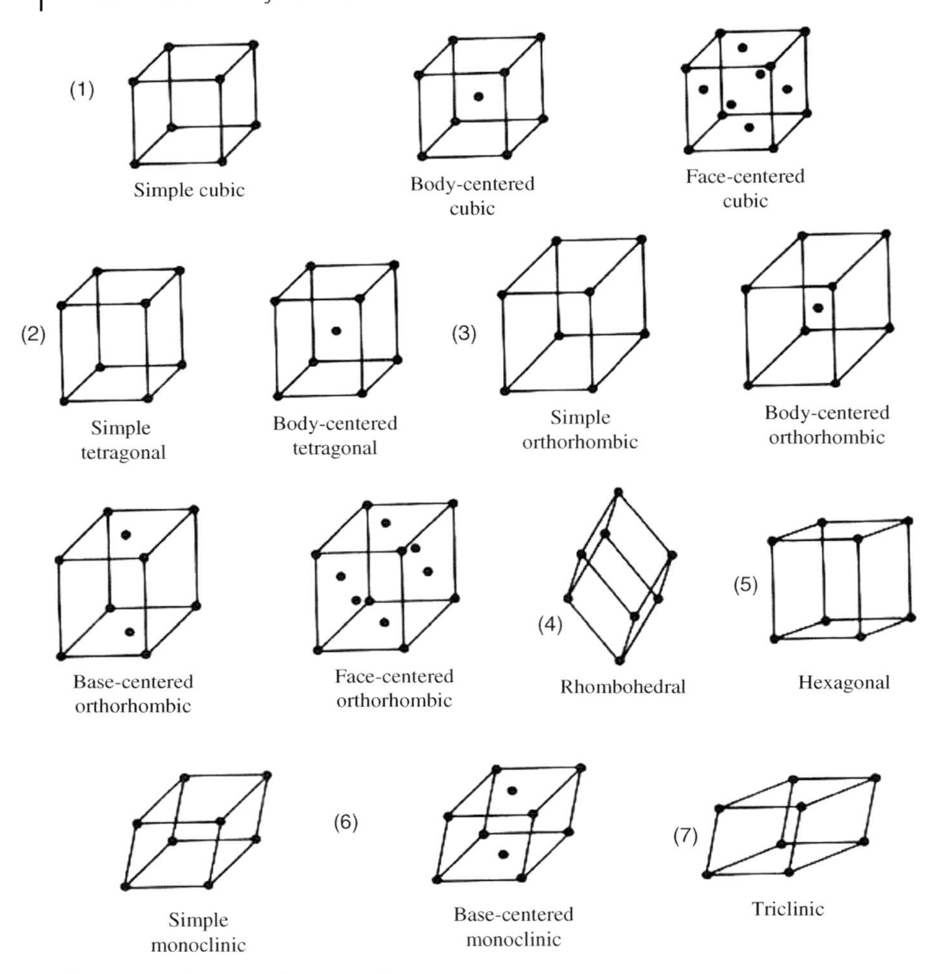

Figure 2.4 Schematics of 14 Bravais lattices.

What is Euler's Rule with Regard to Regular 3D Solids?

$$V - S + F = 2,$$

where V is the number of vertices or corners, S is the number of sides/edges, and F is the number of faces.

For example, for a cube, $V = 8$, $S = 12$, and $F = 6$.

Therefore, $V - S + F = 8 - 12 + 6 = 2$.

Verify the rule by yourself for a regular hexagonal prism. You will be amazed how good it works.

2.1.2
Crystal Structures in Metals

A majority of elements (about three-fourths) in the Periodic Table are metals, and of all the metals, more than two-thirds of them possess relatively simple cubic or hexagonal crystal structures. Metal atoms are bonded by a chemical bond known as the *metallic bond*. The metallic bonds are nondirectional and there is no constraint regarding bond angles (which is not true for covalently bonded materials). Even though a simple cubic system is relatively unknown in metals (or even among elements), α-polonium exhibits a simple cubic crystal structure. The most common crystal structures in metals are discussed in detail in the following sections.

2.1.2.1 Body-Centered Cubic (BCC) Crystal Structure

In the BCC unit cell, an atom is located at the body center of a cubic unit cell in addition to the eight atoms on the cube corners. The coordination number of this type of crystal structure is 8, that is, each atom is surrounded by eight equidistant neighboring atoms. Figure 2.5a shows a schematic of a BCC unit cell. The effective number of atoms in such a unit cell is $(8 \times 1/8) + 1 = 2$, because each corner atom is shared by eight unit cells and the body center atom is fully inside the unit cell to be counted as one. We note that each atom is surrounded by eight nearest neighbors and this factor is known as *coordination number* defined as the number of equivalent nearest neighbors. The *packing efficiency* or *atomic packing factor* is

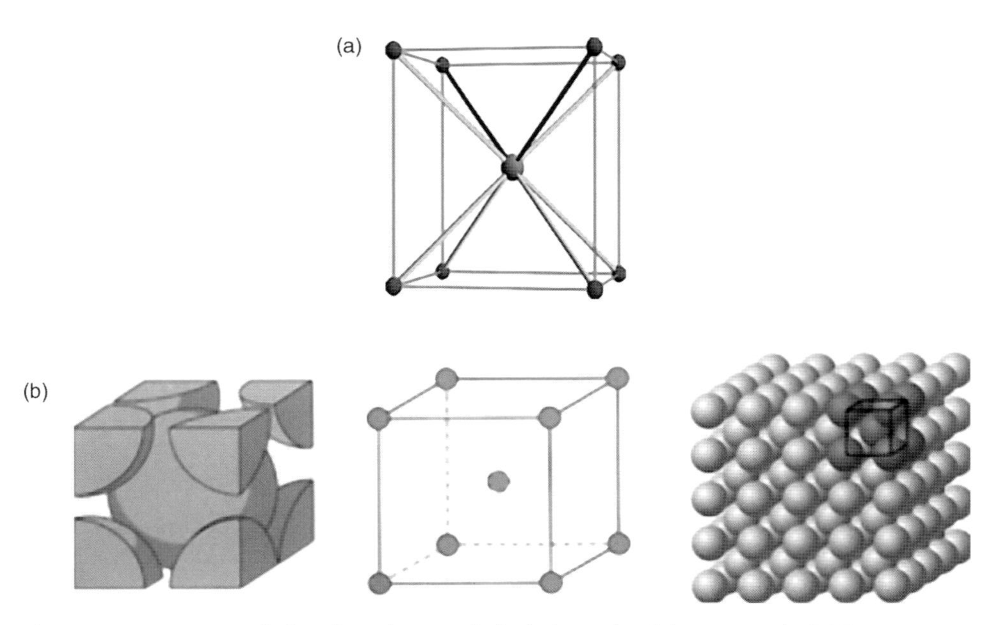

Figure 2.5 (a) BCC unit cell. (b) Relation between the body diagonal and the atomic radii (BCC).

defined as the volume of the space occupied by atoms in a unit cell. Below we present a detailed procedure for calculating the atomic packing factor of a BCC unit cell. Note that the same method can be applied to calculate the packing factors of virtually any kind of known unit cells.

The packing efficiency PE of atoms in the BCC unit cell is derived as follows:

$$\begin{aligned}\mathrm{PE} &= \frac{\text{Volume of atoms in the unit cell}}{\text{Volume of the unit cell}} = \frac{2 \times (4/3)\pi r^3}{a^3} \\ &= \frac{2 \times (4/3)\pi\left((\sqrt{3}/4)a\right)^3}{a^3} = 0.68.\end{aligned}$$

Many metals such as α-Fe, β-Zr, Mo, Ta, Na, K, Cr, and so on have a BCC crystal structure. This structure is relatively loosely packed (only 68% of the crystal volume is occupied by atoms) compared to the best packing that can be achieved using hard incompressible solids spheres. Two corner atoms touch the body center atom and create the body diagonal that is closest-packed direction in the BCC unit cell. This also gives us an opportunity to calculate the relation between the atom radius (r) and the lattice constant (a) using the simple Pythagorean rule. The body diagonal length is given by $(r + 2r + r) = 4r$. The body diagonal makes the hypotenuse of the triangle that a base of a face diagonal ($\sqrt{2}a$) and a cube edge (a), as shown in Figure 2.5b.

Hence, $(4r)^2 = (\sqrt{2}a)^2 + (a)^2 \quad \text{or} \quad (4r)^2 = 3a^2 = (\sqrt{3}a)^2.$

$$\text{Or, } 4r = \sqrt{3}a. \tag{2.1}$$

Example 2.1

Calculate the theoretical density of α-Fe at room temperature.

Solution

The density of any crystal can be calculated from the first principles.

$$\varrho = \frac{\text{Mass of atoms in the unit cell}}{\text{Volume of the unit cell}}.$$

First, we need to know the crystal structure (BCC) and lattice parameter ($a = 0.287$ nm) of α-Fe.

Mass of an Fe atom = 55.85 atomic mass unit (or amu) from the definition of atomic weight.

$$1\ \text{amu} = 1.66 \times 10^{-27}\ \text{kg}.$$

Hence,

$$\varrho\text{Fe} = \frac{2 \times 55.85 \times 1.66 \times 10^{-27}\ \text{kg}}{(0.287 \times 10^{-10}\ \text{m})^3} = 7840\ \text{kg m}^{-3} = 7.84\ \text{g cm}^{-3}.$$

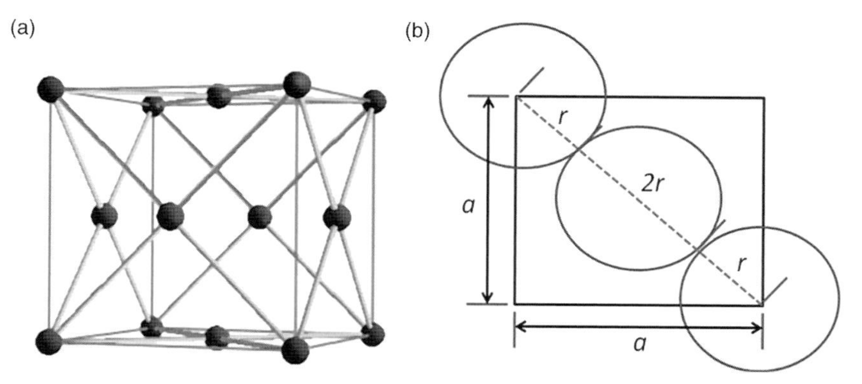

Figure 2.6 Relation between the body diagonal and the atomic radii (FCC).

2.1.2.2 **Face-Centered Cubic (FCC) Crystal Structure**

The largest possible dense packing in a cubic system is achieved in the FCC crystal system. That is why it is sometimes referred to as close-packed cubic (CPC) crystal structure. Each cube face (i.e., six of them in a cube) has one atom at the center of the cube face in addition to eight corner atoms (Figure 2.6a). The effective number of atoms in an FCC unit cell is then given by $(6 \times 1/2) + (8 \times 1/8) = 4$, because each face-centered atom is shared by two unit cells and each corner atom is shared by eight unit cells. This structure has a coordination number of 12. The packing factor can be calculated in much the same way as the BCC crystal structure, and is found to be about 0.74. Following a similar method, the relationship between the atom radius (r) and the lattice constant (a) can be found for an FCC unit cell. Figure 2.6b shows the cube face of an FCC unit cell. The face diagonal is $(r + 2r + r) = 4r$, and now applying the Pythagorean rule, we also know that the face diagonal is $\sqrt{2}a$ when the cube edge is a so that $4r = \sqrt{2}a$.

Metals such as aluminum, γ-Fe, gold, silver, platinum, lead, nickel, and many others have FCC crystal structures. The theoretical density of FCC metals can also be derived from the first principles if the lattice constant and the atomic weight of the metal are known as shown earlier for BCC in Example 2.1.

2.1.2.3 **Hexagonal Close-Packed (HCP) Crystal Structure**

One-third of the hexagonal unit cell shown in Figure 2.2 can be considered as the *primitive* hexagonal unit cell since all crystal structures with a hexagonal symmetry can be created out of it, such that it leads to the creation of a HCP unit cell as well as the graphite crystal structure. A HCP unit cell (Figure 2.7) is quite different from the cubic ones that we discussed until now. Nonetheless, an ideal HCP crystal structure is as close-packed as the FCC one (packing factor of 0.74). This crystal structure is created by alternate layers of atoms on top of each other (see Section 2.1.3). This crystal structure has two characteristic lengths, a, in the base of the unit cell and the height of the unit cell along the c-axis (Figure 2.7). In fact, there are three separate coplanar axes with lattice vectors of a_1, a_2, and a_3 at 120° apart even though they are all equal in length. The upper, middle, and lower

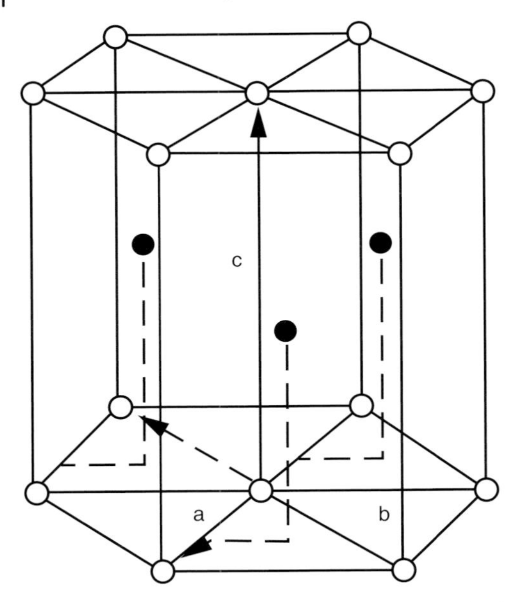

Figure 2.7 A schematic of a HCP unit cell (from Olander).

horizontal planes are called the *basal* planes. Planes (six vertical faces in a HCP unit cell) perpendicular to the basal planes are known as the *prismatic* planes. Planes other than the basal and prismatic planes are known as the *pyramidal* planes. This crystal structure has a coordination number of 12 (note the same as for FCC).

With a close examination of the HCP unit cell, the effective number of atoms can be easily calculated. Two atoms present on the upper and lower bases are shared by 2 HCP unit cells, whereas the corner atoms (12 of them) are shared by 6 unit cells, and 3 atoms reside entirely inside the unit cell. So, the count becomes $(2 \times 1/2) + (12 \times 1/6) + (1 \times 3) = 6$.

The relation between the lattice constant a and the atom radius (r) is simple, $a = 2r$. However, in order to get a relation between the other lattice constant c and r, one needs to know little further. The HCP unit cell is inherently anisotropic. This implies that certain fundamental characteristics along the a- and c-axes are different. One measure of this anisotropy is given by the c/a ratio. An ideal HCP unit cell has a c/a ratio of 1.633 (can be shown from the geometry) based on a hard sphere model. But in reality, HCP metals are hardly ideal in close packing, and thus the packing factors also change from metals to metals although slightly. Table 2.2 lists a number of HCP metals with different c/a ratios. The different c/a ratios are thought to be a result of the specific electronic structures of the atoms. Earlier it was thought that the c/a ratios influence the primary deformation mode, and that is why the corresponding dominant deformation modes on the right-hand side column of Table 2.2 are also included. However, later this conjecture has been proved otherwise noting the glaring exception of Be (low c/a ratio, yet the primary deformation mode is basal). This issue will be further discussed in Chapter 4 when we learn more about the slip deformation mode.

Table 2.2 A list of metals with c/a ratios.

HCP metals	c/a	Primary slip system
Beryllium (Be)	1.568	Basal
Yttrium (Y)	1.571	Prismatic
Hafnium (Hf)	1.581	Prismatic
Ruthenium (Ru)	1.582	Prismatic
Alpha-titanium (α-Ti)	1.588	Prismatic
Alpha-zirconium (α-Zr)	1.593	Prismatic
Rhenium (Re)	1.615	Basal/prismatic
Cobalt (Co)	1.624	Basal
Magnesium (Mg)	1.624	Basal
Zinc (Zn)	1.856	Basal
Cadmium (Cd)	1.886	Basal

Taken from Ref. [1].

Example 2.2

Beryllium (Be) is used in LWRs as reflectors. Be has a HCP crystal structure at temperatures $<1250\,^{\circ}$C. Be has the lattice constants – $a = 0.22858$ nm and $c = 0.35\ 842$ nm (i.e., c/a ratio of 1.568) during normal conditions. The atomic radius of a Be atom is 0.1143 nm. What is the atomic packing efficiency of the unit cell? Discuss the significance of the computed result.

Solution

We know

$$\mathrm{PE} = \frac{\text{Volume of atoms in the unit cell}}{\text{Volume of the unit cell}}.$$

Beryllium has a HCP crystal structure. The volume of a hexagonal prism (unit cell) is given by $V_{\mathrm{HCP}} = (3\sqrt{3}a^2/2)c$, where a is the length of the prism base edge and c is the height of the hexagonal prism.

As $a = 0.22858$ nm and $c = 0.35842$ nm, $V_{\mathrm{HCP}} = 0.04\,865\ \mathrm{nm}^3$ (this is the input value in the denominator of the above expression for the PE).

Now we need to find out how much of the volume of the hexagonal prism is occupied by the atoms. The effective number of atoms in a HCP unit cell is 6. The atomic radius of each Be atom is 0.1143 nm. Therefore, the total volume of the atoms in the unit cell is $6 \times (4/3)\pi[\![(0.1143\ \mathrm{nm})]\!]^3 = 0.03751\ \mathrm{nm}^3$.

Therefore, $\mathrm{PE} = \dfrac{0.03751\ \mathrm{nm}^3}{0.04865\ \mathrm{nm}^3} = 0.77$. Hence, the atoms must be spheroidal rather than spherical if they are assumed to be in contact.

The packing efficiency of a HCP unit cell is 0.74 only when the ideal c/a ratio is maintained. However, in reality, it is not and the packing efficiency of the HCP metals will always be little off from the ideal value. In the case of Be, the packing efficiency is slightly higher than the ideal.

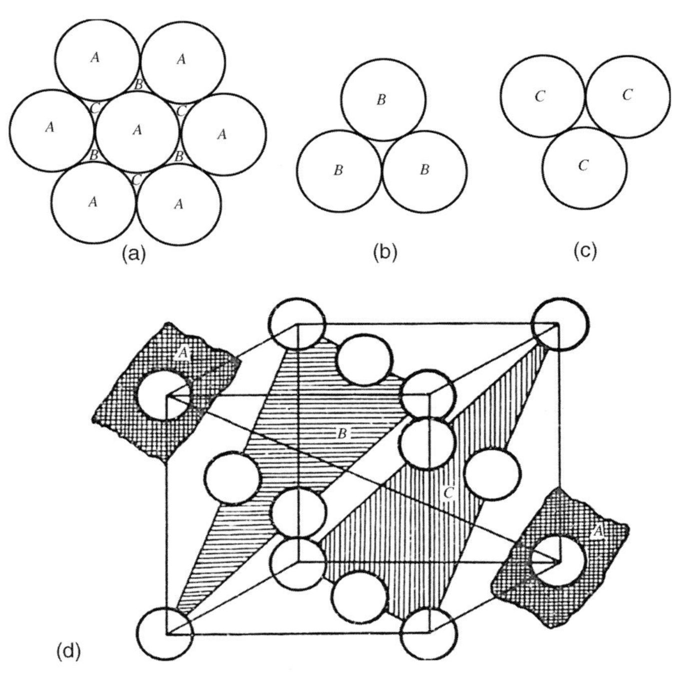

Figure 2.8 Stacking of close-packed planes resulting in an FCC structure. (From V. Raghavan, Physical Metallurgy: Principles and Practices, Prentice-Hall of India Limited)

2.1.3
Close Packing Geometry

So far, we have noted FCC and HCP metals to be close-packed structures with the highest atomic packing factor (0.74), but we have not discussed how a close-packed structure results. Both FCC and HCP crystals can be constructed by stacking close-packed layers of atoms on top of one another. There are two ways the stacking can be achieved leading to the creation of two different stacking sequences and thus two different crystal structures. Figure 2.8a shows a close-packed plane A where six equal-sized spheres surround a central atom. To create a close-packed three-dimensional structure, more spheres need to be placed on top of the A layer to fill up the triangular cavities created by the first layer. There are two sets of cavities on the A layer, B and C positions, as shown in the Figure 2.8a. A second plane of atoms can be placed on top of either B or C position. If one assumes that the second atom layer is over B, there are two ways the third close-packed layer can be stacked either vertically above C positions or directly above A positions. This scheme leads to the possibility of two types of close-packed packing. An FCC crystal shows a stacking sequence of ABCABCABC ... (Figure 2.8d). On the other hand, a HCP crystal shows a stacking sequence of ABABAB ... as shown in Figure 2.9. The lower and upper basal plane layers constitute the A layer and middle layer forms the B layer.

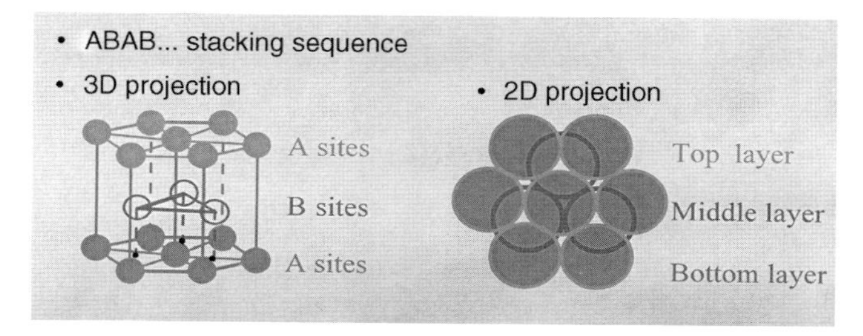

Figure 2.9 Stacking of close-packed planes resulting in a HCP structure.

Due to this kind of stacking, a set of planes known as close-packed planes in FCC and HCP crystals are the densest (i.e., the planar atom density is the greatest). We will discuss this further in a later section.

2.1.4 Polymorphism

Many metals, nonmetals, and minerals exhibit an interesting feature called *polymorphism* (poly means "many" and morphism means "structure"). Polymorphism (sometimes called allotropism) is known as the ability of a material to be present in more than one type of crystal structure as determined by temperature and/or pressure. Many phase transformations are based on this unique feature of the materials. As a general trend, more close-packed crystal structures are favored at lower temperatures, whereas open structure such a BCC crystal structure is most favored at higher temperatures. So, many metals show polymorphic transformation from the HCP to BCC crystal structure as the temperature increases and these phases are commonly referred to by Greek alphabets (α, β, γ, etc.). For example, zirconium (Zr) assumes a HCP structure (α-Zr) at $<865\,^\circ$C, but becomes BCC (β-Zr) above that temperature. Similarly, hafnium (Hf) exhibits HCP structure below 1950 °C, but assumes BCC crystal structure above 1950 °C until its melting point (2233 °C). The example of iron (Fe) is interesting and bit of an exception. α-Fe (BCC) transforms to γ-Fe (FCC) at about 912 °C, and then back to a BCC allotrope known as δ-Fe (with slightly larger lattice constant than that of α-Fe) above about 1394 °C. Thus, with the example of α-Fe and δ-Fe, polymorphism may not necessarily mean the presence of entirely different crystal structures, but the definition needs to be more related to the difference in crystal parameters such as lattice constants. There is also a HCP crystal structure (ε-Fe) that is formed only under very high pressures. Polymorphic transformations of iron as a function of temperature make the heat treatment of steels possible leading to multitude of resulting microstructures and properties. In general, no FCC or BCC metal transforms to a HCP phase without an exception. Nuclear fuels like uranium and plutonium show multiple crystal structures at different temperature regimes as a result of polymorphism, and will be mentioned in more detail in Chapter 7.

2.1.5
Miller Indices for Denoting Crystallographic Planes and Directions

The method devised by the British mineralogist, William H. Miller, in 1839 is still used to denote crystallographic planes and directions. That is why the method named after him uses the *Miller indices* labeling technique. Besides closed-packed planes, there could be a number of crystallographic planes of interest. Their orientation, arrangement, and atom density could be different. It may be cumbersome to name a crystallographic plane as "cube face plane," "octahedral plane," and so on. The same may be applicable to the issue of defining crystallographic directions. One would soon run out of names just to refer to the directions as "cube edge," "face diagonal," "body diagonal," and so on. This type of nomenclature also lacks adequate specificity to be seriously considered. That is why a more convenient and systematic technique, such as Miller indices, is used to denote crystallographic planes and directions. As a matter of fact, each point on a crystal can be reached by the translation vector composed of the sum of the multiples of the crystal lattice vectors.

For labeling a *plane* in a crystal with Miller indices, the following general procedure needs to be followed:

1) Select an origin at a lattice point that is not on the crystallographic plane to be indexed.
2) Fix the three orthogonal axes (a, b, and c or x, y, and z) from the selected origin.
3) Find the intercepts (in multiples of the unit lattice vector) that the plane makes on the three coordinate axes.
4) Take reciprocals of these multiples.
5) Convert the fractions (if any) to a set of integers and reduce the integers by dividing by a common integer factor. However, care should be exercised so that the atom configuration of the original plane remains the same.
6) Enclose the final numbers in parentheses such as (hkl), which is the Miller index of the particular plane. A family of equivalent planes is given by numbers enclosed in curly brackets $\{hkl\}$. For example, the faces of a cube are given by $\{100\}$, whereas an individual plane is denoted by (100), (010), and so on. A negative intercept is denoted by a "bar" on the index such as $(\bar{1}00)$.

Figure 2.10a–d shows four planes in a cubic unit cell. In Figure 2.10a, the hatched plane makes an intercept on the x-axis for a one lattice vector (take it as a unity 1 in place of lattice constant a), however it does not intersect the y- and z-axes making the intercept of ∞. Therefore, the reciprocals of the intercepts become $1/1$, $1/\infty$, and $1/\infty$. Hence, the Miller index of the plane is (100). Similarly, for the plane shown in Figure 2.10b, the intercepts are 1, 1, and ∞ on the x-, y-, and z-axes respectively, and therefore, the reciprocals to the intercepts become 1, 1, and 0. Thus, the Miller index of the plane becomes (110). In Figure 2.10c, the plane with hatch marks portends positive intercepts of 1 each on all the three axes. When the reciprocals of the intercepts are taken, they remain the same, and thus the Miller index of the plane becomes (111). In Figure 2.10d, the plane denoted by the hatch

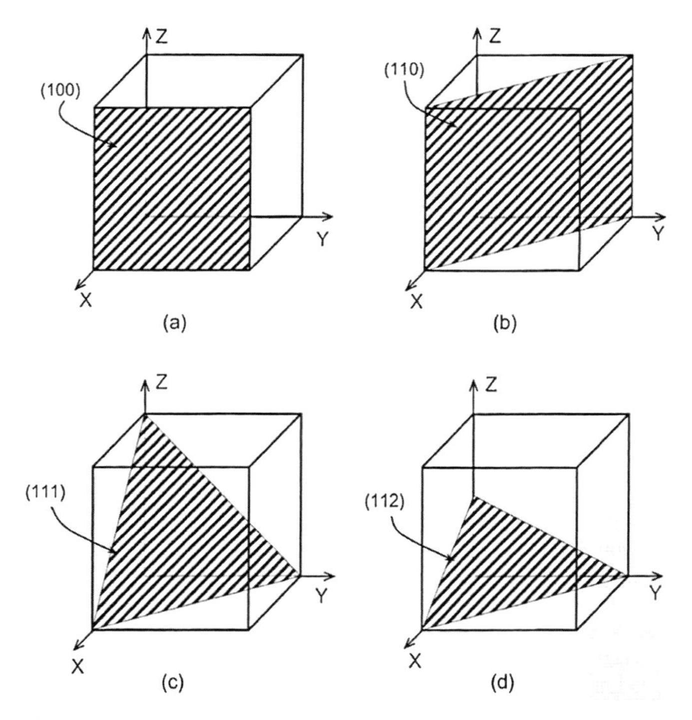

Figure 2.10 Examples of planes in a cubic crystal lattice.

marks cuts the x-, y-, and z-axes creating positive intercepts of 1, 1, and 1/2, respectively. We take the reciprocals of these intercepts to obtain 1, 1, and 2. Thus, the Miller index of the plane would be (112). Even though we have not given a direct example where one needs to reduce the Miller index to lower integers, the issue merits some discussion. This is related to the step 5 in the above-mentioned procedure. The operation needs to be carried out depending on the specific case. For example, (220) and (110) are equivalent planes in an FCC crystal, so be treated so. On the other hand, even though (200) and (100) planes in FCC and BCC crystals are equivalent, that is not the case in a simple cubic crystal. Hence, in these types of cases, individual attention needs to be paid to the atom configurations of the planes in question to ascertain whether further reduction is going to be permitted or not.

Another important issue for denoting crystallographic planes is the situation where negative indices become essential. This can be easily illustrated referring to Figure 2.10a. Let us say one wishes to index the leftmost cube face in Figure 2.10a. As the plane passes through the previous origin, it needs to be shifted to some other position, say shifted to the right side by one lattice spacing. In this case, the intercepts the plane is making to x-, y-, and z-axes can be interpreted as ∞, -1, and ∞, respectively. So by taking reciprocals, we get 0, -1, and 0. This means the Miller

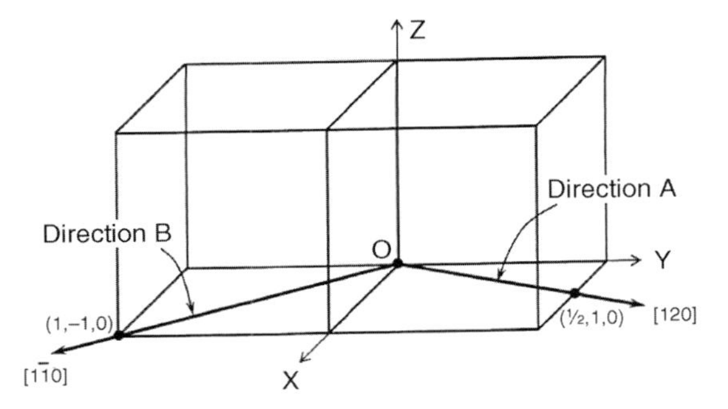

Figure 2.11 Two examples of directions in a cubic crystal structure.

index of the plane can be represented as $(0\bar{1}0)$ where $\bar{1}$ is written in place of −1, and called "bar 1."

So we understand pretty much how to label a plane using Miller indices. We now turn our attention to labeling a crystallographic direction with a Miller index. Let us label the "direction A" in the cubic unit cell on the right-hand side in Figure 2.11. For denoting directions, the direction must pass through the origin (*O*). In a situation, where the original direction does not go through the origin, one needs to a draw a parallel line that passes through the origin. This parallel line will have the same Miller index as the original one. For direction A, the components along the *x*-, *y*-, and *z*-axes need to be resolved. In this case, direction A can be resolved half the lattice constant along *X*-axis, a unit lattice vector distance along *Y*-axis, and no component (i.e., 0) along the *Z*-axis. Now these components need to be converted to the smallest whole integers and then put into the square brackets to obtain the Miller index of the direction. Hence, the Miller index of direction A would be [1/2 1 0], that is, [120]. General notation used for a crystallographic direction is [*uvw*]. The Miller index of "direction B" on the left-hand side unit cell can be found out in the same way. Note that the component of "direction B" can be resolved into a positive unit distance along the *X*-axis (i.e., 1), a lattice vector in the negative *Y*-axis (i.e., −1), and no component (i.e., 0) along the *Z*-axis. Thus, the Miller index of "direction B" would be $[1\bar{1}0]$. Note that the face diagonal of the base face of the cube on the right-hand side unit cell is parallel to "direction B" and will have the same Miller index [110]. A class of equivalent directions is called a "family of directions" and denoted by ⟨*uvw*⟩. In the case of face diagonal of the cubic unit cell, all face diagonals belong to a single family of directions, ⟨110⟩, meaning directions [110], [101], [011], $[1\bar{1}0]$, and so on.

An alternative procedure is quite commonly used where we first determine the coordinates of two points on the direction; the points are specified with respect to a defined set of orthogonal coordinate system (*x*, *y*, *z* or *a*, *b*, *c*), as shown in Figure 2.12a. Next, subtract the coordinates of the "tail" from the "head" and clear the fractions and/or reduce to lowest integers. Enclose the numbers in brackets [− − −] with the negative sign by "bar" above the number. Three examples are

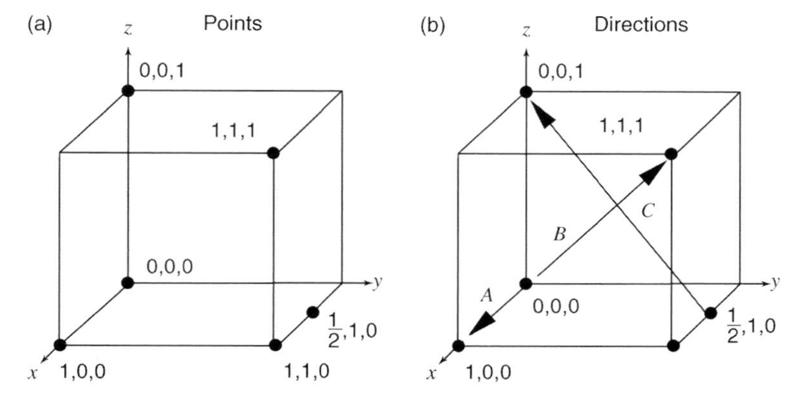

Figure 2.12 Designation of points in a cubic crystal (a) and some directions (b).

illustrated in Figure 2.12b. The Miller indices of the directions A, B, and C in Figure 2.12b are thus [100], [111], and $[\bar{1}\bar{2}2]$, respectively.

We note that the definitions of the Miller indices for planes and directions in cubic structures are such that the direction with Miller index $[hkl]$ is normal to the plane (hkl).

Some simple geometrical results in cubic systems may be useful:

a) When $[uvw] \perp (hkl)$, $u = h$, $v = k$, and $w = l$.
b) When $[uvw] \parallel (hkl)$, $hu + kv + uw = 0$ (i.e., $[uvw] \cdot [hkl] = 0$).
c) The interplanar spacing (d) of $\{hkl\}$ planes is given by

$$d = \frac{a}{\sqrt{h^2 + k^2 + l^2}}.$$

d) The angle (θ) between two planes $(h_1k_1l_1)$ and $(h_2k_2l_2)$ is given by the dot product of the normals to the directions $([h_1k_1l_1] \cdot [h_2k_2l_2])$:

$$\theta = \cos^{-1}\left(\frac{h_1h_2 + k_1k_2 + l_1l_2}{\sqrt{h_1^2 + k_1^2 + l_1^2}\sqrt{h_2^2 + k_2^2 + l_2^2}}\right).$$

The same relation holds valid for finding out the angle between any two directions (say, $[u_1v_1w_1]$ and $[u_2v_2w_2]$) in a cubic crystal structure.

2.1.5.1 Miller–Bravais Indices for Hexagonal Close-Packed Crystals

For hexagonal crystals, it is convenient to use a reference system with four axes (a_1, a_2, a_3, and c), as shown in Figure 2.13, to specify crystallographic planes and

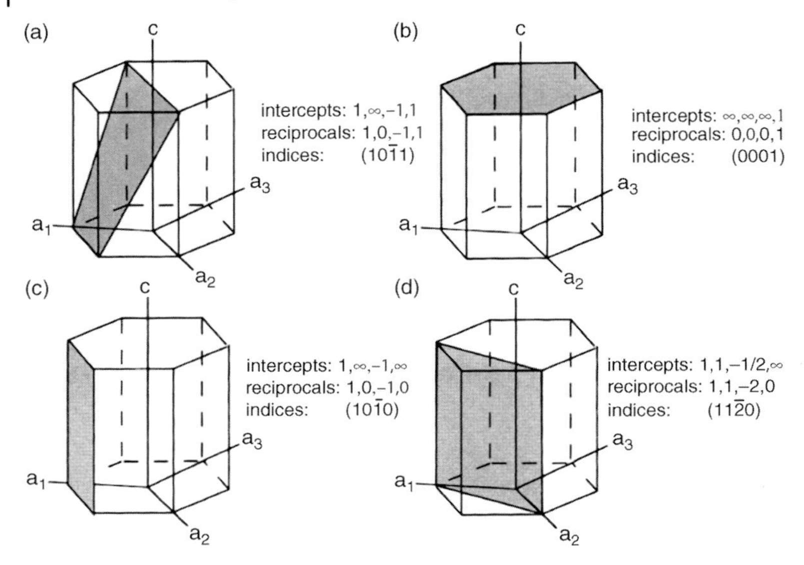

Figure 2.13 Miller–Bravais indices in hexagonal crystals. (From William Hosford, Mechanical Behavior of Materials (2nd Ed.), New York, NY, 2010; with permission)

directions. Three axes (a_1, a_2, and a_3) are coplanar and lie on the base of the hexagonal prism of the unit cell with a 120° angle between them. The fourth axis (c) is perpendicular to the base. Thus, a four-digit notation ($hkil$), known as Miller–Bravais indices, can be used for denoting planes, and [$uvtw$] for directions in a hexagonal crystal. The use of the Miller–Bravais indices enables denoting crystallographically equivalent planes by the same set of indices. However, the basic procedure for the Miller–Bravais indices are the same as that of the Miller indices. Let us take a look at the plane shown in Figure 2.13a. The intercepts created by the four axes are 1, ∞, -1, and 1. Thus, the Miller–Bravais indices of the plane is (1/1 1/∞ -1/1 1/1) or $(10\bar{1}1)$. Similarly, Miller–Bravais indices of the other planes shown in Figure 2.13b–d can be derived. Note that in all the four cases, the condition $h + k = -i$ is satisfied as a_1, a_2, and a_3 axes are coplanar vectors.

For the Miller–Bravais indices of a direction in the hexagonal crystal, the basic procedure again remains the same. However, it is bit different. For example, the direction along or parallel to the axis a_1 can be resolved into components along a_2 and a_3, each component being -1, as shown in Figure 2.14. The first index can be obtained from the relation $u + v = -t$ or $u = -(t + v) = -(-1 - 1) = 2$. The direction does not have any component in the c direction. Thus, the Miller–Bravais index of the direction is $[2\bar{1}\bar{1}0]$.

One can use the alternative method of using points as per the cubic, but in this case we first start from defining a_1, a_2, and c axes as points with the coordinates:

$$a_1 = (100), \quad a_2 = (010), \quad \text{and} \quad c = (001).$$

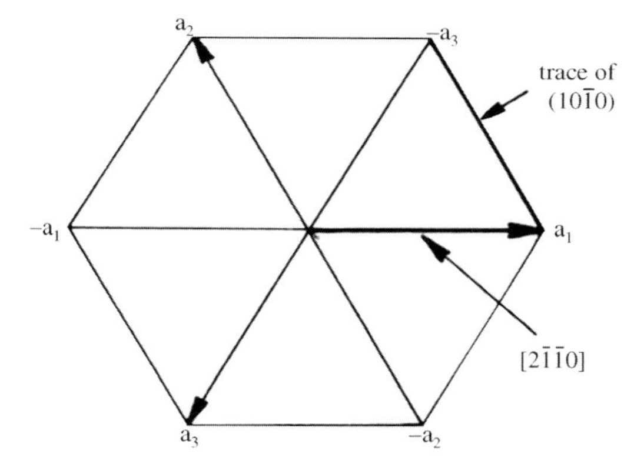

Figure 2.14 Miller indices of directions a_1, a_2, a_3.

As per the cubic system, find the direction as a subtraction of tail from the head and call it $[h'k'l']$. Find the four-axes notation $[hkil]$ by using the following relations and clearing of fractions and reducing to lowest integers:

$$h = \frac{1}{3}(2h' - k'), \quad k = \frac{1}{3}(2k' - h'), \quad i = -\frac{1}{3}(h' + k'), \quad \ell = \ell'.$$

Thus, the Miller index of a_1, for example, is easily found as (100)–(000) so that $h' = 1$, $k' = 0$, and $l' = 0$ or $h = 2/3$, $k = -1/3$, $i = -1/3$, and $l = 0$ or Miller index of a_1 is $[2\bar{1}\bar{1}0]$. This is a very useful procedure in deriving the Miller indices for directions in HCP crystals.

2.1.6
Interstitial Sites in Common Crystal Structures

A lot of space in crystals is empty and not occupied by lattice atoms(recall the definition of atomic packing factor). However, understanding interstitial sites in all the common lattice structures is an important step to understanding more complex crystal structures. Furthermore, interstitial impurity/alloying elements and self-interstitial atoms generally locate themselves in such sites. Figure 2.15a illustrates the octahedral position at the center of an FCC unit cell. Similar octahedral locations can be imagined at the center of each cube edge indicated by an “open” circle. However, to depict fully these octahedral positions just as shown in the center of the cube, one needs to construct additional neighboring unit cells. Tetrahedral interstitial positions (indicated by the “open” circles) of the FCC unit cell are shown in Figure 2.15b. Note that all the tetrahedral positions are entirely within the unit cell unlike the octahedral positions. Figures 2.16 and 2.17 show similar octahedral and tetrahedral sites in both BCC and HCP unit cells.

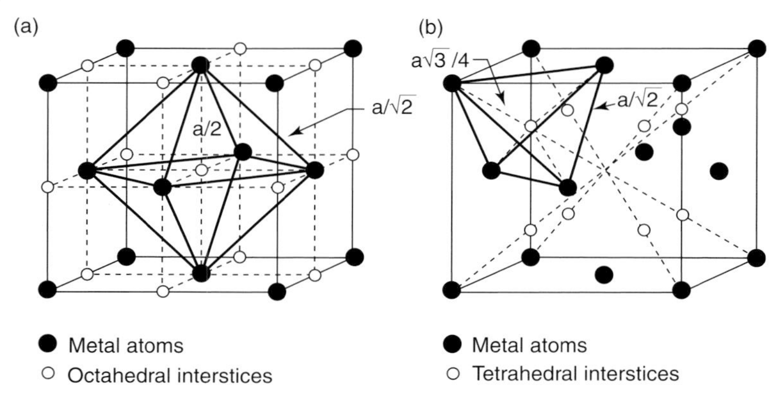

Figure 2.15 FCC unit cell. (a) Octahedral positions. (b) Tetrahedral positions (the "filled" circles represent the lattice atoms and the "open" circles interstitial positions).

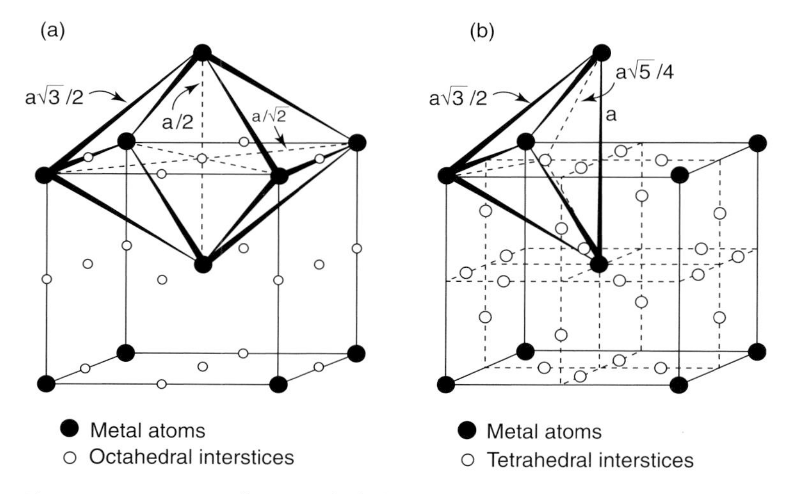

Figure 2.16 BCC unit cell. (a) Octahedral positions. (b) Tetrahedral positions (the "filled" circles represent the lattice atoms and the "open" circles interstitial positions).

2.1.7 Crystal Structure of Carbon: Diamond and Graphite

Carbon exhibits two major polymorphic forms – diamond and graphite. Diamond structure consists of an FCC unit cell of carbon atoms with half of its tetrahedral positions filled by additional carbon atoms (Figure 2.18). Thus, the effective number of atoms in a diamond unit cell is $(4 + 4) = 8$. Due to the presence of strong directional covalent bonding (sp^3 hybridization) between carbon atoms, diamond is the hardest natural material known, although with atomic packing factor of only 0.34.

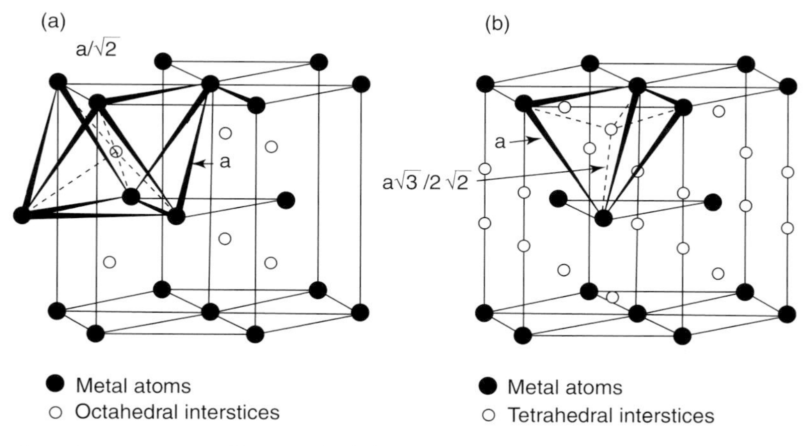

Figure 2.17 HCP unit cell. (a) Octahedral positions. (b) Tetrahedral positions (the "filled" circles represent the lattice atoms and the "open" circles interstitial positions).

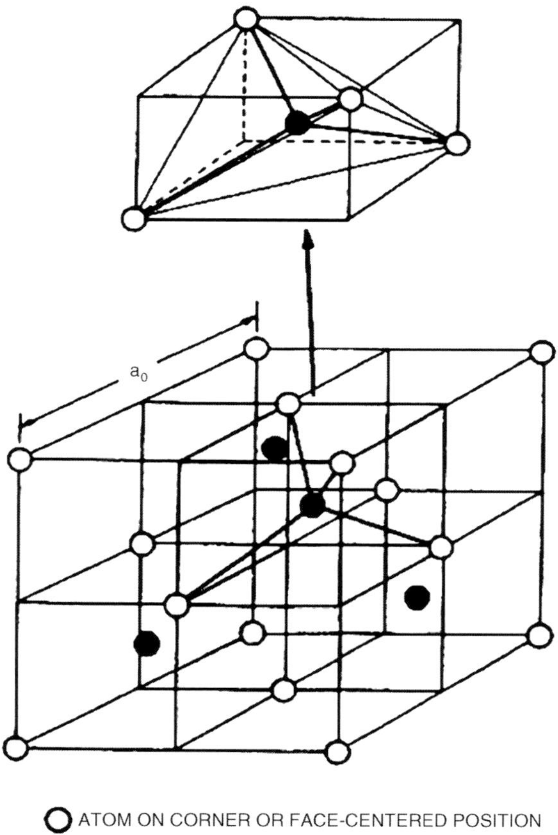

Figure 2.18 Diamond crystal structure (from Olander [15]).

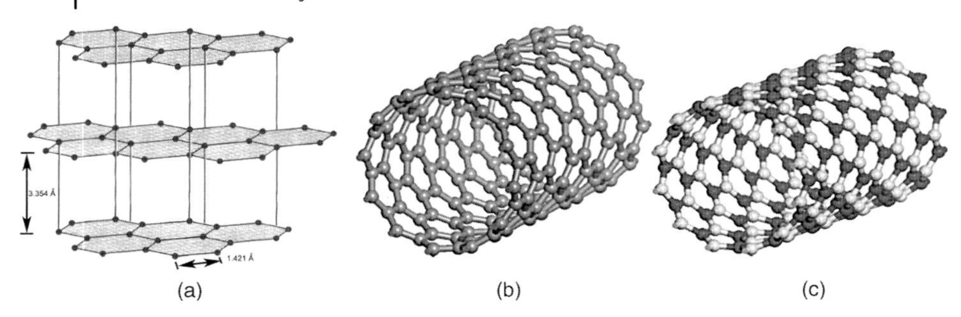

Figure 2.19 Structure of (a) graphite, (b) carbon nanotube, and (c) boron nitride nanotube. D. Golberg et al., Boron Nitride Nanotubes, Advanced Materials, 19 (2007) 2413-2432.

In graphite, carbon atoms are arranged in hexagonal arrays (basal planes) and bonded by strong directional covalent bonding (sp^2 type). However, the bonding between the hexagonal layers is much weaker "van der Waals" interaction. Figure 2.19a illustrates the graphite crystal structure. The presence of highly delocalized electrons makes graphite electrically conductive. Consequently, the graphite structure has very strong directional properties. Hexagonal boron nitride (h-BN) also has a similar structure. Graphite is a known moderator used in many past thermal reactors and high-temperature graphite reactors (HTGRs), and also used as a coating for TRISO (TRI-structural oxide) fuels for the VHTR. A single hexagonal layer of graphite is known as "graphene." When the graphene layer is rolled up, it can form a structure known as "carbon nanotube" (Figure 2.19b). Boron nitride nanotubes have a similar structure (Figure 2.19c).

2.1.8
Crystal Structure in Ceramics

Many of the nuclear fuels are in the form of ceramics. Thus, understanding the crystal structure of ceramics is of paramount importance. Ceramics generally have the ionic bonding between their lattice entities. An ionic solid contains two or more ionic species (positive and negative). The crystal structures of an ionic solid can generally be described as comprising two or more intermingling simpler lattices (called sublattices). The stoichiometry of ionic solids is influenced by the fact that the nearest neighbors of a particular ion should be the ions with the opposite charge in order to maximize the Coulomb energy (attractive) associated with the crystal structure and minimize the repulsive energy that may destabilize the structure. This means a positive ion prefers to be surrounded by negative ions, and a negative ion needs to be surrounded by positive ions as their *first nearest neighbors*. The cations (+ive ions) also tend to keep the maximum separation between the other cations that are their *second nearest neighbors* and vice versa. Usually, the larger of the ions would create an FCC or HCP ion array, and the corresponding interstitial sites are occupied in a regular manner by the opposite type of ions. However, any such arrangement must conform to the rule of local charge neutrality, which can be extended to the entire crystal and still maintains the stoichiometry (i.e., cation/anion ratio) of the crystal. Although some ceramic structures are very complex, we shall limit our discussion to a few common, simpler structures, some of which are of nuclear importance. The crystal structure of a ceramic is generally named after a common compound that shows that particular structure.

2.1.8.1 **Rock Salt Structure**

The rock salt or sodium chloride (NaCl) lattice structure consists of two intermingling cation (Na^+) and anion (Cl^-) FCC sublattices, as illustrated in Figure 2.20. Alternatively, the rock salt structure can also be described as an FCC anion lattice in which all its octahedral interstitial sites are filled up by the cations. Here nearest neighbors to each ion are the six ions of the opposite charge. Coordination numbers of cations and anions are six each in this structure similar to that for simple cubic structure. The number of octahedral sites in an FCC unit cell is the same as the number of atoms (in this case, anions). Therefore, the stoichiometry of this crystal type is MX (M = metal, X = nonmetal). This means that valencies of both cations and anions in this crystal type need to be the same. Some examples of the rock salt structure are KCl, LiF, and KBr (monovalent ions); MgO, CoO, and MnO (divalent ions); and UC (tetravalent ions). Note that UC is a potential nuclear fuel. The *Struckturbericht* notation of a rock salt structure is B1. There are four ion pairs per unit cell similar to an FCC structure. The lattice constant of a NaCl-type crystal structure is given by $2(R^+ + R^-)$, where R^+ and R^- are the cation and anion radii,

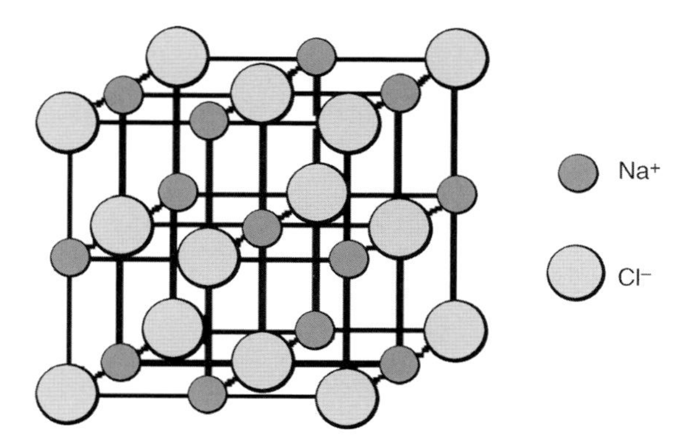

Figure 2.20 A schematic of NaCl-type crystal structure.

respectively. In rock salt structures, the cation to anion *radius ratio* can vary between 0.414 and 0.732.

2.1.8.2 **CsCl Structure**

The CsCl structure contains cations and anions with same valencies similar to those of rock salt-type structure. However, the CsCl crystal structure is essentially different from the NaCl structure. The CsCl structure can be viewed as two interlacing, equal-sized simple cubic cation and anion sublattices. It can also be shown as a simple cubic unit cell of cations (Cs^+) with an anion (Cl^-) in the body center of the cube and vice versa. The lattice constant (*a*) remains the same regardless of the way to view the structure (Figure 2.21). Each ion in this structure is surrounded by eight ions of the opposite type (i.e., coordination number 8 for both cations and anions similar to that of BCC structure) as the first nearest neighbors. There is only one ion pair per unit cell of CsCl as per simple cubic structure. Whether a solid of MX type would assume a NaCl-type structure or CsCl-type structure depends largely on the cation/anion radii ratio. For CsCl, the radius ratio is about 0.92, whereas for NaCl it is 0.61. Structures with the radius ratio higher than 0.732 tend to have eight coordination and may assume CsCl-type structure. Other examples of this type of structure are CsI and CsBr. Note that CsI may be found in the spent fuels among other compounds as Cs and I are generated as fission products in the reactor fuels. However, during the reactor operation, it stays in gaseous form (not crystalline)

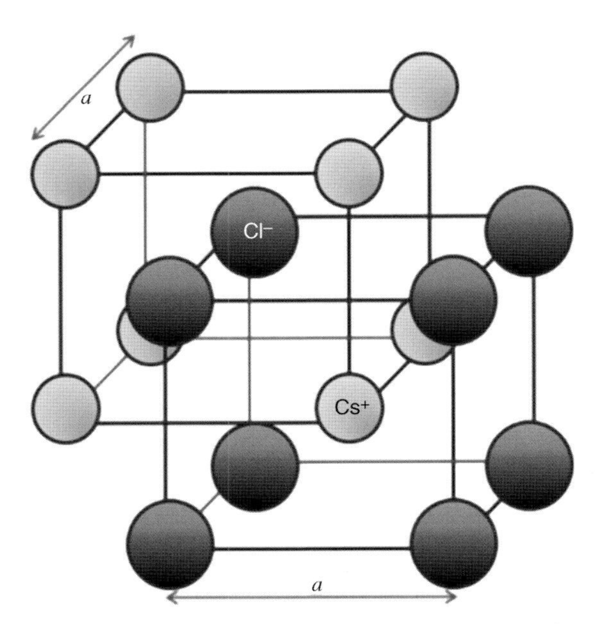

Figure 2.21 CsCl crystal structure (gray circles represent Cs^+ ions, and black circles represent Cl^- ions), not to scale. Note that the body center ion of each cube is shown to illustrate the crystal structure clearly.

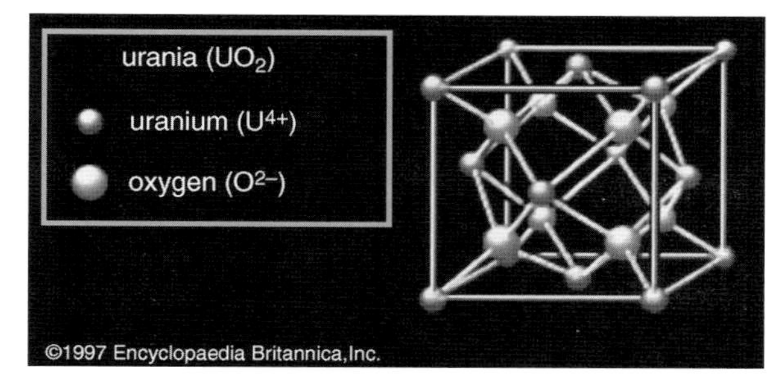

Figure 2.22 A schematic of a fluorite-type crystal structure (shown for UO_2).

inside the fuel. In accident situations, CsI can escape the cladding containment and form solid phases (crystalline) if it comes into contact with the cooling water.

2.1.8.3 **Fluorite Structure**

This structure is named after the mineral *fluorite*, or CaF_2. In this solid, the valency of the cation is twice that of the anion, that is, MX_2. This structure can be described in two ways. Figure 2.22 shows the view of a CaF_2-type unit cell where the FCC cation sublattice is interlaced with a simple cubic anion sublattice. That means all the tetrahedral interstitial positions of the FCC cation sublattice are filled with anions. Thus, each fluorite unit cell contains four cations and eight anions or four CaF_2 molecules. This type of structure has the following relation between the lattice constant and ionic radii: $a = (4/\sqrt{3})\ (R^+ + R^-)$, which can be derived from purely geometrical aspects noting that the F ion is at $a/4$, $a/4$, $a/4$ position, thus, $R^+ + R^-$ becomes the body diagonal of a cube with side equal to $a/4$. Some examples of the fluorite-type crystal structure are ZrO_2, HfO_2, UO_2, ThO_2, PuO_2, and CeO_2. UO_2 is the most widely used nuclear fuel in most commercial power reactors. One of the beneficial effects of using UO_2 is its high melting point (2860 °C), thus giving it excellent stability. UO_2 structure has more empty space than a UC structure due to the simple cubic structure of the interior F^- sublattice. Even though this feature aids in providing more space for the fission products to accumulate inside the structure, it also reduces the fissile atom density. Notably, when the cations and anions exchange their positions in the fluorite structure (i.e., becomes M_2X type), the resulting structure is called an "antifluorite" structure. Examples of such crystal structure are alkali metal oxides, that is, Li_2O, Na_2O, K_2O, and Rb_2O.

Special Note

Polymorphism in ceramics is also possible. One example is zirconium dioxide or zirconia (ZrO_2) that can change its crystal structure from monoclinic to tetragonal to cubic.

Example 2.3

Calculate the density of uranium dioxide (UO_2) from the first principles given: atomic weights of uranium and oxygen are 238 and 16, respectively. The atomic radius data are $U^{4+} = 0.105$ nm and $O^{2-} = 0.132$ nm.

Solution

The theoretical density of UO_2 can be obtained from the mass of the atoms in the unit cell divided by the volume of the unit cell.

The lattice constant of a UO_2 unit cell (a) is given by $(4/\sqrt{3})\ (R^+ + R^-) = (4/\sqrt{3})\ (0.105 + 0.132)\ \text{nm} = 0.547\ \text{nm}$.

The volume of the UO_2 unit cell (a^3) is given by $(0.547\ \text{nm})^3 = 0.164\ \text{nm}^3 = 0.164 \times (10^{-9})^3\ \text{m}^3 = 0.164 \times 10^{-27}\ \text{m}^3$.

UO_2 has a fluorite-type crystal structure, that is, the effective number of U^{4+} ions is 4 and of O^{2-} ions is 8.

The atomic weight of the uranium is 238. This means the mass of each U^{4+} ion is 238 amu or $(238 \times 1.66 \times 10^{-27})$ kg (since 1 amu $= 1.66 \times 10^{-27}$ kg). There are four U^{4+} ions of a combined mass of $(4 \times 238 \times 1.66 \times 10^{-27})$ kg $= 1.58 \times 10^{-24}$ kg.

Similarly, the mass of eight oxygen ions in the unit cell is given by 0.213×10^{-24} kg.

The total mass of the UO_2 unit cell is $(1.58 + 0.212) \times 10^{-24}\ \text{kg} = 1.792 \times 10^{-24}\ \text{kg}$.

The density of UO_2 is then given by $(1.792 \times 10^{-24}\ \text{kg}/0.164 \times 10^{-27}\ \text{m}^3) \approx 10\,926\ \text{kg m}^{-3} \approx 10.93\ \text{g cm}^{-3}$.

2.1.8.4 Zincblende Structure

Oxides and sulfides (such as ZnO, ZnS, and BeO) containing smaller cations tend to have tetrahedral coordination (i.e., the coordination number of both the cation and the anion is 4). The same structure is also shown by the covalent compounds, such as SiC, BN, and GaAs. This structure class is named after the mineral zincblende in which the anions (S^{2-}) form the FCC sublattice, and only half of the available tetrahedral sites are filled with the cations (Zn^{2+}), as depicted in Figure 2.23. Note the similarity between the zincblende and diamond structures (refer back to Figure 2.19).

2.1.8.5 Corundum Structure

One of the most common oxide crystal structures is of corundum or α-alumina (Al_2O_3). The stoichiometry of the corundum-type structures is M_2X_3. These compounds are also known as "sesquioxides." The basic structure contains hexagonal close-packed layers of the anion (i.e., O^{2-}) sublattice. If the cations need to fill in all the octahedral sites, the stoichiometric ratio becomes MX as the number of the octahedral sites is equal to the number of the regular lattice sites in a HCP structure. Because of the particular stoichiometry of Al_2O_3 (i.e., cation/anion = 2 : 3), the cations only fill in two-thirds of the available octahedral interstitial sites, however,

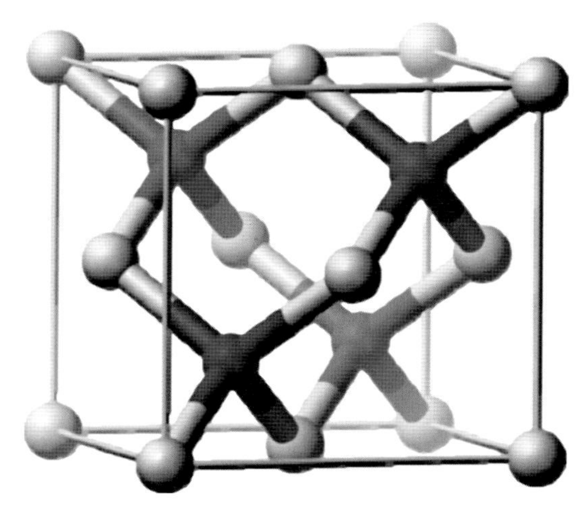

Figure 2.23 The crystal structure of zincblende (the gray circles represent S^{2-} and the black circles represent Zn^{2+}).

maintaining maximum cation separation in an orderly manner. A simpler way to understand the corundum structure is to see the ion positions from both the top and a vertical section. A basal plane of a corundum structure is shown in Figure 2.24a. The picture shows a layer of cations (indicated by filled circles) forming a hexagonal network with the same ion spacing between the two anion layers (only one layer is delineated by the larger open circles in this figure). Note that x positions indicate the unfilled octahedral sites. The next cation layer would have the same hexagonal configuration, but shifted to one atom spacing in the direction of "vector 1," as shown in Figure 2.24a. When an another close-packed oxygen layer is added on top of this cation layer, an another cation layer will sit on it, although shifted by yet another atom spacing in the direction of "vector 2" while maintaining the similar hexagonal configuration. The structure becomes clearer if we take a vertical plane $\{10\bar{1}0\}$ across the corundum structure, as shown in Figure 2.24b. This shows that each octahedral column of cations contains two cations in a row followed by one missing cation and so on. If one takes into account the periodic spacing of both the cations and anions at the same time, we find that the structure repeats itself after six such layers (shown as ABABAB), thus making the lattice constant in the vertical direction (c) as 1.299 nm. As the octahedral sites in the corundum share faces and two out of every three octahedral sites are occupied, the electrostatic repulsion between cations helps move them slightly into the unfilled octahedral site, whereas oxygen ions also shift slightly from their ideal close-packed configuration, thus making the corundum structure bit inherently distorted.

The corundum structure is exhibited by a multiple number of oxides, such as Fe_2O_3, Cr_2O_3, Ti_2O_3, V_2O_3, Ga_2O_3, Rh_2O_3, La_2O_3, Nd_2O_3, and so on. There are many ternary compounds such as ilmenite ($FeTiO_3$) and lithium niobate ($LiNbO_3$)

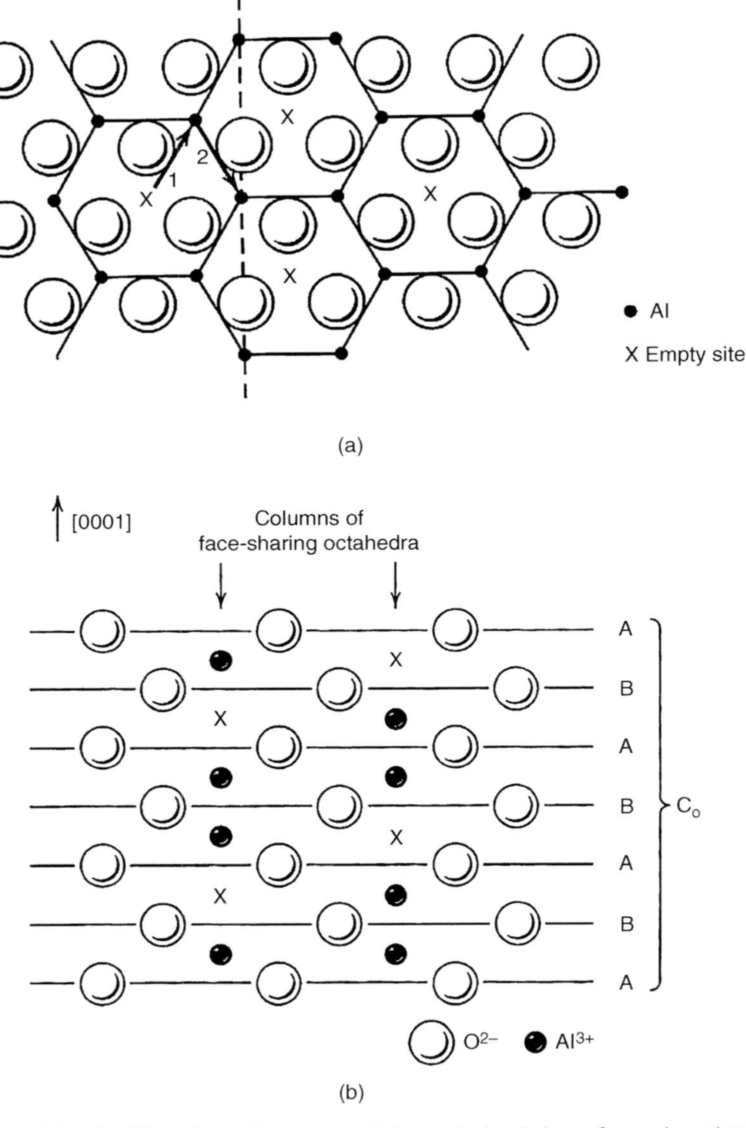

Figure 2.24 (a) Filling of two-third octahedral sites in the basal plane of corundum. (b) Two-third occupancy of octahedral site columns (the plane shown is the dashed line in part (a)) (Ref. 10)

that assume "derivative" corundum structures. The Cr-containing steels in contact with water generally have corrosion layers composed of Fe_2O_3 and Cr_2O_3, both of which have a corundum-type crystal structure. Furthermore, the rare earth-based fission products, such as Nd_2O_3 and La_2O_3, have corundum-type crystal structure.

Table 2.3 A summary of important characteristics in some crystal structures.

Important parameters	SC	BCC	FCC	HCP
Effective number of atoms per unit cell	1	2	4	6
Coordination number	6	8	12	12
Atomic radius (r) versus lattice constant (a)	$a = 2r$	$\sqrt{3}a = 4r$	$\sqrt{2}a = 4r$	$a = 2r$
Atomic packing fraction	0.52	0.68	0.74	0.74
Closest-packed direction	$\langle 100 \rangle$	$\langle 111 \rangle$	$\langle 110 \rangle$	$\langle 11\bar{2}0 \rangle$
Closest-packed plane	$\{100\}$	$\{110\}$	$\{111\}$	$\{0002\}$
Close-packed stacking sequence	—	—	ABCABC ...	ABABAB ...

2.1.9 Summary

There are 7 basic crystal structures and 14 Bravais lattices. Most metals assume one of three Bravais lattices, FCC, BCC, or HCP. A comparison between simple SC, FCC, BCC, and HCP crystal structures is given in Table 2.3.

The crystal structures of ceramics are typically more complex than those of the metals. The presence of two or more types of ionic species makes the structures complex. Only a few ceramic crystal structures (NaCl, CsCl, CaF_2, Al_2O_3, and ZnS type) are discussed, but others like Perovskite and Spinels are omitted. Readers are referred to Refs [2, 3] for details.

2.2 Crystal Defects

As described in Section 2.1, crystals are hardly perfect even when there is no radiation damage. The deviations from the ideal crystal structure are instrumental in influencing various structure-sensitive properties of crystals. There could be several types of these defects, and collectively they are called *crystal or lattice defects*. Interestingly, a perfect crystal could be composed of atoms at rest with only zero-point oscillation at the absolute zero temperature. However, as the temperature increases, the amplitude of the lattice vibration also increases. This lattice vibration basically manifests itself as elastic waves and can influence some very important physical properties (such as thermal conductivity). This type of lattice vibration is called *phonon* because of its similarity in behavior with the light *photons* (mainly because of the relationship between their frequency expressions). Electrons can jump to higher orbits creating electron holes. This can also affect electronic properties (recall the semiconductor theories). However, electronic properties are not of pressing importance in the context of nuclear reactor materials. Henceforth, in this section, our focus would be to give the readers an introduction to various types of crystal defects like point defects (vacancies, self-interstitials, substitutional, or interstitial impurity atoms), line defects (dislocations), surface defects (grain boundaries), and volume defects (voids, cavities, and precipitates).

2.2.1 Point Defects

2.2.1.1 Point Defects in Metals/Alloys

Point (or zero-dimensional) defects are associated with imperfections involving an atom or only a few atoms in a localized region. They are often described as "zero-dimensional" defects. There are many types of point defects that one should be aware of. They are schematically shown in Figure 2.25:

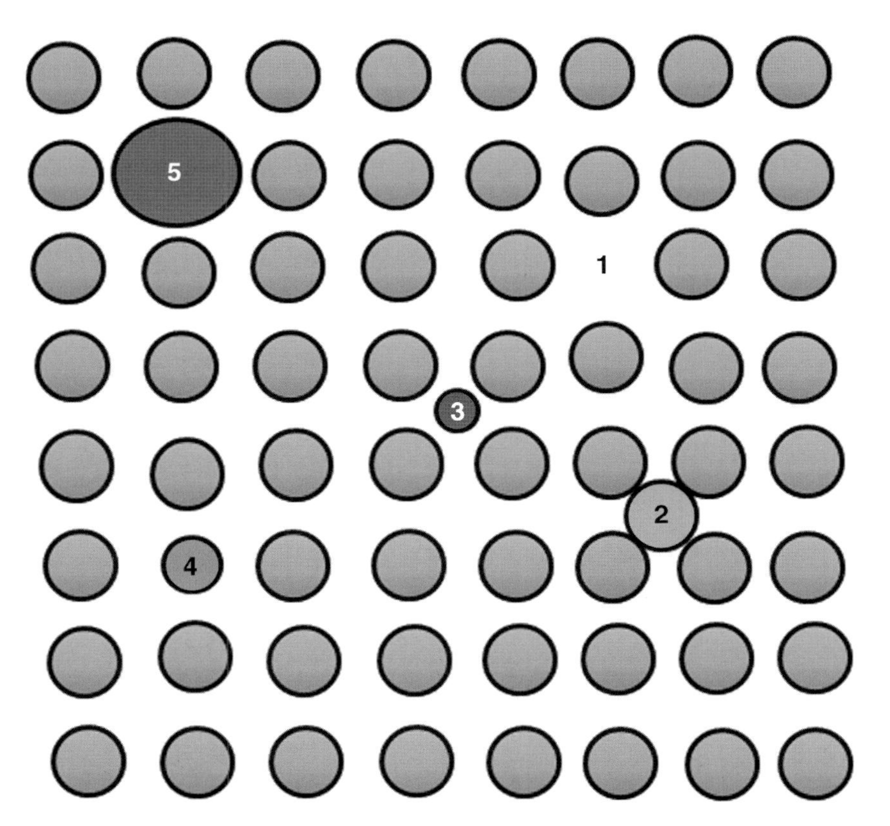

Figure 2.25 Various point defects schematically shown on a 2D crystal lattice (1 – monovacancy, 2 – self-interstitial, 3 – interstitial impurity atom, 4 – undersize substitutional atom, and 5 – oversize substitutional atom).

Vacancies

Vacancy is simply a vacant lattice site. Vacancies are present in all crystalline solids, pure or impure, under almost all conditions, predictable by the laws of thermodynamics. Vacancies were first imagined to explain the diffusion phenomenon in solids (see Section 2.3). Here we develop a derivation for calculating the equilibrium vacancy concentration. A basic knowledge of thermodynamics is required to understand this approach. We assume that the process of vacancy formation must obey $dG = 0$ for maintaining a thermodynamic equilibrium condition. One can clearly note here that due to the existence of vacancies, the enthalpy and the entropy of the crystal would be greater than it would have been without the vacancies. Thus, we can write from the definition of the Gibbs free energy,

$$\Delta G_{\mathrm{v}} = \Delta H_{\mathrm{v}} - T\Delta S_{\mathrm{v}}, \tag{2.2}$$

where ΔG_{v} is the Gibbs free energy change, ΔH_{v} is the enthalpy increase, and ΔS_{v} is the entropy increase due to the presence of vacancies in the crystal.

Therefore, if $G(o)$ is the Gibbs free energy of a perfect lattice and $G(n)$ is the Gibbs free energy of the lattice with n number of vacancies, we can simply write

$$G(n) - G(o) = \Delta G_{\mathrm{v}}. \tag{2.3}$$

Now we definitely need some discussion on the entropy term. The entropy term is broadly categorized in two ways: vibrational entropy and configurational entropy. The vibrational entropy term is readily intuitional in that the atoms present in the neighborhood of a vacancy are less restrained than the atoms in the perfect portion of the crystal. Thus, each vacancy can provide a very small contribution to the total entropy of the crystal due to the more irregular or random *vibration*. Although a detailed theoretical treatment of vacancies would require consideration of the small vibrational entropy contribution, it is generally not considered in this type of derivation as it is of secondary importance. Instead, we should take into account the effect of configurational entropy that arises due to the probabilistic nature of the vacancy creation process. If N is the total number of lattice atoms, there are W different ways of arranging the N atoms and n vacancies on $(N+n)$ lattice sites. Hence, W can be given by the following expression following the *combinatorial* rules:

$$W = (N+n)_{C_n} = \frac{(N+n)!}{n!(N+n-n)!} = \frac{(N+n)!}{n!N!}, \tag{2.4}$$

where $N! = N(N-1)(N-2)\ldots 3{\cdot}2{\cdot}1$, and so forth. On the other hand, the configurational entropy term is given by $S_{\mathrm{conf}} = k \ln W$, where k is the Boltzmann's constant. This constant appears in the science equations quite a bit and has the value on the order of the thermal energy per atom.

Thus, Eq. (2.2) becomes

$$\Delta G_{\mathrm{v}} = ng_{\mathrm{v}} - kT \ln W, \tag{2.5}$$

where g_v is the Gibbs free energy associated with forming a vacancy. For thermal equilibrium, the following relation should hold:

$$\frac{\partial(\Delta G_v)}{\partial n} = 0 = g_v - kT\frac{\partial(\ln W)}{\partial n}. \tag{2.6}$$

For large numbers of N and n (in reality their values are in millions), Stirling's approximation $\ln(N!) = N\ln(N) - N$ can be used to show that $\partial(\ln W)/\partial n = \ln((N+n)/n)$.

Thus, Eq. (2.6) becomes

$$g_v = kT\ln\left(\frac{N+n}{\nu}\right), \quad \text{or} \quad g_v = -kT\ln\left(\frac{n}{N+n}\right), \quad \text{or} \quad \ln\left(\frac{n}{N+n}\right) = -\frac{g_v}{kT}, \quad \text{or} \quad \left(\frac{n}{N+n}\right) = \exp\left(-\frac{g_v}{kT}\right).$$

Since $N \gg n$, we can also write

$$\frac{n}{N} = C_v = \exp\left(-\frac{g_v}{kT}\right), \tag{2.7}$$

where C_v is defined as the equilibrium vacancy concentration.

We know from the definition of Gibbs free energy,

$$g_v = H_v - TS_v.$$

So, Eq. (2.7) becomes

$$C_v = \exp\left(-\frac{H_v}{kT}\right)\exp\left(\frac{S_v}{k}\right). \tag{2.8}$$

The above equation presents the compromise between the enthalpy (i.e., energy, E_v, in this case as P_V term is negligible, that is, $H_v \approx E_v$) and entropy. The vacancy formation energy can also be defined as the energy needed to remove one atom from the lattice and place it on the crystal surface. However, it does not give us any indication of how much time it would take to accomplish it. That is why the thermal vacancy formation is guided by the thermodynamic principles, not kinetic ones. It is clear from this that a large vacancy concentration is favored with a decrease in the vacancy formation energy, whereas a large entropy of vacancy formation tends to increase the vacancy concentration. Furthermore, it is not only the vacancy formation energy alone that is solely important, but the ratio of E_v to thermal energy is also important. At higher temperatures, the thermal energy is high causing a significant probability of strong thermal fluctuations leading to the formation of vacancies. Conversely, at low temperatures, the probability of large thermal fluctuations is so low that less number of vacancies is created. So, the vacancy concentration strongly depends on the temperature and would increase exponentially. In other words, the term $\exp(-E_v/kT)$ represents a probability term giving the chances that a crystal with thermal energy (kT) has to create thermal fluctuations sufficient enough to provide the energy needed to produce vacancies. It has been found that the contribution of $\exp(S_v/k)$ is very small at all temperatures compared to the $\exp(-E_v/kT)$ contribution, and hence will not be considered further. Hence, Eq. (2.7) can also be written as

$$C_v = \exp\left(-\frac{E_v}{kT}\right). \tag{2.9}$$

The results of experimental and theoretical studies have shown that the vacancy formation energies are typically on the order of 1 eV. There are several methods of measuring vacancy concentration – one being the electrical resistivity measurement. The electrical resistivity of a metal generally increases because of the presence of vacancies, and the change in resistivity is proportional to the vacancy concentration. The vacancy formation energy is generally obtained from the slope of the semilog plot of C_v versus $1/T$. Vacancies play a major role in the diffusion processes and thus affect various phase transformation, deformation, and physical processes. Sometimes, the nonequilibrium concentration of vacancies can be sustained at room temperature by heating a metal followed by quenching (i.e., fast cooling). Quenching ensures that the high concentration of vacancies characteristic of higher temperature is retained at room temperature without being depleted (as the migration of the vacancies become slower at lower temperatures).

The type of vacancies that we discussed previously should be called "monovacancy" since only one (i.e., *mono*) lattice atom is missing (Figure 2.26a). At very high temperatures, many vacant lattice sites find the neighboring sites vacant, too. If two vacancies come side by side, a "divacancy" is formed (Figure 2.26b). The divacancy formation energy ($E_v^{(2)}$) can be expressed as ($2E_v - B$), where B is the binding energy of a divacancy (this energy is basically the energy required to separate a divacancy into two isolated monovacancies). It is generally hard to measure or calculate the binding energy. One estimate for copper has shown values of the

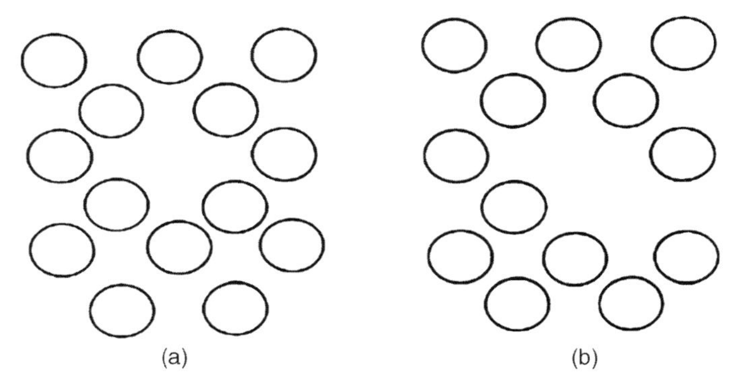

Figure 2.26 Configuration of (a) a monovacancy and (b) a divacancy.

binding energy of a divacancy in the range of 0.3–0.4 eV. The equilibrium divacancy concentration ($C_v^{(2)}$) is given by the following equation:

$$(C_v^{(2)}) = \frac{\beta}{2} \exp\left(-\frac{2E_v - B}{kT}\right), \tag{2.10}$$

where β is the coordination number. Equation (2.10) can also be expressed in the following form:

$$\frac{C_v^{(2)}}{C_v^2} = \frac{\beta}{2} \exp\left(\frac{B}{kT}\right). \tag{2.11}$$

Self-Interstitial Atom (SIA)

A self-interstitial atom is a type of point defect where a lattice atom occupies an interstitial site instead of its regular position (recall the interstitial space as discussed in Section 2.1). For example, if a copper atom is in one of the interstitial positions, a self-interstitial type of point defect would be created. It can also be thought of as removing one atom from the crystal surface and moving it to an interstitial site. Any interstitial site is smaller than its own atom size, and the presence of such an interstitial atom could badly strain the lattice surrounding it. As a result, the formation energy of an SIA (E_i) in copper even under equilibrium conditions is quite high (~4 eV) compared to the similar quantity for a monovacancy (~1 eV). The equilibrium concentration of SIA (C_i) is given by

$$C_i = \frac{n_i}{N_i} = \exp\left(-\frac{E_i}{kT}\right), \tag{2.12}$$

where n_i is the number of interstitial atoms and N_i is the number of interstitial sites. Thus, thermal energy (kT) is not sufficient to create self-interstitials. The self-interstitials can be generated more readily under energetic particle radiation, such as fast neutron irradiation. Furthermore, the actual configuration of SIAs could be much different from the simple model that we just discussed (see Chapter 3).

Example 2.4

Calculate the concentrations of thermal monovacancies, divacancies, and self-interstitials in copper (FCC crystal structure, coordination number 12) at 20 °C, 500 °C, and 1073 °C. Comment on the results. Assume $E_v = 20$ kcal mol^{-1}, $B = 7$ kcal mol^{-1}, and $E_i = 90$ kcal mol^{-1}.

Solution

Since the activation energies are given as per mole, it is convenient to use the gas constant R (=1.987 cal mol^{-1} K^{-1}) rather than Boltzmann's constant. We use the above formulations to evaluate the concentration of monovacancies (C_v), divacancies ($C_v^{(2)}$), and self-interstitials (C_i). It will be easier to use a standard spreadsheet software in which you can write in the equations, and then calculate for all three temperatures. The results are summarized in the table below.

T (in °C/K)	C_v	$C_v^{(2)}$	C_i
20/293	1.20×10^{-15}	1.45×10^{-24}	7.30×10^{-68}
500/773	2.21×10^{-6}	2.80×10^{-9}	3.57×10^{-26}
1073/1346	5.65×10^{-4}	2.63×10^{-5}	2.43×10^{-15}

Note that the melting temperature of copper is 1083 °C. So, if the temperature rises to that level or above, the concentration of point defects will lose their physical meaning as the liquid state of copper would not contain any crystal point defects.

Interstitial Solutes

The solute atoms whose sizes are much smaller than the parent atom size can occupy the interstitial spaces easily. These solute atoms could be in the alloy and result in increased strength. They are called interstitial impurity atoms (i.e., IIA). However, the same type of point defect would be created if the interstitial atoms act as alloying constituents. In this case, they are more likely to be called as "interstitial alloying atoms." The presence of interstitial atoms in the host lattice may lead to the creation of "interstitial solid solutions." These are the homogeneous mixture of two or more kinds of atoms, of which at least one is dissolved in the host lattice by occupying the interstitial spaces. Elements with relatively small atomic radii (less than 100 pm) like hydrogen, oxygen, nitrogen, boron, and carbon are the most common interstitial atoms. As per Hume–Rothery, extensive interstitial solid solubility results if the apparent solute atom diameter is 0.59 smaller than that of the solvent. It has been noted that atomic size factor is not the only sufficient factor to determine the feasibility of interstitial solid solution formation. Transition elements such as iron, titanium, tungsten, molybdenum, thorium, uranium, and so on have preferred solubility for the interstitial atoms than the nontransition metals. It is postulated that the unusual electronic structure of the transition metals (incomplete d subshell) is the reason for higher interstitial solid solubility, which is however almost always smaller than the typical substitutional solute solubility (see the next section).

In plain carbon steels, different solid solution phases can be formed based on the type of the host iron lattice structure. Here, we give an example of the formation of two different phases, austenite and α-ferrite, in steel. In the FCC unit cells, octahedral interstitial sites are located at the midpoints of the cube edges and at the body center. The maximum size of an atom that can be accommodated in the confine of this type of octahedral site without causing distortion is $0.414r$, where r is the parent atom radius surrounding the interstitial space. On the other hand, smaller tetrahedral sites are located on the body diagonals (Figure 2.15a). The size of an atom that fits the tetrahedral space is only $0.225r$. We know that the iron atom radius is 140 pm. Thus, the maximum interstitial atom size that can be accommodated in the octahedral and tetrahedral sites without distortion are ~58.0 and 31.5 pm, respectively. The carbon atom radius is only 29.0 pm. So, the carbon atoms can easily occupy the octahedral sites as well as tetrahedral sites without causing any lattice distortion, and hence it becomes energetically stable configuration. Thus, the austenite phase can be defined as an interstitial solid solution of carbon atoms located at both octahedral and tetrahedral sites of the FCC iron lattice (i.e., γ-Fe). Conversely, in a BCC unit cell, the tetrahedral spaces are slightly larger ($0.29r$) compared to the octahedral spaces ($0.15r$). The maximum interstitial atom sizes to be fitted inside the octahedral and tetrahedral sites in a BCC iron would be 21.0 and 40.6 pm, respectively. Given the carbon atom radius (29 pm), only the tetrahedral sites could be filled without causing any lattice distortion. Thus, the α-ferrite phase can be defined as the interstitial solid solution of carbon atoms (located at the tetrahedral sites) in a BCC iron (α-Fe). In general, interstitial solutes cannot occupy all the interstitial spaces available in the host lattice because of the possibility of much greater lattice distortion.

Substitutional Atoms

When solute atoms substitute the parent lattice atoms from their original sites, the solid solution is called a substitutional solid solution. There could be two possibilities: First, solute atoms can substitute the host lattice atoms randomly, forming a random (or disordered) substitutional solid solution. An overwhelming majority of substitutional alloys are of this type. Second, an ordered (substitutional) solid solution results when the solute and solvent atoms are arranged in a regular fashion on the lattice sites (Figure 2.27). Perfect order becomes possible when the two metals are mixed in some fixed proportions such as 3 : 1, 1 : 1, and so on and at under certain temperature. For example, Cu_3Au (75 at% of Cu and 25 at% of Au) alloy can exhibit an ordered structure at lower temperature range, but above a certain temperature range the ordered alloy loses perfect order, thus becoming disordered. However, the Cu–Ni alloy system at all compositions and in all conditions are disordered, which is the norm most times.

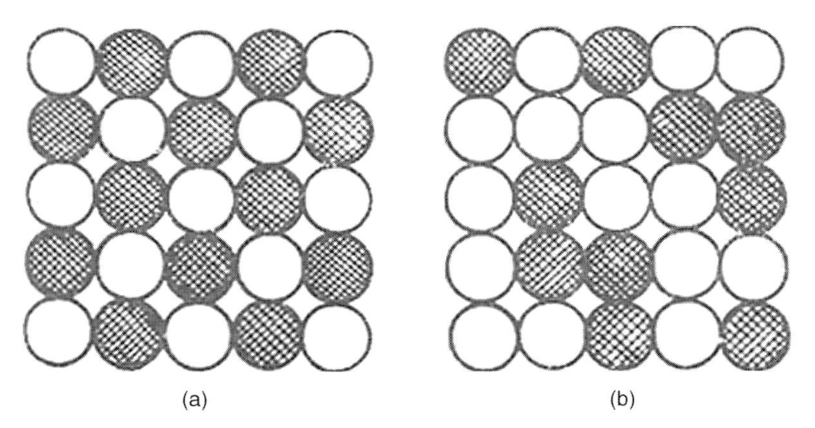

Figure 2.27 Schematic structures of the (a) ordered and (b) disordered alloys.

Hume–Rothery studied a number of substitutional alloys and developed a set of basic rules for substitutional solid solubility. The principles derived from his study are known as the Hume–Rothery's rules:

1) "The atom size difference between the host atom and the solute atom should be less than 15%." If the size difference becomes higher, more intense stress fields are created. This observation is true for both large and small solute atoms. Generally, a large atom induces a compressive stress field around itself, while a small atom tends to introduce a tensile stress field. These stress fields can increase the potential energy of the crystal. Therefore, the solid solubility becomes more limited as the size difference increases. This is known as the *size factor effect*. For example, the atomic radii of copper and nickel are 145 and 149 pm., respectively. The percentage size difference is only about 2.8%. They have wide solid solubility in each other.
2) "The electronegativity difference between the solvent and the solute should be small." Electronegativity is defined as the ability of an atom to attract electrons to itself. The electronegativity difference between two elements can be calculated from Pauling's electronegativity scale. The difference is generally quite small between typical metallic elements. If the electronegativity difference is more, the tendency would be to create compounds rather than alloys. This is called *chemical affinity effect*. For example, the Pauling electronegativity values of copper and nickel are 1.9 and 1.91, respectively, thus giving the difference of only 0.01.
3) "The valency of the atoms constituting the alloy must be the same for extensive solid solubility." However, when the valencies are different, a metal with lower valency tends to dissolve in a metal of higher valency more readily than vice versa. It is found that an excess of electrons is more readily tolerated rather than a deficiency of bonding electrons. This is known as *relative valency effect*. For instance, zinc (two valence electrons) dissolves appreciable amount of copper (up to 38%), whereas copper (one valence electron) dissolves only about 3% in zinc.
4) "The crystal structures of the solvent and solute should be the same for achieving extensive solid solubility, known as *crystal structure effect*." This implies that solute atoms can substitute the host lattice atoms continuously, forming a series of solid solutions. A nice example of this is the isomorphous system of copper and nickel (both have FCC crystal structure).

2.2.1.2 **Point Defects in Ionic Crystals**

There are two types of point defects, Frenkel and Schottky defects that can be found in ionic crystals.

a) *Frenkel defect or disorder* forms when an ion leaves its original place in the lattice creating a vacancy, and becomes an interstitial by moving into a nearby interstitial space not usually occupied by another ion. This type of defect was first discovered by the Soviet physicist Yakov Frenkel (1926). This defect is actually a

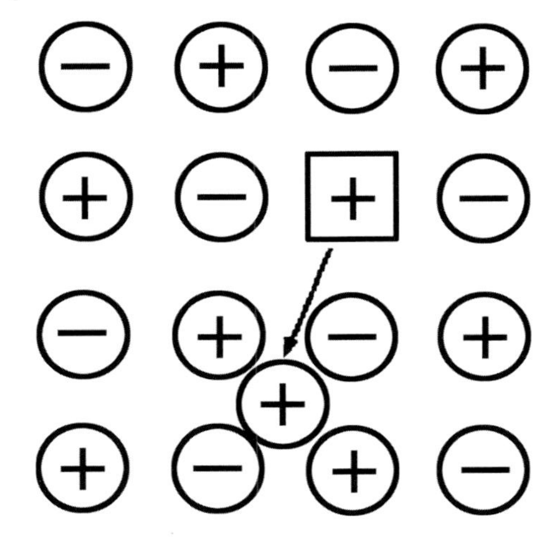

Figure 2.28 A schematic of a Frenkel disorder. After Olander

pair of vacancy and interstitial (Figure 2.28). Only a negligible volume expansion occurs when the Frenkel defects are formed. Frenkel defects can form on both the cation sublattice and the anion sublattice. Anion Frenkel defects are also known as anti-Frenkel. It is to be noted that a vacancy–interstitial pair (known as Frenkel pairs) can be formed in metals under irradiation damage. For an MX-type of ionic crystal, the concentrations of these defects are given by the following:

$$C_{\mathrm{vM}} = C_{\mathrm{iM}} = \exp\left(-E_{\mathrm{F}}/2kT\right) \tag{2.13}$$

and

$$C_{\mathrm{vX}} = C_{\mathrm{iX}} = \exp\left(-E_{\mathrm{F}}/2kT\right), \tag{2.14}$$

where C_{vM} and C_{iM} are the concentrations of the cation Frenkel, C_{vX} and C_{iM} are the anti-Frenkel concentrations, E_{F} is the formation energy for a Frenkel defect, and k, T have the usual meaning. In a given ionic crystal, either Frenkel or anti-Frenkel defects are created but never both kinds.

b) *Schottky defect or disorder* was named after Walter H. Schottky. A Schottky defect is composed of differently charged pairs of vacancies, that is, missing Na^+ and Cl^- ions in the NaCl crystal (Figure 2.29). However, in TiO_2, the Schottky defect consists of one titanium ion vacancy and two oxygen ion vacancies in order to maintain the electrical neutrality of the crystal. Unlike Frenkel defect, Schottky defect is unique to ionic compounds only. Since the number of ions has to stay constant, no matter how many Schottky defects are present, the surplus of the ions must be thought of as sitting on the crystal surface. That is why the crystal

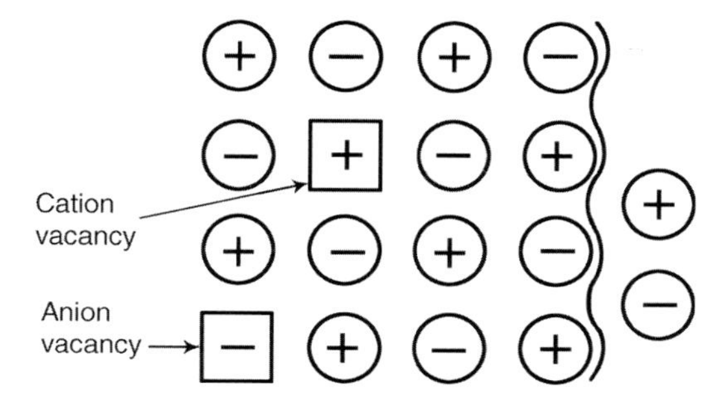

Figure 2.29 A schematic of a Schottky defect in an MX-type ionic lattice. After Olander

expands measurably when Schottky defects are formed. The concentration of Schottky defect pair is expressed as the following:

$$C_{vX} = C_{vM} = \exp(-E_S/2kT), \tag{2.15}$$

where E_S is the Schottky defect formation energy.

Note

Dopants in the ionic crystals can also introduce defects. When the valence of solute is different from that of the host ion, the solute is called *aliovalent*. A divalent substitutional cation impurity in an MX-type crystal creates a cation vacancy to maintain charge neutrality. For example, cation vacancies are created due to the addition of Ca^{2+} in NaCl, and thus maintain the charge neutrality. Hence, q ppm of Ca^{2+} creates $q \times 10^{-6}$ cation vacancies. These are athermal and not dependent on temperature, but equal to the divalent impurity concentration.

Nonstoichiometry is often observed in ceramic compounds where the cation/anion ratio deviates from the ideal stoichiometry (based on the structure of the compound), and as a result multiple ion valence states exist. For example, ferrous oxide has a NaCl-type crystal structure. But it is cation deficient due to the existence of iron (more than 5%) as Fe^{3+}. This implies that there should be cation vacancies or oxygen interstitials. We also know that cation vacancies are created more easily than the oxygen interstitials.

There are more details on defects in ionic crystals. The readers may refer to ceramic books enlisted in Bibliography if needed.

2.2.2
Line Defects

The example of a line defect is dislocation, and it has a very important role to play in plastic deformation of crystalline materials. In this section, we introduce the concept of dislocations; however, the majority of the dislocation theories will be

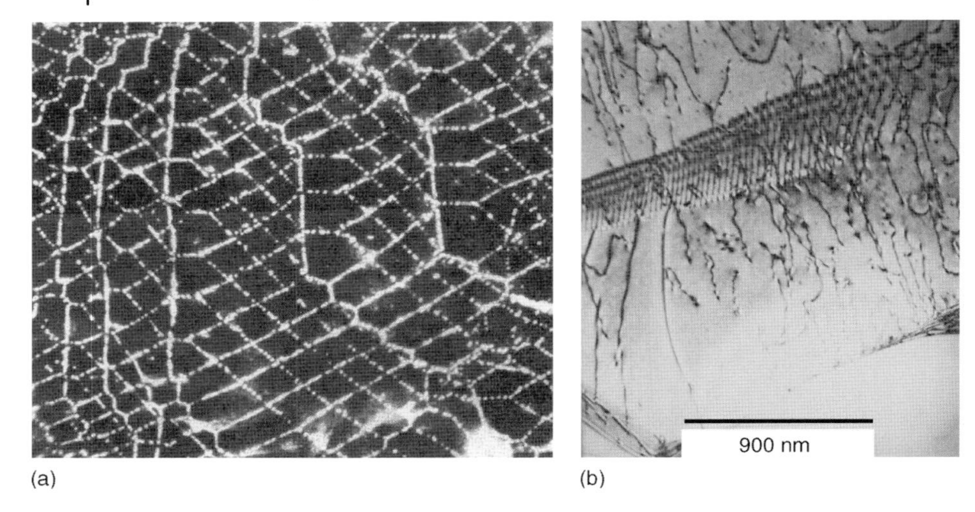

Figure 2.30 (a) Dislocations in a thin KCl crystal are revealed by adding AgCl to the melt prior to crystal growth using the "decoration" method using optical microscopy. *Courtesy:* Taken from Ref. [4]. (b) An array of dislocations seen in a Ti–3Al–2.5V alloy after creep deformation, revealed using transmission electron microscopy [13].

discussed in detail in Chapter 4. Dislocations are not equilibrium defects like point defects because the associated energy is higher than the increase in the enthalpy. They are generally created during solidification, cooling, and mechanical working, and sometimes just by handling. Hence, they are introduced in the crystal in a non-equilibrium way due to the action of mechanical stresses, thermal stresses, collapse of vacancies, or during the precipitate growth and different other events such as exposure to high-energy radiation.

Orowan, Taylor, and Polanyi first conceptualized crystal dislocations during 1930s without directly observing them. The direct evidence of the presence of dislocations was obtained later (during 1950s) using X-ray topography and transmission electron microscopy (TEM) techniques. Some other indirect techniques (such as etch pit method and decoration method) were also used, but none was viable in comparison to TEM (Figure 2.30).

Under normal conditions, plastic deformation occurs through the relative shearing of two crystal parts on a particular plane, called slip plane, along certain crystallographic direction (recall close-packed planes and directions in Section 2.1). It would have taken a lot of energy to create the deformation if the atoms needed to jump all at the same time. It has been estimated from the theories of rigid body shear and some other simple assumptions that the shear stress required to initiate plastic deformation (shear yield stress) in a crystal should be $\sim G/2\pi$, where G is the shear modulus of the crystal. This leads to a huge yield strength number for real metals, which is physically never observed. That is why it led to the belief that the deformation remains localized in a narrow region and propagates through the

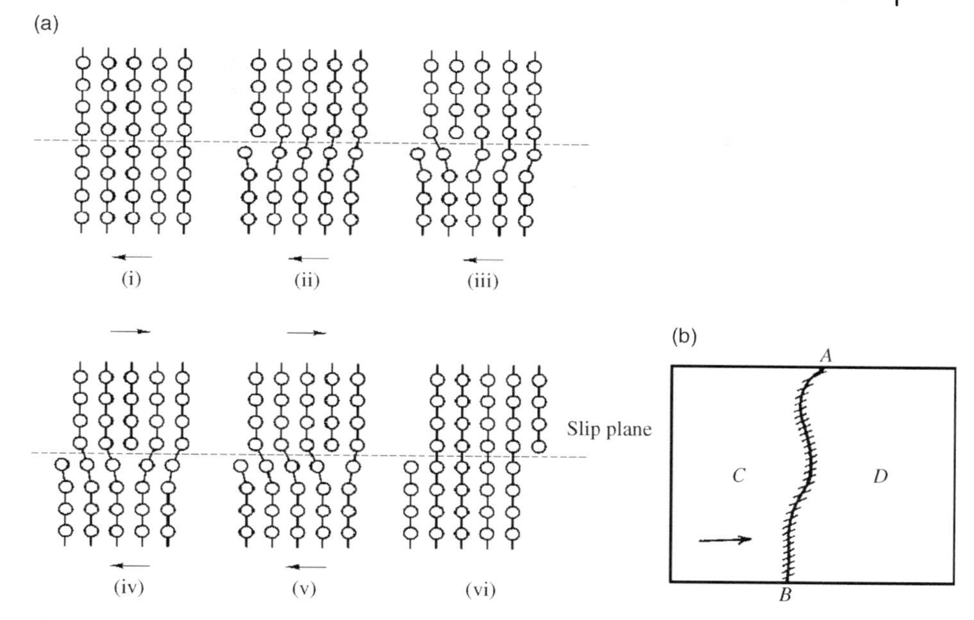

Figure 2.31 (a) Spread of slip through dislocation motion. (b) Dislocation defined as the demarcation line between the unslipped and slipped portions of a crystal.

crystal in a wave-like fashion (Figure 2.31a). Thus, there are some defects that aid in the plastic deformation such that the stress required to initiate plastic deformation becomes quite less. Let us try the case of aluminum. Aluminum has a shear modulus of ~27 GPa; thus, a shear yield stress of ~4.3 GPa. But we know that the shear yield stress of pure aluminum is in the range of 3–10 MPa. There is a difference of several orders of magnitude between the predicted and observed values. This dichotomy can be solved by the inclusion of dislocations in the conversation. Dislocations can be defined as the line defect (AB) that demarcates the unslipped (D) and slipped (C) regions of a crystal (Figure 2.31b).

The dislocation line perpendicular to the slip direction is called *edge or Orowan–Taylor dislocation* (Figure 2.32a), and that parallel to the slip direction is called *screw* or *Burgers dislocation* (Figure 2.32b). But most dislocations remain in a *mixed* configuration as the dislocation line is typically curved. An important characteristic of dislocation is *Burgers vector* (b) that represents the unit slip distance and is always along the slip direction. An edge dislocation is illustrated by inserting an extra half-plane of atoms, thus creating a large disturbance in the atomic configuration in a region just below the extra half-plane. If the half-plane is above the slip plane, the dislocation is called a *positive edge dislocation* (represented by $\perp$), and if it is situated below the slip plane, it is called a *negative edge dislocation* (represented by $\top$). On the other hand, the situation of screw dislocation is little different. The screw dislocation moves in a single surface helicoid, much like a spiral staircase. If we look

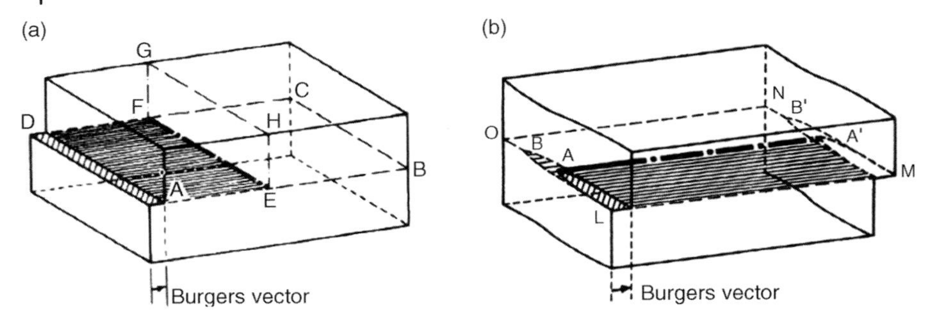

Figure 2.32 (a) A pure edge dislocation at EF. (b) A pure screw dislocation at AA'. *Source:* Taken from Ref. [5].

down on the dislocation and the helix appears to advance in a clockwise circuit, the dislocation is called a *right-handed screw dislocation* (or *positive screw*), and if it is anticlockwise, it is called a *left-handed screw dislocation* (or *negative screw*).

The Burgers vector can be found out by constructing a *Burgers circuit* around the dislocation. Burgers circuit is any atom-to-atom path taken in a crystal that forms a closed path, while the circuit passes through the good part of the crystal. In the presence of a dislocation, the vector needed to close the circuit is the Burgers vector, as illustrated for a schematic edge dislocation configuration in Figure 2.33a. Construction of Burgers circuit for a screw dislocation is shown in Figure 2.33b. Dislocations can end at the crystal surface, internal interfaces (grain boundaries), and so on, but never within the grain unless it forms a node (the sum of the dislocation Burgers vectors is zero at the node, $\Sigma b = 0$) or a closed loop.

A pure edge dislocation can glide or slip perpendicular to its line vector (t) on the slip plane. As the edge dislocation line is normal to its Burgers vector, it remains confined to a specific plane along with its Burgers vector, and hence its glide is limited to a specific plane that contains both the Burgers and line vectors. But it can move vertically leaving the slip plane via a process known as *climb*. The climb process requires addition or subtraction of atoms from the end edge of the half-plane of atoms that characterize the edge dislocation. It implies that atomic diffusion must take place, and most often it occurs through vacancy diffusion

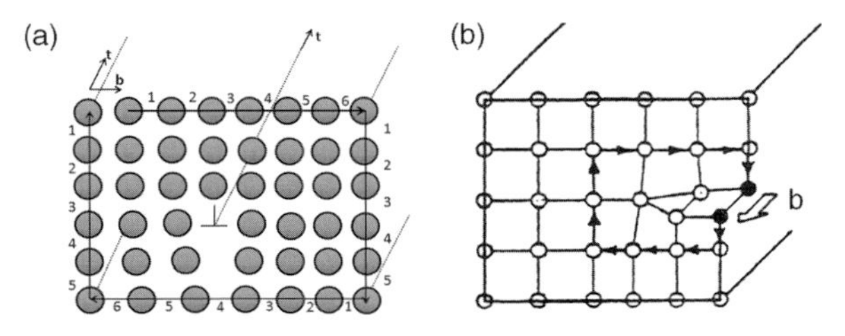

Figure 2.33 Burgers circuit in (a) an edge dislocation and (b) a screw dislocation [5].

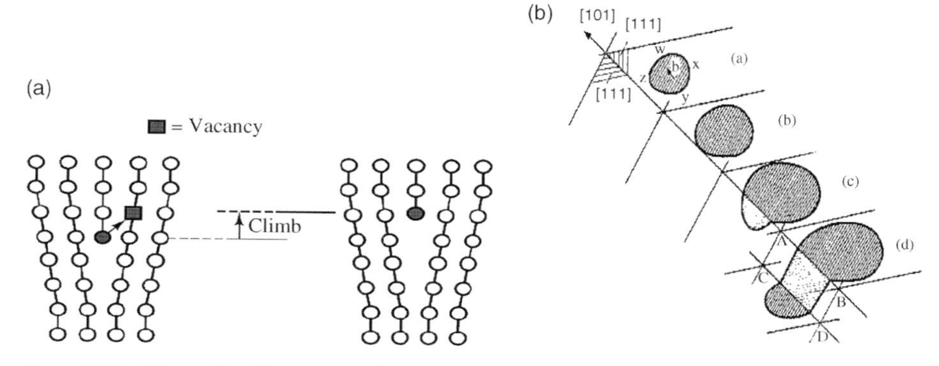

Figure 2.34 Schematics of (a) positive climb of an edge dislocation, and (b) cross-slip in an FCC crystal Ref. [5].

mechanism. Since the climb process requires diffusion, the process is likely only at a higher homologous temperature, and it is generally slower than the glide. Because the climb process requires new atoms coming into or going out of the region, the local mass is not conserved, and that is why it is called nonconservative movement (unlike glide that is called conservative movement). If the dislocation moves vertically upward, the process is called *positive climb* (Figure 2.34a); if the dislocation moves vertically down, it is called *negative climb*. On the other hand, if a screw dislocation having the Burgers vector is parallel to its line vector, it does not have a preferred slip plane and thus the glide of screw dislocation is much less restricted. It is worth noting that screw dislocation cannot climb, but it can leave its slip plane by a process known as *cross-slip* or *cross-glide* (Figure 2.34b). The segments of z and x are the screw components of the dislocation loop shown in Figure 2.34b, and thus are able to cross-slip. Table 2.4 summarizes some interesting features of the edge and screw dislocations in a comparative way. More discussion on this topic will be initiated in Chapter 4.

An important aspect of dislocation microstructure is the *dislocation density* (ϱ_d) so that the number of dislocations in a given volume can be quantified to be used in a variety of relations describing mechanical behavior of crystalline materials. Thus,

Table 2.4 An edge dislocation versus a screw dislocation.

Dislocation feature	Edge	Screw
Relationship between the dislocation line vector (t) and Burgers vector (b)	Perpendicular	Parallel
Slip direction	Parallel to b	Parallel to b
Direction of dislocation line relative to b	Parallel	Perpendicular
Process by which dislocation may leave slip plane	Climb	Cross-slip

dislocation density is primarily defined as "the total line length of dislocations per unit volume." Hence, the unit of dislocation density would be cm cm^{-3}, that is, cm^{-2}. Based on the derived unit, there is another way to define dislocation density: the number of dislocation lines that intersect a unit area. Carefully prepared crystal tends to have a low dislocation density, $\sim 10^2\,cm^{-2}$. Some single crystal whiskers can be made nearly free of any dislocation. On the other hand, heavily deformed metals (cold worked) may contain dislocation density in the range of 10^{10}–$10^{12}\,cm^{-2}$ or more, whereas an annealed crystal may contain 10^6–$10^8\,cm^{-2}$.

2.2.3 Surface Defects

Surface defects are two-dimensional defects, also known as *planar defects*. There are a variety of surface defects such as grain boundaries, twin boundaries, stacking faults, interphase boundaries (coherent, semicoherent, and incoherent), and antiphase boundaries (specifically in intermetallic compounds). Each of these planar defects has an important role to play affecting various properties of crystalline materials.

2.2.3.1 Grain Boundaries

Here, we will define and discuss *grain boundaries*. Grain boundaries play an important role in strengthening in that finer grain sizes lead to higher strength and vice versa, also popularly known as Hall–Petch strengthening. Their presence may lower the thermal/electrical conductivity of the material. They may act as the preferred sites for corrosion (intergranular corrosion), for precipitation of new phases to occur, or may contribute to the plastic deformation or failure at higher temperatures (grain boundary sliding) and many other phenomena. A grain boundary can be defined as the interface boundary between two neighboring grains. In a polycrystalline material, each grain is a single crystal with a particular orientation (Figure 2.35a).

The grain boundaries are the regions of misfit where the atoms are *confused*. When the misorientation angle (θ) between the grains is small ($\sim 10^\circ$), the boundary is called a *low-angle boundary*. Low-angle boundaries can be described as an array of dislocations and are of two types: *tilt* and *twist*. Tilt boundaries can be generated by bending a single crystal with the rotation axis being parallel to the boundary plane (Figure 2.36a), while the twist boundary is created when the rotation axis is normal to the boundary plane (Figure 2.36b). Tilt boundaries can be described as an array of parallel edge dislocations, as illustrated in Figure 2.37. The tilt angle (θ) is given by

$$\tan(\theta) = \theta = b/h, \tag{2.16}$$

where b is the Burgers vector of the dislocation and h is the vertical distance of separation between two neighboring edge dislocations at the boundary. The tilt boundaries are generated during a metallurgical process known as recovery when excess dislocations of the same type arrange themselves one below the other. The

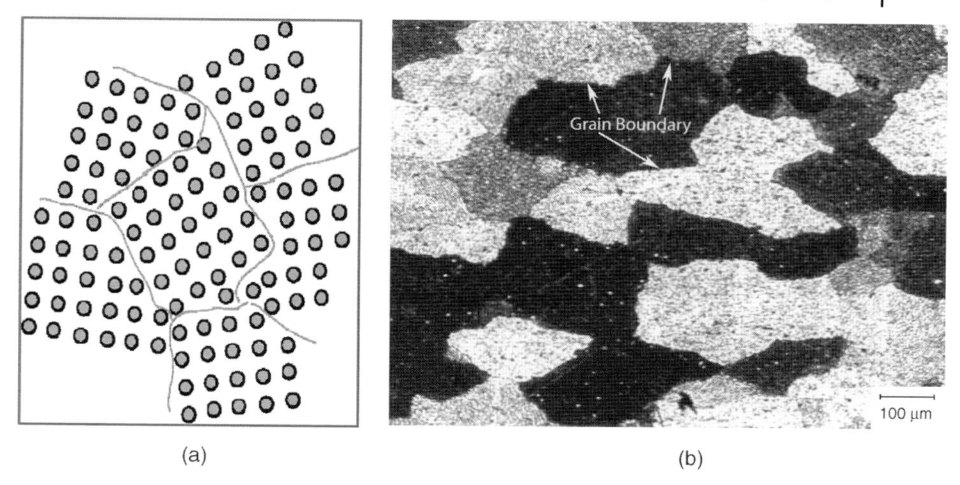

Figure 2.35 (a) A schematic of a polycrystal in which grain boundaries developed as a result of different orientations between adjacent grains. (b) An optical micrograph of a coarser grained 2024 Al alloy with arrows showing three grain boundaries.

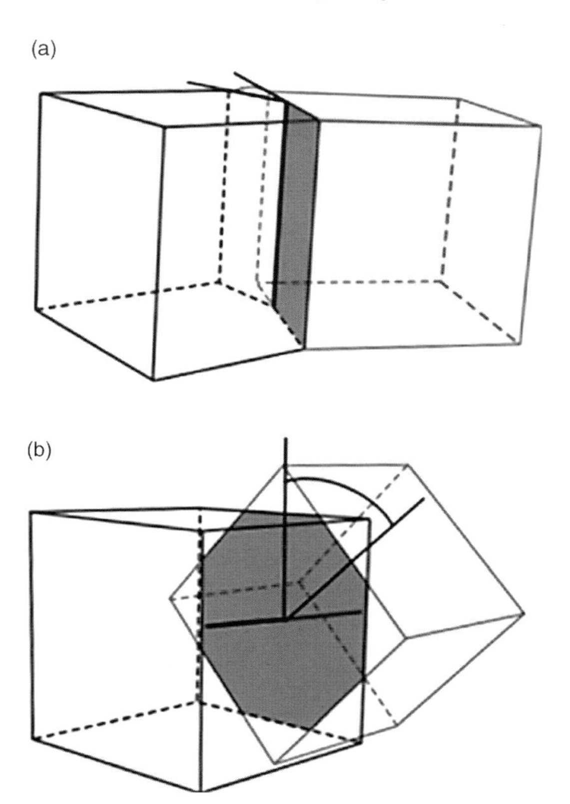

Figure 2.36 The elemental geometrical processes for creating (a) a tilt boundary and (b) a twist boundary.

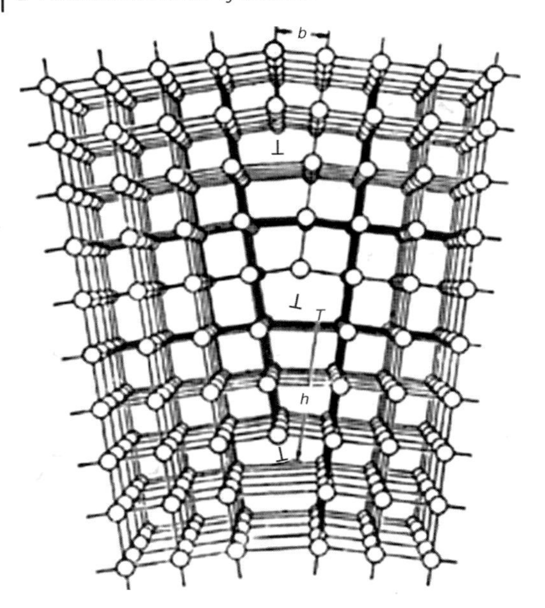

Figure 2.37 The dislocation array model of tilt boundary [12].

twist boundaries refer to similar low-angle boundaries formed by arrays of screw dislocations.

High-angle boundaries ($\theta > 10$–$15°$) can be extrapolated from the simple theory of low-angle grain boundaries. But the dislocation model becomes invalid when there are too many dislocations at the boundary such that the adjacent dislocation cores start overlapping. To overcome this difficulty, grain boundaries are often described by the *coincident site lattice model*; however, a complete discussion is outside the scope of the chapter. The high-angle boundaries generally have a "free space" or "free volume," whereby solutes can collect and a *solute-drag effect* can be generated. The energy of high-angle grain boundary varies from 0.5 to $1.0\,\mathrm{J\,m^{-2}}$ for most metals. Migration of high-angle boundaries occurs due to atom jumps across the boundary during the grain growth, which can be influenced by the grain boundary crystallography, presence of impurities, and temperature.

2.2.3.2 **Twin Boundaries**

In addition to the slip, another method by which the crystal can plastically deform is "twinning." When the arrangement of atoms on one side of a boundary is a mirror image of that on the other side due to a homogeneous shear action, the boundary is known as a twin boundary (Figure 2.38). Twin boundaries appear in pairs where the orientation difference introduced by one boundary is nullified by the other. Twin boundaries are formed during annealing (called annealing twins) or during deformation (deformation twins). Annealing twins generally appear as

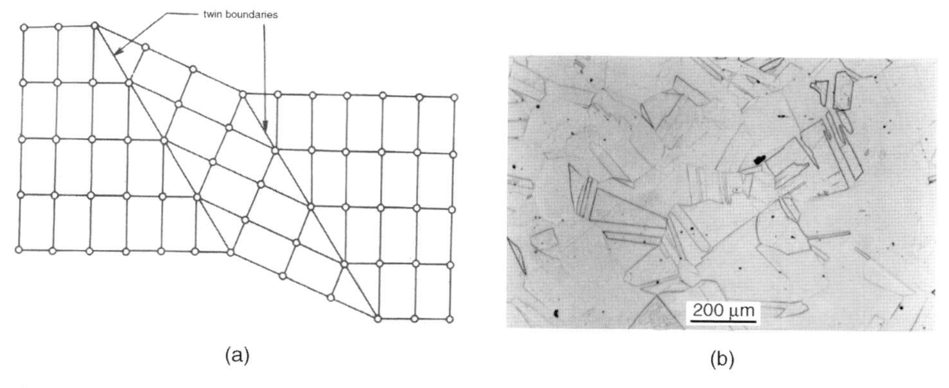

Figure 2.38 (a) A schematic configuration of a twin bound by twin boundaries. (b) An optical micrograph showing the presence of annealing twins in a heat-treated 70–30 brass.

straight edged band, whereas deformation twins appear with a lenticular shape with jagged edges. The energy of twin boundaries is in the range of 0.01–0.05 $\mathrm{J\,m^{-2}}$, which is clearly less than those of the high-angle grain boundaries.

2.2.3.3 Stacking Faults

The close-packed lattices have a specific atom stacking sequence. For example, FCC crystal has ABCABC ... stacking sequence, as shown in Figure 2.39a. Any disruption in this stacking sequence causes local region to get out of perfect sequence, as shown in Figure 2.39b. Locally, this region is improper or faulty, and so this region is called a *stacking fault*. Stacking faults possess surface energy. The surface tension due to this tends to minimize the area of the faulted region. Stacking fault energies of some typical metals/alloys are as follows: Al: 0.2 $\mathrm{J\,m^{-2}}$, Cu: 0.04 $\mathrm{J\,m^{-2}}$, Ni: 0.03 $\mathrm{J\,m^{-2}}$, Cu–25Zn: 0.007 $\mathrm{J\,m^{-2}}$, Fe–18Cr–8Ni: 0.007 $\mathrm{J\,m^{-2}}$. This topic will be revisited in detail in Chapter 4.

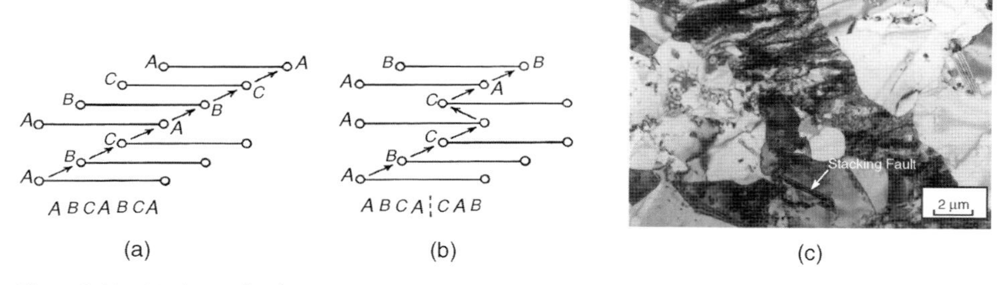

Figure 2.39 (a) The perfect lattice sequence (ABCABC ...) in an FCC lattice [14]. (b) A stacking fault configuration in the same FCC lattice. (c) A TEM image showing the presence of stacking fault in an austenitic stainless steel.

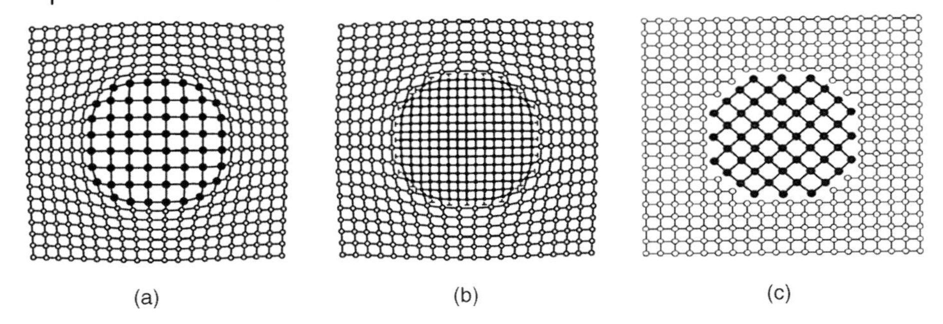

Figure 2.40 (a) A coherent interface. (b) A semicoherent interface. (c) An incoherent interface. The inner shaded region in (a)–(c) represents the particle and rest area is the matrix lattice, and the interface lies in the region where they meet [5].

2.2.3.4 **Other Boundaries**

When the crystal structures of the two contiguous phases are similar and the lattice parameters are nearly equal, the boundary between the two crystals is called a *coherent interface* with one-to-one correspondence of atoms at the interface, as illustrated in Figure 2.40a. The surface energy of the coherent interface is 0.01–0.05 J m^{-2}. When the phase partially loses the coherence with the matrix, the interface is called a *semicoherent interface* (Figure 2.40b). When there is no such similarity or no matching exists, the interface is called *incoherent*, which is quite similar in structure and energy to high-angle grain boundaries. An incoherent interface is shown in Figure 2.40c.

2.2.4 **Volume Defects**

Volume defects are three-dimensional in nature, and include precipitates, dispersoids, inclusions, voids, bubbles, and pores that can occur in materials under different environmental conditions or processing conditions. They do have various important effects on the properties of materials. A TEM picture is shown in Figure 2.41 showing various types of precipitates present in a 2024 Al alloy. We will discuss these vital roles in Chapters 4 and 7 in detail.

2.2.5 **Summary**

This section introduced the concept of crystal defects. Defects are always present in engineering materials in some or the other form. Four types of crystal defects, namely, point, line, surface, and volume defects, are discussed. Each of these defects has profound influence on properties and performance of these materials. This discussion was limited to defects that are observed in crystalline materials under conventional conditions. We will continue the discussion on crystal defects

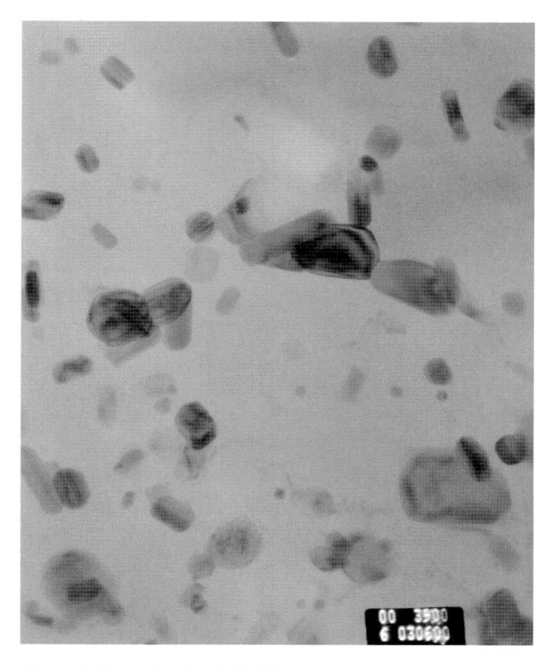

Figure 2.41 A bright field TEM image showing various types of precipitates in a 2024 Al alloy.

in Chapter 3 where we will learn about the defect formation due to the radiation damage.

2.3 Diffusion

No kinetic aspect in materials is as fundamental and important as *diffusion*! Diffusion can be defined as the effective movement of atoms/molecules relative to their neighbors under the influence of a gradient. The process is assisted by the intrinsic thermal or kinetic energy of atoms. The driving force or the gradient can be of various types. It can be chemical potentials arising from the concentration gradient or gradients in electrical field, mechanical stresses, or even gravitational field. The movement of atoms could be over a large number of interatomic distances (i.e., long-range diffusion) or over one or two interatomic distances (i.e., short-range diffusion). Although diffusion in liquid and gaseous states is easier to visualize, diffusion of atoms in solids is not so. Diffusion is regarded as one of the most important mechanisms of mass transport in materials. It is sometimes difficult to assume that atoms remain "diffusible" until the temperature of the solid is brought down to the absolute zero (still a hypothetical situation though)! It is again amazing to know how many well-known materials phenomena are influenced by diffusion. Here are some examples: phase transformations, precipitation, high-temperature creep, high-temperature oxidation of metals, metal joining by diffusion bonding, impurity transistors, grain growth, and *radiation damage defects and their migration.*

There are two general ways by which diffusion can be categorized. If one considers diffusion of atoms in a pure metal, the diffusion happens basically between its own lattice atoms. This diffusion is called *self-diffusion*. On the other hand, diffusion of alloying elements or impurities may well be occurring in the parent lattice and then the diffusion is termed as *heterodiffusion*. In the following section, the macroscopic diffusion theories are first dealt with and then the topics of atomic diffusion.

2.3.1 Phenomenological Theories of Diffusion

If we consider random walk phenomenon as a general event in atomic diffusion, the net movement of atoms (or net motion of point defects) through a homogeneous crystal will be zero. However, if one assumes that the impurity concentration is more in region A than in region B, a net movement of atoms occurs from region A to region B until the concentrations of impurities in both the regions become uniform. In a simple kind of picture, a net movement of atoms occurs along a concentration gradient. However, without knowing the details of atom movement, one can come up with the phenomenological diffusion theories as apparent in Fick's laws for diffusion.

2.3.1.1 Fick's First Law

A German physiologist (not a materials scientist!), Adolf Fick first proposed the laws of diffusion in 1855. These laws are applicable to all three general states (gas, liquid, and solid) of matter. These "macroscopic" laws are basically continuum diffusion in nature, and are very akin to equations of heat conduction (Fourier's laws) or electrical conduction (Ohm's law), and do not explicitly take into account any defect-assisted atomistic or microscopic mechanisms involved in diffusion processes.

The first law states that the net flux of solute atoms in a solution will occur from the regions of high solute concentrations to the regions of low concentrations, and the solute flux along a particular coordinate axis is directly proportional to the concentration gradient. In a unidirectional (say, along x-axis) diffusion event, it is expressed as

$$J = -D\frac{\partial c}{\partial x}, \tag{2.17}$$

where J is the diffusion flux, D is the proportionality factor known as diffusion coefficient (or simply diffusivity), c is the concentration of the diffusing species, and x is the diffusion distance. The negative sign on the right-hand side of Eq. (2.17) implies that the diffusion or the mass flow occurs down the concentration gradient. A schematic representation of Fick's law is shown in Figure 2.42. One assumption is inbuilt with Fick's first law. It applies to a *steady-state* (or nearly steady-state) condition only. It is important to understand what the steady-state flow

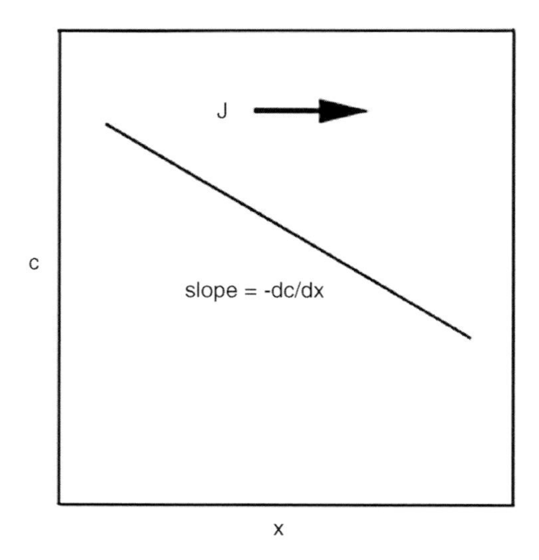

Figure 2.42 The concentration gradient ($dc/dx < 0$) is shown with the direction of diffusive flux (J) in the case of $D \neq f(c)$. The concentration gradient is not a straight line if $D = f(c)$.

means. In such a state, the diffusion flux at any cross-sectional plane across the diffusion distance remains constant (i.e., time-independent).

Flux (J) is the net amount atoms diffusing normal to an imaginary plane per unit area in a unit time (much like the definition of heat flux or neutron flux). That is why the unit of diffusion flux is generally given in atoms per m^2 s (or mol m^{-2} s^{-1}). The concentration c is given in unit of atoms per m^3 (or mol m^{-3}). If the dimension of diffusion distance (x) is given in m, the unit of diffusivity (m^2 s^{-1}) can then be found through dimensional analysis. Diffusivity (D) is a significant parameter that is directly linked to the atomistic and defect mechanisms. D depends on factors such as temperature and composition. Figure 2.43 shows the variation of diffusion coefficients in three gold alloys as a function of composition. The variation in diffusivity with composition is especially marked in Au–Ni alloy, followed by Au–Pd and Au–Pt. However, the diffusivity of species would be practically independent of concentration if present in a trace quantity. Sometimes, D is considered to be independent of composition to preserve the simplicity of the mathematical solutions.

2.3.1.2 **Fick's Second Law**

Although Fick's first law is helpful, it cannot express concentration profile at a point as a function of time. It is because Fick's first law cannot link the concentration gradient at a point to the rate at which the concentration changes within a fixed volume element of a material. That is why Fick's second law was derived from embodying Fick's first law and the law of mass conservation. The resulting law can deal with *nonsteady-state diffusion* phenomena satisfactorily. Nonsteady-state

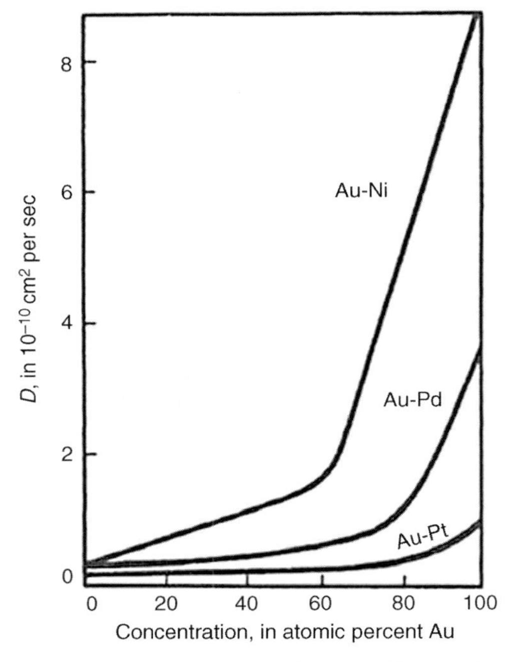

Figure 2.43 The variation of diffusion coefficient with composition in Au–Ni, Au–Pd, and Au–Pt alloys. *Courtesy:* Taken from Ref. [6].

diffusion refers to a situation when the diffusion fluxes at various cross-sectional planes across the diffusion distance are not constant; they vary with time. Thus, the concentration profile changes over time. In fact, this situation is found in most practical cases.

Let us derive Fick's second law using a simple method. At the onset, one needs to assume that the concentration change in a volume element can be calculated if one knows how much flux is getting in (J_{in}) and how much is getting out (J_{out}) in a given volume element. Also, consider two parallel planes in a volume element of thickness dx. Then, the flux through the first plane is

$$J_{in} = -D\frac{\partial c}{\partial x}. \tag{2.18}$$

On the other hand, the flux from the second plane is

$$J_{out} = J_{in} + \frac{\partial J}{\partial x}\mathrm{d}x = -D\frac{\partial c}{\partial x} + \frac{\partial}{\partial x}\left(-D\frac{\partial c}{\partial x}\right)\mathrm{d}x. \tag{2.19}$$

By subtracting Eq. (2.18) from Eq. (2.19), we get the following relation:

$$\frac{\partial J}{\partial x} = -\frac{\partial}{\partial x}\left(D\frac{\partial c}{\partial x}\right). \tag{2.20}$$

It can also be shown that

$$\frac{\partial c}{\partial t}\mathrm{d}x = -(J_{\text{out}} - J_{\text{in}}) = -\frac{\partial J}{\partial x}\mathrm{d}x.$$

$$\text{Therefore,} \quad \frac{\partial J}{\partial x} = -\frac{\partial c}{\partial t}. \tag{2.21}$$

Comparing Eqs (2.20) and (2.21), we get

$$\frac{\partial c}{\partial t} = \frac{\partial}{\partial x}\left(D\frac{\partial c}{\partial x}\right) = D\frac{\partial^2 c}{\partial x^2}. \tag{2.22}$$

One can only take diffusivity (*D*) term out of the del operator as constant if it is assumed that the diffusivity does not change with concentration, and thus with distance. One of the great significances of Fick's second law is that the equation can be solved to describe the concentration profile *c*(*x*,*t*) as a function of time and position given the appropriate initial and boundary conditions. A couple of examples are given in the following sections. More cases can be found in standard diffusion textbooks.

Thin-Film Solution

If a finite thin-film source contains an initial amount of solute *M* (in mol m^{-2}), the concentration profile of an infinite volume at both sides is given by the following relation (see Figure 2.44 for details):

$$c(x,t) = \frac{M}{2\sqrt{\pi D t}}\exp\left(-\frac{x^2}{4Dt}\right). \tag{2.23}$$

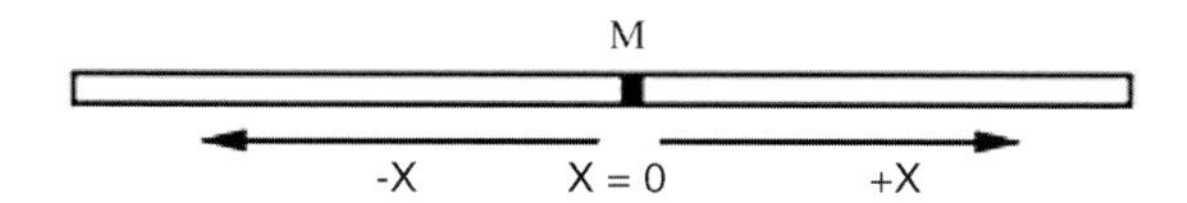

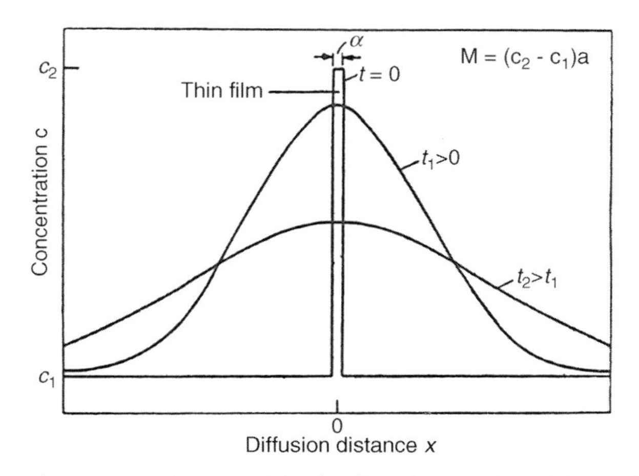

Figure 2.44 A depiction of the thin-film solution. Adapted from Ref. [7].

This has significance in determining D. If the solute present in concentration M in the thin film is a radioactive isotope of the matrix element, the diffusion equation can help in determining self-diffusion characteristics by taking logarithm on both sides of Eq. (2.23):

$$\ln c = \frac{-x^2}{4Dt} + \ln\left(\frac{M}{2\sqrt{\pi Dt}}\right). \tag{2.24}$$

If one plots ln (c) against x^2 using experimental data, a straight line with a slope of $(4Dt)^{-1}$ is easily obtained and the diffusivity D is calculated since t would be known. Remember that D does depend on temperature, and suitable experimental data collection at different temperatures will give more details like activation energy for diffusion.

Carburizing/Decarburizing

If the composition of the surface is changed from the initial composition c_0 to c_s and is maintained at that level, the concentration profile at any point in the semi-infinite volume is given by

$$c(x,t) = c_s - (c_s - c_0)\,\text{erf}\left(\frac{x}{2\sqrt{Dt}}\right), \tag{2.25}$$

where "erf" or the error function is an integral defined by the following expression:

$$\text{erf}\,(z) = \frac{2}{\sqrt{\pi}}\int_0^z \exp\,(-\xi^2)\,\text{d}\xi. \tag{2.26}$$

Equation (2.25) is applicable for describing the concentration–distance profile during carburization (surface hardening through addition of carbon atoms) and decarburization (taking away carbon atoms from the surface layer of a material) of steels.

For $z \geq 1$, $\text{erf}\,(z) =\sim 1$; for $1 \geq z \geq -1$, $\text{erf}\,(z) = z$; and for $z \leq -1$, $\text{erf}\,(z) =\sim -1$.

Furthermore, relations such as $\text{erf}(0) = 0$, $\text{erf}(\infty) = 1$, $\text{erf}(-z) = -\text{erf}(z)$, and, for very small values of x, $\text{erf}\,(z) = 2z/\sqrt{\pi}$ are helpful in determining the concentration profiles. Table 2.5 lists the error function values for different x values. Detailed error function tables are generally available in the books that deal exclusively with diffusion and can be found in the references cited at the end of this chapter.

One extended case would be the determination of the concentration profile of two semi-infinite plates of different solute concentrations of c_1 and c_2 joined together (also known as *diffusion couple*). In this case, the appropriate equation is

$$c(x,t) = [(c_2 + c_1)/2] - [(c_2 - c_1)/2]\,\text{erf}\left(\frac{x}{2\sqrt{Dt}}\right). \tag{2.27}$$

Table 2.5 The error function.

z	erf(z)	z	erf(z)
0	0	0.6	0.6039
0.05	0.0564	0.7	0.6778
0.10	0.1125	0.8	0.7421
0.2	0.2227	0.9	0.7970
0.3	0.3268	1.0	0.8427
0.4	0.4284	1.1	0.8802
0.5	0.5205	1.2	0.9103

Note

In many applications of Fick's second law, the solution is of the following form:

$$f(\text{concentrations}) = \frac{x}{\sqrt{Dt}},$$

where the function involving concentration at a specific point depends on the initial concentration and surface concentration. But when the concentration does not change, $x/\sqrt{Dt}$ is essentially constant. In many instances, this relation can be used to solve diffusion distance/time problems in a much simpler way.

2.3.2 Atomic Theories of Diffusion

Random walk diffusion refers to the situation where an atom can jump from one site to another neighboring site with equal probability. Interestingly, one can derive Fick's first law from the theories of random walk diffusion. Note that the following derivation does not assume any particular micromechanism. Consider two adjacent crystal planes (A and B) that are λ distance apart, and there are n_A number of diffusing species per unit area in plane A and n_B in plane B with $n_A > n_B$ in x-direction only. Γ is the number of atoms making jumps from one plane to the other neighboring plane per second (i.e., jump frequency). If we consider the fact that the atomic jump may be either along the forward direction (A to B) or along the backward direction (B to A), the following equations can be written:

$$J_{A\to B} = \frac{1}{2} n_A \Gamma. \tag{2.28}$$

$$J_{B\to A} = \frac{1}{2} n_B \Gamma. \tag{2.29}$$

Then, the net flux crossing over from plane A to plane B is

$$J_{\text{net}} = \frac{1}{2}(n_A - n_B)\Gamma. \tag{2.30}$$

In terms of concentration (number of diffusing species per unit volume), one can write $n_A = c_A\lambda$ and $n_B = c_B\lambda$; Eq. (2.30) then becomes

$$J_{net} = \frac{1}{2}\lambda(c_A - c_B)\Gamma. \tag{2.31}$$

However, $$c_A - c_B = -\lambda\frac{\partial c}{\partial x}. \tag{2.32}$$

Then, Eq. (2.31) can be reduced to

$$J_{net} = -\frac{1}{2}\lambda^2\Gamma\frac{\partial c}{\partial x}. \tag{2.33}$$

Comparing Eq. (2.33) with Eq. (2.18), that is, Fick's first law, we can write

$$D = \frac{1}{2}\lambda^2\Gamma. \tag{2.34}$$

Hence, a general form of diffusivity is obtained from this atomic theory, which could not be obtained from the continuum theory of diffusion. Diffusivity is the product of a geometric factor, square of diffusion jump distance, and the jump frequency. The geometrical factor for a simple cubic lattice would be 1/6 because of the atomic jump that can take place in three orthogonal directions and thus

$$D = \frac{1}{6}\lambda^2\Gamma. \tag{2.35}$$

If we wish to describe the geometric factor in a more general form, it can be described by $\chi = 1/N_c$, where N_c is the number of adjacent sites to which the atom may jump. So, λ in general is the distance between such sites. We can take an example from UC that has a NaCl-type structure (see Section 2.1). In ionic crystals, cation can jump to equivalent cation sites and anion to equivalent anion sites, and these sites need to be neighboring sites. Now the coordination number for U cation and C anion in the UC crystal structure is 6 and the jump distance is the length associated with the vector, $(1/2)\langle 110\rangle$. The distance between the two jump sites along $\langle 110\rangle$ would be $a_0/\sqrt{2}$. Hence, the diffusivity term can be expressed by

$$D = \chi\lambda^2\Gamma = \frac{1}{12}\left(\frac{a_0}{\sqrt{2}}\right)^2\Gamma, \tag{2.36}$$

where a_0 is the lattice constant.

Note that the geometric factor and the jump distance do not vary much; however, diffusivity value varies a lot among between different materials. This variability is attributed to the dependence of jump frequency on the energy of migration, temperature, and probability of adequate jump sites, which in turn depends on the defect concentration.

Einstein arrived at the similar expression by directly using random walk theory. He derived the following relations for mean square displacement $(\overline{r^2})$:

$$\overline{r^2} = \Gamma\tau\lambda^2. \tag{2.37}$$

$$\overline{r^2} = 6Dt, \tag{2.38}$$

where t is the time interval between jumps and other terms are as defined before. Comparing Eqs. (2.37) and (2.38), the relation $D = (1/6)\lambda^2\Gamma$ is obtained (similar to the relation obtained in Eq. (2.34)). A more detailed account of this approach can be found in the books listed in Bibliography.

2.3.3
Atomic Diffusion Mechanisms

Here, a number of diffusion mechanisms from an atomistic viewpoint are discussed. The simplest picture of diffusion can be visualized through diffusion of interstitial atoms through a lattice. In this mechanism, interstitial atoms move from one interstitial site to another. In the dilute interstitial solid solutions, the probability of finding an interstitial site is very high and close to unity. Note that Figure 2.45a shows a two-dimensional schematic illustration of a monatomic crystal with a very few number of impurity atoms that are of much smaller size than the interstitial sites themselves. So the impurity atom can squeeze past through the host lattice atoms to fall into another interstitial site. While it tries to go past the host atoms, repulsive forces would act on the impurity atom and so energy would be needed to surmount the barrier. That energy is supplied by the thermal energy of the interstitial atom. Figure 2.45b shows such an *activated state*. Following this state, the interstitial atom falls into a new interstitial position completing one jump (Figure 2.45c). If we calculate the energy of an atom as a function of position, we would see that energy is minimum when the impurity atom is at normal position (as shown in Figure 2.45a and b) and the maximum is at the midway between the two positions (i.e., at the activated state). The situation is shown in Figure 2.45d. The amount of this energy barrier is given by the difference between the energy at the activated state and that at the normal state, and is referred to as the activation energy for interstitial diffusion. The real event may consist of a series of such unit atomic jumps. Examples of such *interstitial* mechanism may comprise diffusion of C inside any allotropic form of iron (alpha iron, gamma iron, or delta iron) or hydrogen diffusion in zirconium, and so on.

Now, let us consider the atomic mechanism by which self-diffusion may occur. In self-diffusion, like atoms exchange lattice positions leaving the lattice identical before and after diffusion. One of the simplest modes of this is the *direct exchange* mechanism. In this mode, atom X can move to the site of lattice atom Y and at the same time, atom Y moves to the site of atom X (Figure 2.46). But such a direct exchange of atoms is not at all energetically favorable. This may seem implausible even in a very open structure as there are other mechanisms that can actually achieve the same result without expending that much energy. *Ring mechanism* is one such example. A four-ring mechanism is also depicted in Figure 2.46 (right). As is evident, a greater coordination between atoms is essential for this mechanism to have any consequence. Some evidences suggest that self-diffusion in chromium and sodium may occur by ring mechanism. However, most metals and other engineering materials are, in general, too close-packed for that mechanism to occur. That is to say that self-diffusion activation energy associated with direct exchange and four-ring mechanisms will be always higher than what has commonly been observed such as through vacancy mechanism.

The predominant way by which the self-diffusion or self-substitutional atoms can diffuse is by changing its position with a neighboring vacant site. Figure 2.47a–c shows the different steps involved in substitutional atom diffusion. The same applies to self-diffusion, but for clarity of the process, it is shown in terms of substitutional atoms. Repetition of this process can result in the transfer of matter over

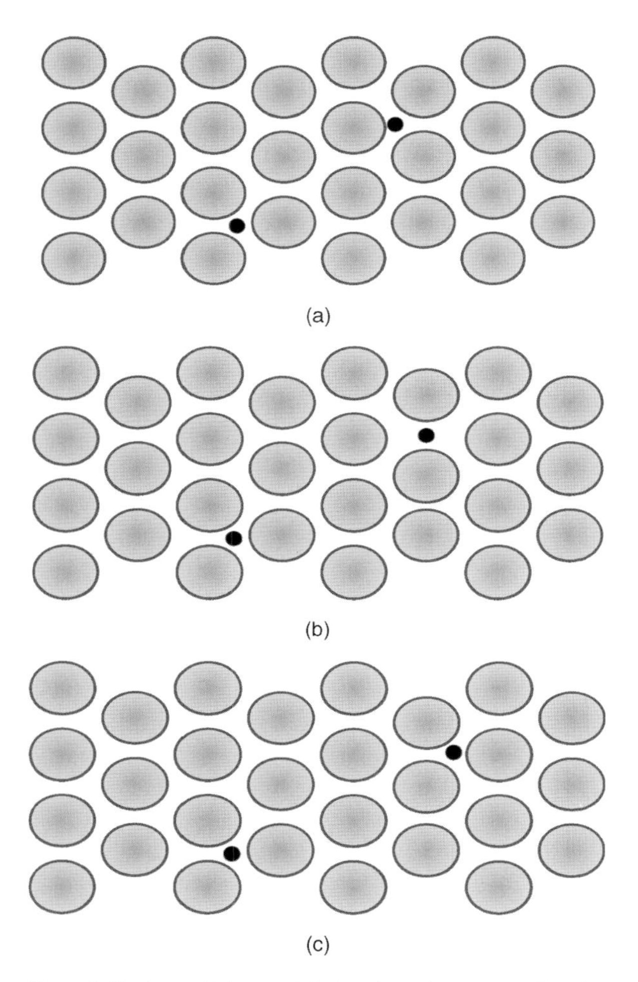

Figure 2.45 Interstitial atom diffusion through a 2D crystal lattice (host atom – larger filled circle; interstitial atom – smaller filled circle). (a) Initial configuration of atoms. (b) One interstitial atom in "activated state" squeezing past two host lattice atoms to the neighboring interstitial site. (c) Configuration of the interstitial atom after diffusion. (d) Energy of an impurity atom as a function of position (E^* is the activation barrier).

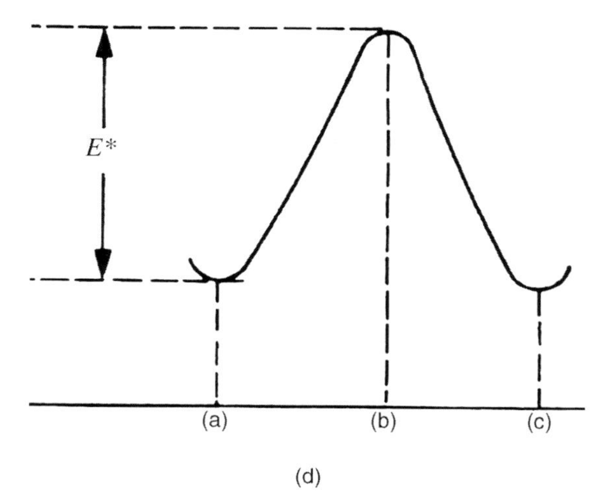

Figure 2.45 *(Continued)*

large atomic distances. This is known as the "vacancy mechanism of diffusion." There is always an equilibrium number of vacant lattice sites (thermal vacancies) present at any particular temperature and their concentration increases with increasing temperature, as seen in Section 2.2 (Eq. (2.7)). Likewise, self-diffusion also takes place through this mechanism. It is also instructive to note that atom jump can occur into divacancies. However, larger vacancy agglomerates like trivacancy and quadrivacancy are relatively immobile, and do not take part in general diffusion.

In the *interstitialcy* mechanism, an atom from a regular lattice site jumps into a neighboring interstitial site that is too small to accommodate it fully. As a result, it displaces another atom from a regular lattice site. Hence, both the atoms share a common site, although displaced from their original lattice sites.

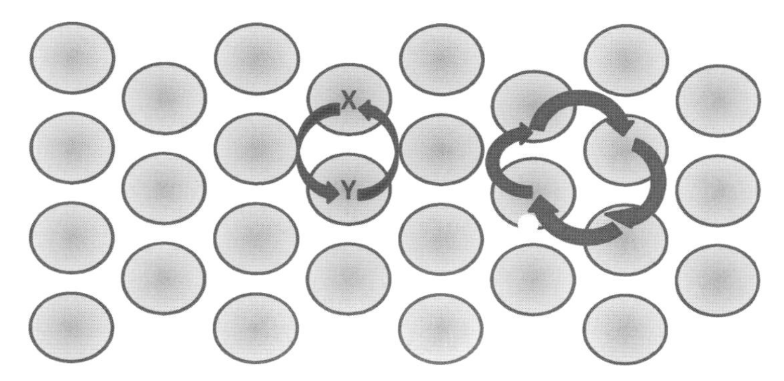

Figure 2.46 Direct exchange mechanism and ring mechanism in a 2D lattice.

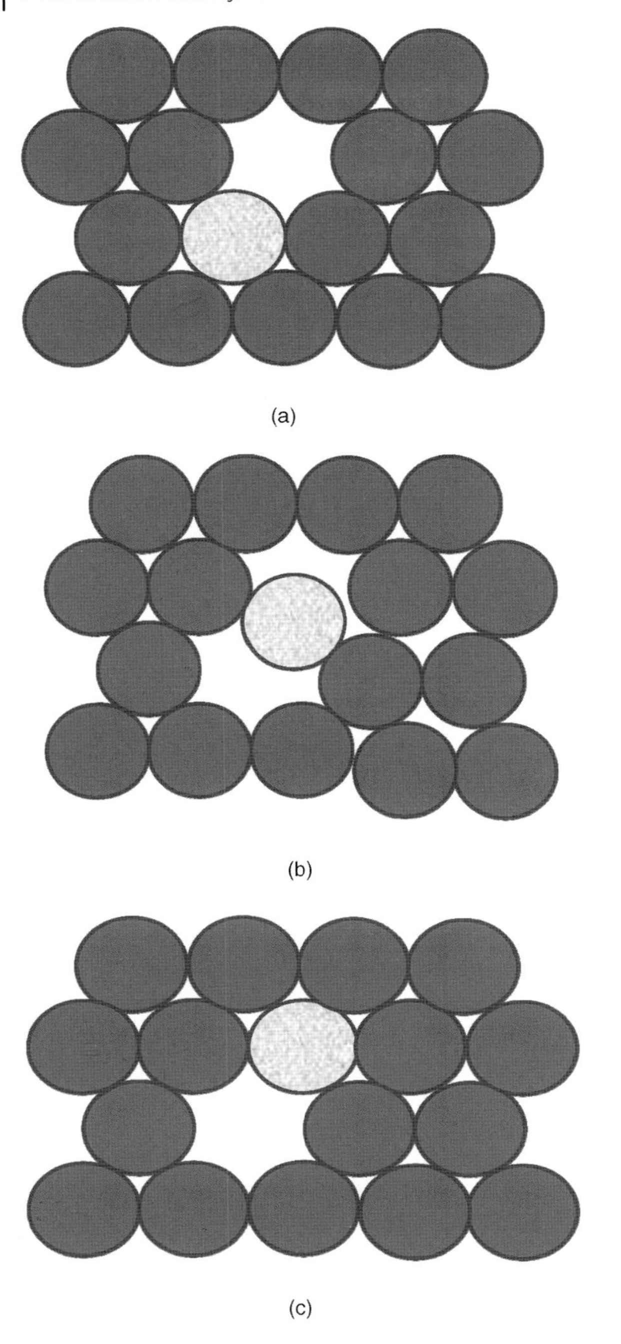

Figure 2.47 Schematic illustration of diffusion of a substitutional atom by a vacancy mechanism. (a) Initial configuration. (b) Activated state. (c) Final configuration.

2.3.4
Diffusion as a Thermally Activated Process

It is, in general, intuitive that diffusion will enhance with increasing temperature. Here, we attempt to derive diffusivity as a function of temperature from atomic theories as applicable to the vacancy mechanism of diffusion. For diffusion of atoms to occur, there should be availability of vacant neighboring sites around the atom. If Γ (as defined before) is proportional to the number of nearest neighbors (β) and the probability of finding neighboring lattice sites vacant or vacancy concentration (C_v), the following mathematical relation can be written:

$$\Gamma = \beta C_v \omega, \tag{2.39}$$

where ω is the atomic jump frequency. Hence, the number of successful jumps by an atom can be given by

$$\omega = \nu_D \exp\left(-\Delta G_m / RT\right), \tag{2.40}$$

where ν_D is the lattice vibration frequency (same as the Debye frequency, typically $\sim 10^{13}\,s^{-1}$, which is defined as the theoretical maximum frequency of vibration that make up the diffusion medium crystal), ΔG_m is the free energy maximum (in calories per mole) along the diffusion path or activation barrier to vacancy migration (also known as free energy for migration), and T is the temperature (in K).

From Eq. (2.35), we know $D = (1/6)\lambda^2\Gamma$. By replacing the Γ expression using Eqs. (2.39) and (2.40), we obtain

$$D = \frac{1}{6}\lambda^2 \beta C_v \nu_D \exp(-\Delta G_m / RT). \tag{2.41}$$

We know from earlier chapters

$C_v = \exp(-\Delta G_f / RT)$, where ΔG_f is the free energy of vacancy formation (see Section 2.2).

Using the general relation from the second law of thermodynamics, $\Delta G = \Delta H - T\Delta S$, we can write

$$D = \frac{1}{6}\lambda^2 \beta \exp\left(-\frac{\Delta H_f}{RT}\right)\exp\left(\frac{\Delta S_f}{R}\right)\nu_D \exp\left(-\frac{\Delta H_m}{RT}\right)\exp\left(\frac{\Delta S_m}{R}\right). \tag{2.42}$$

Rearranging Eq. (2.42), we get

$$D = \frac{1}{6}\lambda^2 \beta \nu_D \exp\left(\frac{\Delta S_f + \Delta S_m}{R}\right)\exp\left(-\frac{\Delta H_f + \Delta H_m}{RT}\right). \tag{2.43}$$

The above equation can also be written as

$$D = D_0 \exp\left(-Q_{SD}/RT\right), \tag{2.44}$$

$$\text{where} \quad D_0 = \frac{1}{6}\lambda^2 \beta \nu_D \exp\left(\frac{\Delta S_f + \Delta S_m}{R}\right) \tag{2.45}$$

$$\text{and} \quad Q_{SD} = \Delta H_m + \Delta H_f. \tag{2.46}$$

D_0 is called the frequency factor and Q_{SD} is called the activation energy for self-diffusion. D_0 and Q_{SD} can be obtained by measuring D at different temperatures from experiments. A plot of ln (D) versus $1/T$ yields a straight line and the slope equals $-Q/R$ and the intercept on the y-axis is ln (D_0). It is interesting to note that the final diffusivity term does not contain the defect concentration term, rather it has the activation enthalpy for vacancy formation. The derivation is equally applicable for substitutional vacancy diffusion and interstitial diffusion mechanisms. On assigning approximate values to the terms in Eq. (2.45) and considering a small positive value of the entropy term, D_0 is generally found to be between 10^{-3} and $10\,\text{cm}^2\,\text{s}^{-1}$.

Example Problem

Let us assume the following relation is applicable for determining the lattice self-diffusivity value in copper (FCC, lattice parameter $a_0 = 0.3615$ nm):

$$D_L = \frac{b}{2\delta}\lambda^2 \nu_D \exp\left(-\frac{Q_L}{RT}\right),$$

where β is the number of positions an atom can jump to, δ is the dimension (if for one-dimensional flow, $\delta = 1$; for two-dimensional flow, $\delta = 2$; and for three-dimensional flow, $\delta = 3$), and other terms are already defined in Eq. (2.43).

Determine β, λ, and D_L, given that diffusion takes place along $\langle 110\rangle$ direction at 500 °C (773 K) and $Q_L = 209\,\text{kJ mol}^{-1}$.

Solution

For an FCC along $\langle 110\rangle$ direction, $\beta = 12$, and λ is given by half of the face diagonal length ($\sqrt{2}\,a_0/2$), that is, 0.2566 nm or 2.566×10^{-8} cm.

Therefore, using the given equation, we obtain

$$D_L = \frac{12}{2 \times 3}(2.556 \times 10^{-8}\,\text{cm})^2(10^{13}\,\text{s}^{-1})\exp\left(-\frac{209\,000\,\text{J mol}^{-1}}{8.314\,\text{J mol}^{-1}\,\text{K}^{-1} \times 773\,\text{K}}\right)$$
$$= 9.83 \times 10^{-14}\,\text{cm}^2\,\text{s}^{-1}.$$

The activation energy for vacancy diffusion is composed of two terms, activation enthalpy (or energy) for migration and activation enthalpy for vacancy formation. Calculated and experimental activation energies for diffusion in gold and silver through vacancy mechanism are shown in Table 2.6. The activation energy for diffusion of self-interstitials appears in the form similar to the vacancy diffusion (i.e., both formation energy and migration energy are included). Although for interstitial impurity diffusion the formulation is almost the same, for vacancy diffusion, it does not involve any probability factor similar to the vacancy formation energy; instead, it contains only the migration energy term. That is why the substitutional diffusion (including self-diffusion) occurring through vacancy mechanism is much slower than the interstitial impurity diffusion. See the example in Figure 2.48

Table 2.6 Calculated and experimental activation energies (in kJmol^{-1}) for diffusion by the vacancy mechanism.

Element	ΔH_m	ΔH_m	$\Delta H_m + \Delta H_f$	Q
Au	79	95	174	184
Ag	80	97	177	174

Courtesy: Taken from Ref. [7].

showing carbon (interstitial) diffusion in γ-iron, Cr substitutional diffusion, and self-diffusion in γ-iron. Substitutional impurity diffusion is also influenced by the atom size and charge effects of the impurities. Generally, oversized (compared to the host atom) substitutional impurities have a higher migration energy than that of the undersized substitutional impurities. Increase in the valence of the substitutional impurity atom has been found to reduce the activation energy. When solute–vacancy complexes are created, they would also affect the diffusion.

For a given crystal structure and bond type, Q_{self}/RT_m is more or less constant, where T_m is the melting temperature (K). It has been found that most close-packed metals tend to possess a Q_{self}/RT_m of ~18. The activation energy for self-diffusion is proportional to the melting temperature. For example, Figure 2.49 gives the activation energy for self-diffusion of various FCC metals plotted against their melting temperatures. Correlations by Sherby and Simnad [8] revealed the following relation between the activation energy for self-diffusion, melting point, and valence:

$$Q_{self} = R(K_0 + V)T_m, \tag{2.47}$$

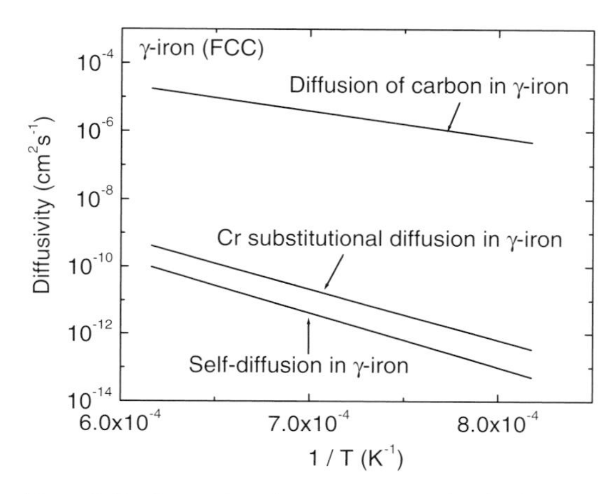

Figure 2.48 Comparison between interstitial and substitutional impurity diffusion; and self-diffusion in γ-iron.

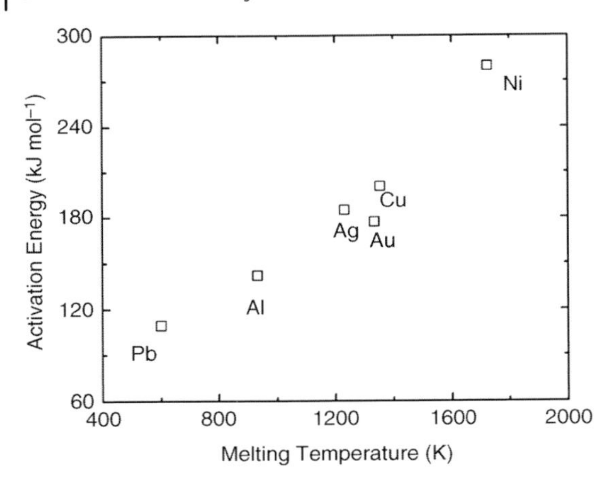

Figure 2.49 Activation energy for lattice self-diffusion versus melting temperature for various FCC metals [8].

where $R = 1.987\ \text{cal mol}^{-1}\,\text{K}^{-1}$, V is the valence, T_m is the melting point in K, and K_0 depends on the crystal structure,

$$K_0 = \begin{cases} 14 \text{ for BCC} \\ 16 \text{ for HCP} \\ 18 \text{ for FCC} \\ 20 \text{ for diamond structure} \end{cases} \quad (2.48)$$

While this formulation worked well for many metals, some metals such as Zr, Ti, Hf, U, and Pu were noted to deviate from the predictions.

Special Note

Direct measurement of self-diffusivities is not easy because of the difficulty in tracking identical, individual atoms. That is why tracers (such as radioisotopes) that are chemically the same as the host atoms but detectable by analytical methods are often used. However, tracer diffusivity is quite close to the self-diffusivity, yet it is little less. It occurs because the tracer jumps could be *correlated* in that they can jump back to the sites wherefrom they originally started the jump. Because of these basically "wasted" jumps, the tracer diffusivity (D_{tracer}) values are less than the self-diffusivity (D_{self}), that is, $D_{tracer} = f\, D_{self}$, where f is the correlation factor. Based on exhaustive geometric principles, the correlation factors can be found out. For vacancy diffusion, the following are the correlation factors for different crystal structures (0.781 for FCC, 0.721 for HCP, and 0.655 for BCC). In general, we can see that the variation due to the correlation factors is mostly quite small since the accuracy with which diffusivity can be determined is limited. Furthermore, consideration of correlated jumps is required in the case of associated defects (such as the solute–vacancy complex).

2.3.5
Diffusion in Multicomponent Systems

Solute diffusion in a dilute alloy can be treated with a simple assumption that the environment the solute sees while diffusing almost entirely consists of host lattice atoms. The same is not true for a diffusion couple, say metal A and metal B brought together and held at elevated temperatures for longer time. Diffusion across the interface (A/B) will take place, and diffusion parameters such as the jump frequency and vacancy concentration will depend on the position and time. For explaining such a case, Darken defined a diffusivity term, chemical interdiffusion coefficient ($\tilde{D}$), to describe the diffusion that takes place in the diffusion couple. It is given by the following relation:

$$\tilde{D} = x_A D_B + x_B D_A, \tag{2.49}$$

where x_A and x_B are the atom fractions of A and B, respectively, at the point the interdiffusion coefficient is measured, and the intrinsic diffusion coefficients of A and B are D_A and D_B at the same point, and are not necessarily constant. More refinement of this model has been done by incorporating activity coefficients, known as the Darken–Manning relation. Readers are referred to Refs [3, 7, 9–11] for more information.

Intuitively, it is clear now that the diffusion rate of A into B is in general different from the diffusion rate of B into A. Kirkendall has conducted a famous experiment to elucidate the operation of vacancy diffusion in metals. A number of experimental and theoretical research studies have since then followed and expanded the understanding of diffusion in a significant way. In this experiment, molybdenum wires were wound around an alpha-brass (70Cu–30Zn, wt%) block and then plated with a copper coating of appreciable thickness (Figure 2.50). The molybdenum wires act

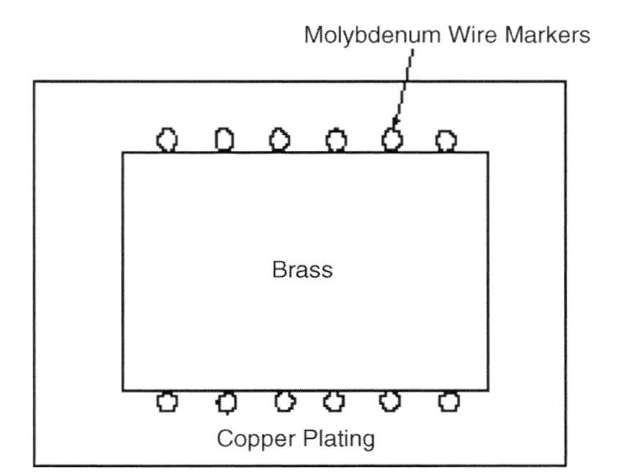

Figure 2.50 The initial configuration of the Kirkendall couple.

as inert marker to locate the original interface. When the sample is kept in a furnace and sufficient diffusion is allowed, Kirkendall noticed that the wire markers present on the opposite sides of brass moved toward each another. This observation implied that more material has moved away from brass to copper than what entered from copper to brass. Now let us think about the atomic picture of the situation. Direct interchange mechanism and ring mechanism of diffusion require that the net number of atoms crossing the interface is zero. But this is clearly not the case. The vacancy mechanism of diffusion is the only plausible explanation for this behavior. When zinc diffuses by a vacancy mechanism, there is a net flux of zinc atoms going in the opposite direction. That is, an equal number of vacancies are entering the brass block. However, this enhanced concentration of vacancies is thermodynamically unstable. There are many vacancy sinks (such as grain boundaries and dislocations) in the material, so the vacancy concentration does not go above the equilibrium vacancy concentration. Thus, it means that zinc leaves brass and the excess vacancies in brass get annihilated at the preexisting sinks in brass. So the natural result of the event is that the volume of brass decreases and the wire markers move closer together. It is true that similar event is also happening in copper plating, but because the diffusion rate of zinc is higher than that of copper, the net effect of diffusion of the latter does not show up.

2.3.6 Diffusion in Different Microstructural Paths

A microstructure is not as homogeneous as we think from a larger length scale. If it is possible to delve into the microscale, we can encounter various features like grain boundaries and dislocations. Diffusion through these features would be different from the diffusion that takes place through the lattice interior or the bulk of the crystal.

2.3.6.1 Grain Boundary Diffusion

We have discussed the characteristics of grain boundaries in Section 2.2. Even though we have not fully covered grain boundary models, we can easily develop a picture of grain boundary where atoms are more loosely packed compared to the crystal interior. Grain boundary in itself is typically only a few atomic diameters in thickness. Naturally, it leads us to believe that atomic migration rates are greater in grain boundaries than in the grain interior or single crystals. In order to understand the effect of grain boundaries, one needs to compare the diffusion results between a single crystal and a polycrystalline material. Let us take an example involving the diffusion in single crystal and polycrystalline silver. The data were plotted as $\log(D)$ versus $1/T$ (a schematic plot is shown in Figure 2.51 without actual experimental data). The plot is a straight line for the single crystal across the temperature range studied, but the curve for the polycrystal coincides with the single crystal at the higher temperature range. However, with decreasing temperature, the diffusivity in the polycrystal silver becomes higher. It can only happen if the

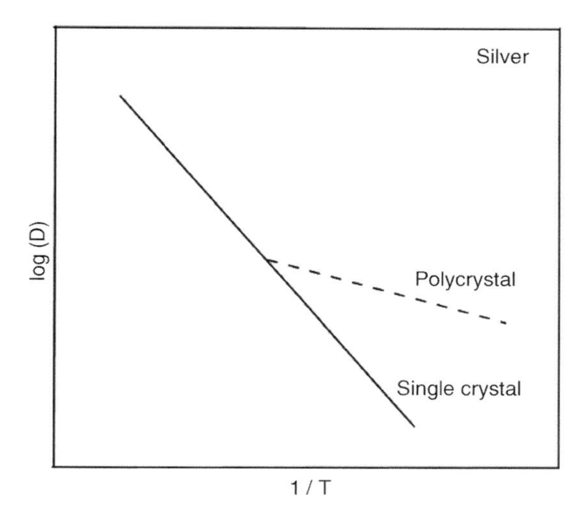

Figure 2.51 Effect of grain boundaries on the diffusivity [9].

grain boundary activation energy is lower than that for bulk diffusion. Equation (2.50) expresses the diffusion data in single-crystal silver crystal, whereas Eq. (2.51) shows the diffusivity data in polycrystalline silver.

$$D = 0.9\exp(-1.99\,\text{eV}/kT)\,\text{cm}^2\,\text{s}^{-1} \tag{2.50}$$

and

$$D = 0.9\exp(-1.01\,\text{eV}/kT)\,\text{cm}^2\,\text{s}^{-1}. \tag{2.51}$$

The equations above reflect the fact that at lower temperatures, the lower activation energy for grain boundary diffusion lowers the overall activation energy and thus grain boundary diffusion becomes more dominant at lower temperatures. However, as the temperature is increased, the contribution of bulk diffusion becomes more dominant. That is why at higher temperatures, the grain boundary diffusion contribution becomes negligible compared to the bulk diffusion. Thus, one cannot see much change at all in the position of the curves (Figure 2.49) in the higher temperature range. For example, the grain boundary activation energy for bulk diffusion in aluminum is taken as 142 kJ mol^{-1}, whereas the grain boundary activation energy for aluminum is only 84 kJ mol^{-1}. However, grain boundary activation energy has been variously taken as 0.35–0.60 times the activation energy for lattice diffusion. It is worth noting that polycrystals are generally composed of small grains oriented randomly with one another. Thus, when the diffusion rates are measured, it gives an average value of measurement over several grains, thus simulating a *macroscopic isotropy* of diffusion, even though diffusion in each grain (single crystal) in essence is highly anisotropic.

2.3.6.2 **Dislocation Core Diffusion**

Dislocation core structure is quite different from the lattice crystal structure. Diffusion along the dislocation core may provide a faster diffusion path and contribute to the overall diffusion, especially at lower temperatures. This type of diffusion is also known as *pipe diffusion*. Activation energy for the dislocation core diffusion is generally close to the activation energy for grain boundary diffusion.

2.3.6.3 **Surface Diffusion**

At the outset, surface diffusion is generally assumed to be simple. On the contrary to the belief, it is rather difficult to precisely estimate. This problem arises from the difficulty in describing the surface as "atomically flat," as surfaces also have grooves, scratches, steps, and so on. Moreover, surfaces are easily susceptible to contamination. The general view is that the activation energy for surface diffusion is considerably less than that for the bulk diffusion as a surface atom has almost half the nearest-neighbor atoms compared to an atom in the bulk. So, it is relatively easier for a surface atom to hop from one lattice position to another equivalent atom position.

2.3.7
Summary

This section discussed a very important kinetic phenomenon in materials known as diffusion. Diffusion has a vital role to play in several nuclear-specific materials, which will be elucidated in later chapters. Both phenomenological and atomistic theories of diffusion are discussed. Microstructural paths of diffusion have an important bearing on the activation energy and thus on diffusion rates; activation energy values in a material would be sequenced from smaller to higher magnitudes as shown in the following: surface (Q_s) $<$ grain boundary (Q_{gb}) $<$ dislocation core (Q_c) $<$ lattice self-diffusion (Q_l).

Problems

2.1 Given the following information on tantalum (Ta): atomic number = 73, atomic mass = 180.95 amu, atomic radius = 0.1429 nm, density = 16.6 g/cc m^{-3}. (a) Find out the number of atoms per mm^3. (b) What is the atomic packing factor? (c) If it is cubic, what is its Bravais lattice?

2.2 What is polymorphism? Give four examples.

2.3 The unit cell of uranium has orthorhombic symmetry with *a*, *b*, and *c* lattice parameters of 0.286, 0.587, and 0.495 nm, respectively. If its density, atomic weight, and atomic radius are 19.05 g cm^{-3}, 238.03 g mol^{-1}, and 0.1385 nm, respectively. Calculate the atomic packing factor.

2.4 Calculate the density of uranium carbide (UC) given the atomic radii of U and C species.

2.5 Copper has an FCC structure with an atomic radius of 0.128 nm and atomic weight of 63.5 g mol^{-1}.
   a) Calculate its density in g cm^{-3}?
   b) Draw a neat sketch of a unit cell with axes appropriately shown and depict the planes and directions: (100), (110) and (111); [110] and [111].
   c) Calculate the planar density (per cm^2) of atoms on the above planes?
   d) Calculate the linear density (per cm) of atoms along these directions?

2.6 What is the angle between the planes (011) and (001) in a cubic crystal?

2.7 a) What are the crystal structures of UC and UO_2 (draw a neat sketch of a unit cell and show the ionic positions) and what are the advantages of UO_2 over UC as a nuclear fuel?
   b) In UO_2, what are the coordination numbers of U and O ions?
   c) Given that the ionic radii of U and O are 0.97 and 1.32 Å, respectively, calculate the lattice constant of UO_2.

2.8 Draw a neat sketch of a unit cell of Fe (BCC) and depict a close-packed plane and close-packed direction (i.e., the densest plane and direction) showing clearly the choice of coordinates.
   a) What are the Miller indices of the specific plane and direction you chose?
   b) Calculate the planar and linear atomic densities for the plane and direction above?

2.9 a) Compute and compare the linear densities of the [110] and [111] directions for BCC.
   b) Calculate and compare the planar densities of the (100) and (111) planes for FCC.

2.10 Show that the *ideal* c/a ratio for HCP crystals is 1.633?

2.11 On a neat sketch of a HCP crystal, show the following (clearly depict the axes):
   a) Prism plane $(10\bar{1}0)$ and a close-packed direction in that plane – What are the Miller indices of the direction chosen and what is the interatomic distance?
   b) Pyramidal plane $(11\bar{2}2)$ and a close-packed direction in that plane – What are the Miller indices of the direction chosen?
   c) A basal plane and a close-packed direction in that plane – What are the Miller indices of the plane and direction chosen?

2.12 a) If the activation energy for vacancy formation in Fe is 35 kcal mol^{-1}, what is the vacancy concentration at (i) 900 °C and (ii) 400 °C?
   b) Calculate the number of vacancies per unit cell at 900 °C in the above problem.

2.13 In alkali halides, such as NaCl, one finds intrinsic and extrinsic vacancies (cation vacancies) due to divalent impurity such as Ca. On an Arrhenius plot show (schematically) the temperature variation of vacancy (cation) concentration indicating the extrinsic and intrinsic regions. Note down the equations relevant to these regions.

2.14 Compare and contrast the edge and screw dislocations.

2.15 Show how the dislocation model of low-angle grain boundaries can explain the creation of high-angle grain boundaries.

2.16 Show that Hume–Rothery's rules apply to copper–nickel alloy system. Why is the solubility of carbon in FCC iron more than in BCC iron?

2.17 Steel surfaces can be hardened by *carburization*, the diffusion of carbon into the steel from a carbon-rich atmosphere. During one such treatment at 1000 °C, there is a drop in carbon concentration from 5.0% to 4.0% carbon between 1 and 2 mm from the surface of the steel. Estimate the flux of carbon into the steel in this near-surface region to be 2.45×10^{19} atoms per m^2 s (The density of γ-Fe at 1000 °C is 7.63 g cm^{-3}.).

2.18 A steel with 0.2% C is to be carburized in a carburizing atmosphere to reach a carbon concentration of 1.1% at the surface. After 10 h at 890 °C, at what depth below the surface one would find 0.4% C concentration? (For diffusion of C in austenite, $D_0 = 2.0 \times 10^{-5}\ s^{-1}$, and $Q = 140$ kJ mol^{-1})

2.19 a) Given the diffusion data (D in $cm^2\ s^{-1}$) for yttrium in chromium oxide at different temperatures, find the activation energy and diffusion coefficient (D_0): 1.2×10^{-13} @800 °C; 5.4×10^{-13} @850 °C; 6.7×10^{-13} @900 °C; 1.8×10^{-12} @950 °C; and 4.6×10^{-12} @1000 °C.
b) Find D at 925 °C.

Bibliography

1 Poirier, J.P. and Price, G.D. (1999) Primary slip system of epsilon-iron and anisotropy of Earth's inner core *Physics of the Earth and Planetary Interiors*, **110**, 147–156.

2 Callister, W.D. and Rethwisch, D.G. (2007) *Materials Science and Engineering*: An Introduction (ed. 7e), John Wiley & Sons, New York.

3 Chiang, Y.-M., Birnie, D.P., III, and Kingery, W.D. (1997) *Physical Ceramics: Principles for Ceramic Science and Engineering*, John Wiley & Sons, Inc., New York.

4 Amelinckx, S. (1958), Dislocation patterns in potassium chloride, *Acta Metallurgica*, **6**, 34–58.

5 Hull, D. and Bacon, D.J. (1984) *Introduction to Dislocations*, 3rd edn, Butterworth-Heinemann.

6 Matano, C. (1933) *Japanese Journal of Physics*, On the relation between the diffusion-coefficients and concentrations of solid metals, **8**, 109–113.

7 Raghavan, V. (1992) *Solid State Phase Transformations*, Prentice Hall, New Delhi, India.

8 Sherby, O.D. and Simnad, M.T. (1961) Prediction of atomic mobility in metallic systems. *Transactions of the American Society for Metals*, **54**, 227–240.

9 Girifalco, L.A. (1964) *Atomic Migration in Crystals*, Blaisdell Publishing Company, New York.

10 Chiang, Y.-M., Birnie, D.P., III, and Kingery, W.D. (1997) *Physical Ceramics*, John Wiley & Sons, Inc., New York, NY.

11 Darken, L.S. and Gurry, R.W. (1953) *Physical Chemistry of Metals*, McGraw-Hill, Tokyo, Japan.

12 Raghavan, V. (2006) *Physical Metallurgy: Principles and Practice*, 2nd edn, Prentice Hall, New Delhi, India.

13 S. Gollapudi, I. Charit and K.L. Murty, Acta Materialia, **56** (2008) 2406–2419.

14 Dieter, G.E. (1986) *Mechanical Metallurgy*, 3rd edn, McGraw-Hill.

15 Olander, D.R. (1976) *Fundamental Aspects of Nuclear Fuel Elements*, University of Michigan Library.

3
Fundamentals of Radiation Damage

"Nothing in Life is to be Feared. It is Only to be Understood."
—Marie Curie

Interactions of high-energy radiation such as α-, β-, and, γ-rays as well as subatomic particles such as electrons, protons, and neutrons with crystal lattices give rise to defects/imperfections such as vacancies, self-interstitials, ionization, electron excitation, and so on. Fission fragments and neutrons cause the bulk of the radiation damage. Other types of radiation either do not have enough energy or are not produced in sufficient number density to cause any major radiation damage. In a nuclear reactor scenario, the microscopic defects produced in materials due to irradiation are referred to as *radiation damage*. These defects result in changes in physical, mechanical, and chemical properties, and these macroscopic material property changes in aggregate are referred to as *radiation effects*. Before discussing the effects of radiation on various properties, we need to know how to describe the radiation damage in a quantitative fashion. The timescales in which the damage and effects take place are quite different. While radiation damage events take place within a short time period of around 10^{-11} s or less, the radiation effects occur in a relatively large timescale ranging from milliseconds to months. Radiation effects range from the migration of defects to sinks that takes place in milliseconds to changes in physical dimensions due to swelling and so on with much longer duration. Quantitative characterization of radiation damage is covered in this chapter, while radiation effects are discussed in Chapter 6 after descriptions of various properties of materials. We mainly consider neutron irradiation here and the reader is referred to other monographs for damage calculations when charged particles (such as heavy ions and protons) and photon (such as γ-ray) irradiations are involved.

The binding energy of lattice atoms is very small (~10–60 eV) compared to the energy of the impinging particles so that a scattering event between them results in the lattice atom getting knocked off from its position. The atom will generally have such a high energy that it can interact with another lattice atom that will also get knocked off from the lattice position. The atom that was knocked off by the incoming high-energy particle is known as "primary knock-on atom" or PKA, which in turn knocks off a large number of atoms before it comes to rest in an interstitial position, thereby creating a Frenkel pair in which case the atom is

An Introduction to Nuclear Materials: Fundamentals and Applications, First Edition.
K. Linga Murty and Indrajit Charit.
© 2013 Wiley-VCH Verlag GmbH & Co. KGaA. Published 2013 by Wiley-VCH Verlag GmbH & Co. KGaA.

considered to have been "displaced." However, if the atom is in proximity to a vacancy, it would occupy the vacant lattice position in which case it becomes a "replacement collision." Thus, in general, a PKA can lead to a large number of higher order knock-on atoms (also known as recoil atoms and/or secondary knock-on atoms) resulting in many vacant lattice sites and this conglomeration of point defects is known as "displacement cascade." If during these collision processes many nuclei go into higher energy states at their lattice position, thermal spike is created. Before the particle–lattice atom interactions, the incoming particle may interact with electrons leading to ionization. As will be seen later, there is an electron energy cutoff above which no additional atomic displacements take place until the particle energy becomes lower than this cutoff value, as envisioned in the Kinchin–Pease (K–P) model.

Brinkman [1] first came up with the *displacement spike* model, as shown in Figure 3.1a. In this model, PKA motion creates a core consisting of several vacancies surrounded by a periphery rich in interstitials. Later, Seeger [2] further refined the concept and showed that vacant lattice sites in proximity lead to zones devoid of atoms commonly referred to as *depleted zones* (Figure 3.1b). However, if these regions are large enough, they form voids that lead to decreased density or volume increase known as "swelling." In cases where elements such as B, Ni, and Fe are present, (n, α)[1] reactions will lead to the production of He that will stabilize these voids, in which case they are referred to as "cavities". These cavities once formed are stable and cannot be removed by thermal annealing.

The following are the radiation defects induced by intense nuclear radiation, in particular high-energy $\{E \geq 0.1\,\text{MeV}\}$ neutrons:

- Vacancies.
- Interstitials.
- Impurity atoms – produced by transmutation.
- Thermal spikes – regions with atoms in high-energy states.
- Displacement spikes – regions with displaced atoms, vacancies, self-interstitials (Frenkel pairs) produced by *primary* and *secondary knock-on* atoms.
- Depleted zones – regions with vacancy clusters (depleted of atoms).
- Voids – large regions devoid of atoms.
- Bubbles – voids stabilized by filled gases such as He produced from (n, α) reactions with B, Ni, Fe, and so on.
- Replacement collisions – scattered (self) interstitial atoms falling into vacant sites after collisions between moving interstitial and stationary atoms and dissipating their energies through lattice vibrations.

A flux of neutrons then results in a large number of PKAs, which in turn produce higher order knock-on atoms, as illustrated in Figure 3.1. Our goal is to first calculate the number of PKAs produced due to a flux of neutrons with a range of energies comprising the neutron spectrum. Next step is to find the number of knocked-on atoms due to these PKAs with varied energies. Integration through the

1) For example, $B^{10} + n^1 \rightarrow Li^7 + He^4$.

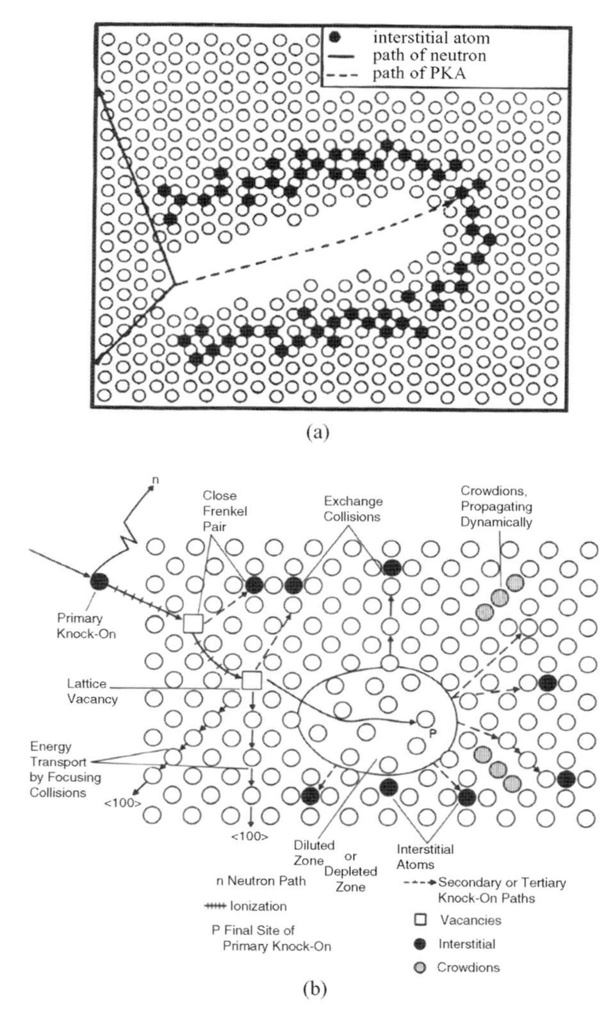

Figure 3.1 (a) Displacement spike as described by Brinkman [1]. (b) Seeger's [2] refined concept of primary damage events in an FCC metal.

whole neutron spectrum then yields the number of atoms displaced. The displacements per atom or dpa will give us a measure of quantitative radiation damage that can be later related to changes in the macroscopic properties of materials due to the given neutron spectrum. Earlier, the total fluence or dose (flux $\times$ time) in units of n cm^{-2} was commonly used, but this does not take into account the different spectral variations, and so dpa is a far better unit. It is commonly observed that the properties of irradiated materials depend on the specific neutron spectrum to which they are exposed and thus will be different for the same total neutron

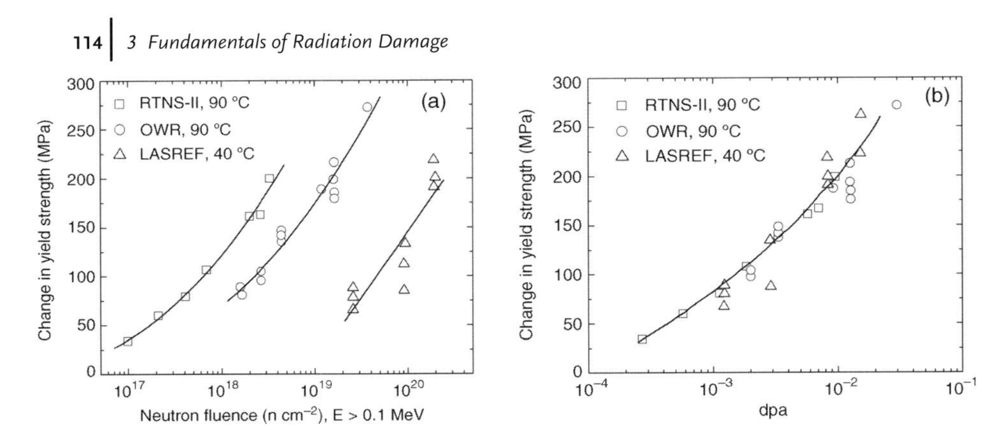

Figure 3.2 (a) and (b) The effectiveness of the use of dpa as a measure of radiation damage as opposed to the neutron fluence illustrated through an example of 316-type stainless steel irradiated at three different reactor facilities. After Was and Greenwood.

dose. Figure 3.2 clearly demonstrates this behavior. The example here shows the correlation for mechanical properties, while other such relations can be found for other properties such as physical, thermal, electrical, and so forth, which are generally referred to as radiation effects. It is important to note that many atomic displacements occur due to neutron–lattice interactions, but only a small fraction ($\sim$1%) survive since most of the defects anneal out *in situ* during irradiation mainly due to the proximity of the defects to the appropriate sinks and mutual recombination of vacancies and interstitials.

3.1 Displacement Threshold

Before starting discussions on various radiation damage models, let us first understand the concept of displacement threshold. While describing the primary damage events, it is essential to develop a clear understanding of the displacement energy or displacement threshold (denoted by E_d), which is defined as the minimum energy that must be transferred to a lattice atom in order for it to be dislodged from its lattice site. Generally, average displacement energy of 25 eV is used. However, the fixed value of 25 eV is only an average of all the possible displacement energies calculated along different crystallographic directions in a given material. This value also agrees well with the experimentally measured displacement energy values. The specific values of displacement energy depend on the nature of the momentum transfer, trajectory of the knock-ons, crystallographic structure, and thermal energy of the atoms. It has been noted that higher melting point metals tend to have higher displacement energies (Figure 3.3). Even though all elements do not follow the trend as there are many other factors that influence displacement

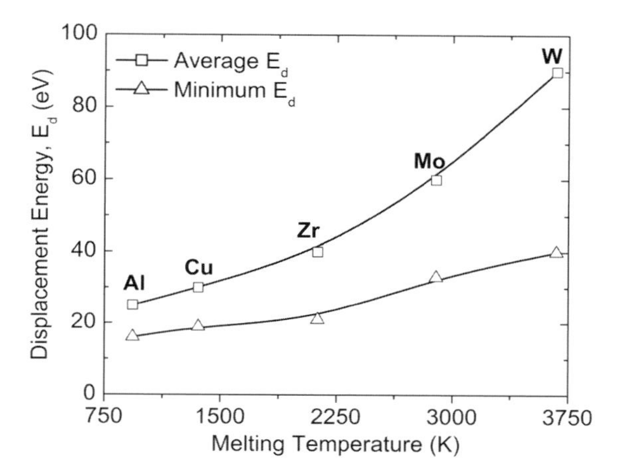

Figure 3.3 The variation of displacement energy as a function of melting temperature of metals.

energy, it is clear from Figure 3.3 that the refractory metals (such as Mo and W) tend to have much higher displacement energies compared to those of lower melting point ones. It is most possibly related to the stronger binding energy present in higher melting point metals.

As regards to the lower limiting value of the displacement energy, it is close to the energy needed to produce Frenkel pairs (3–6 eV). If the energy transferred by the knock-on atom to the struck lattice atom is less than the displacement energy, the atom will not dislodge from its site. Rather it will vibrate around an equilibrium position, transfer the energy through the neighboring lattice atoms, and eventually dissipate as heat. In order to calculate the displacement energy, it is essential to know the description of the interatomic potential fields since this is the energy barrier that the struck atom needs to surmount to eject successfully from its regular lattice site. A simple example is shown below.

Example 3.1

Calculate the displacement energy along $\langle 110 \rangle$ direction (i.e., a face-centered position to an adjacent face-centered position) in an FCC metal unit cell (as shown in Figure 3.4a) in which the interatomic potential is given by a simple repulsion potential as described below:

$$V(r) = -U + \tfrac{1}{2}\, k(r_{\text{eq}} - r)^2 \quad \text{at} \quad r < r_{\text{eq}}, \tag{3.1a}$$

$$V(r) = 0 \quad \text{at} \quad r > r_{\text{eq}}, \tag{3.1b}$$

where U is the binding energy of atom (energy per atomic bond), k is a force constant indicative of the repulsive portion of the potential, r_{eq} is the equilibrium atom separation, and r is the general separation distance. The force constant is ka_0^2 and U values can be taken as 60 and 1 eV, respectively. Note that a_0 is the lattice constant of the crystal.

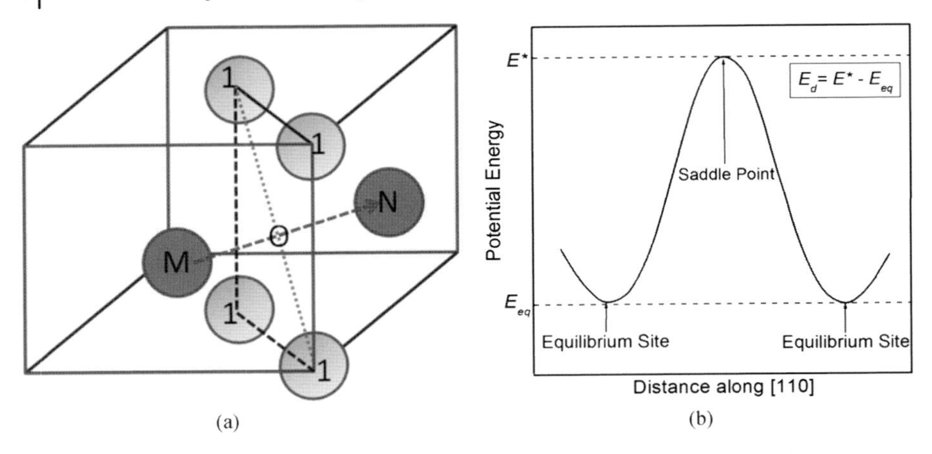

Figure 3.4 (a) An FCC unit cell is shown with only the atoms forming the barrier, and at positions M and N (the dashed arrow designates the struck atom trajectory along ⟨110⟩). (b) The variation of potential energy as a function of position along the trajectory of the knock-on atom motion.

Solution

The atom at position "M" after being dislodged from the site needs to follow the direction of the arrow (parallel to ⟨110⟩ direction) to the position "N." During the trajectory motion of the atom M, it needs to go through a potential field created by the four-atom barrier (1-1-1-1). The center of this four-atom barrier represents the saddle point (at O) along the trajectory of M atom with the maximum potential energy; the variation of potential energy as a function of position is shown in Figure 3.4b.

The energy of a single atom in the FCC crystal can then be given by $E_{eq} = -12U$, considering that an FCC atom is surrounded by 12 equidistant nearest neighbors (due to the coordination number of FCC crystal being 12). In turn, we can say that when M atom is at the equilibrium site, the energy of M atom is given by as shown above.

Now when the M atom moves to the center of four-atom barrier, it achieves the highest potential energy following the potential described in Eq. (3.1a). Therefore, the potential energy at the saddle point "O" is given by

$$E^* = 4V(r) = 4[-U + {}^1\!/_2\, k(r_{eq} - r)^2].$$

The displacement energy (E_d) is given by

$$\begin{aligned} E_d &= E^* - E_{eq} = 4[-U + {}^1\!/_2\, k(r_{eq} - r)^2] - (-12U) \\ &= 8U + 2k(r_{eq} - r)^2. \end{aligned} \tag{3.2}$$

Now we need to express r_{eq} and r in terms of lattice constant (a_0). This can be accomplished from the geometrical relations in the FCC unit cell, as shown in Figure 3.4a. In this case, r_{eq} is the minimum distance from M to O. A simple geometric construction can show that the distance can be determined by calculating half the hypotenuse of a triangle with other two sides being $a/2$ and $a/2$.

Hence, we get

$$r_{\text{eq}}\ (\text{i.e., MO}) = \frac{1}{2}\sqrt{\left(\frac{a_0}{2}\right)^2 + \left(\frac{a_0}{2}\right)^2} = \frac{\sqrt{2}a_0}{4}, \tag{3.3}$$

and r (the impact parameter), the distance of each atom-1 of the four-atom barrier from the point O, is given by

$$r\ (\text{i.e., O1}) = \frac{1}{2}\sqrt{(a_0)^2 + \left(\frac{\sqrt{2}a_0}{2}\right)^2} = \sqrt{\frac{3}{8}}a_0. \tag{3.4}$$

Now we take r_{eq} and r relations from Eqs (3.3) and (3.4), respectively, and use them in Eq. (3.2). Thus, we obtain

E_{d} along $\langle 110\rangle = 8U + 2k(r_{\text{eq}} - r)^2 = 8U + 2k\left(\frac{\sqrt{2}a_0}{4} - \sqrt{\frac{3}{8}}a_0\right)^2 = 8U + 2\times k$ $(0.213)^2 a_0^2 = (8U) + (2\times 0.045)(k\ a_0^2) = (8\times 1\ \text{eV}) + (0.09\times 60\ \text{eV}) = {\sim}13.4\ \text{eV}$.

As additional exercises, determine the displacement energy along $\langle 100\rangle$ and $\langle 111\rangle$ in FCC yourself using the same potential expression given in Eq. (3.1). It is very clear that accurate knowledge of interatomic potentials is quite important in the accuracy of the calculated displacement energy values.

Special Note: The use of displacement energy in ceramics is similar but bit complicated because of the presence of multiple atomic species (cations and anions). Table 3.1 summarizes some displacement energy values of cations and anions in some well-known ceramics.

In case of multicomponent ceramics, an effective displacement energy ($E_{\text{d}}^{\text{eff}}$) is often used to calculate the extent of radiation damage. This is given by

$$E_{\text{d}}^{\text{eff}} = \left(\sum_i \frac{S_i}{E_{\text{d}}^i}\right)^{-1}, \tag{3.5}$$

where S_i is the stoichiometric fraction and E_{d}^i is the displacement energy of the ith atomic species. Taking into account the nature of Coulombic interactions in

Table 3.1 A summary of displacement energies in some ceramic materials.

Material	Threshold displacement energy (eV)
Al_2O_3	$E_{\text{d}}^{\text{Al}} \sim 20,\ E_{\text{d}}^{\text{O}} = 50$
MgO	$E_{\text{d}}^{\text{Mg}} = 55,\ E_{\text{d}}^{\text{O}} = 55$
ZnO	$E_{\text{d}}^{\text{Zn}} \sim 50,\ E_{\text{d}}^{\text{O}} = 55$
UO_2	$E_{\text{d}}^{\text{U}} = 40,\ E_{\text{d}}^{\text{O}} = 20$

collision cascades, the scaling parameter is more appropriately expressed as $(S_i Z_i^2)/A_i$, where Z_i and A_i are the atomic number and atomic mass of the ith species.

3.2
Radiation Damage Models

A simple model for calculating the atomic displacements is due to Kinchin and Pease [3], known as the Kinchin–Pease model. Before discussing this model, let us consider a case of collision between a high-energy neutron of mass M_1 and a lattice atom of mass M_2 (Figure 3.5). At certain energy ranges, neutron–nucleus interaction can be described by an elastic scattering process, where both kinetic energy and momentum of the particles are conserved before and after the collision. In the majority of the events like this, binary (two-body) collision is an appropriate approximation. A neutron has no electrical charge (i.e., neutral particle), and hence does not get perturbed by electrical fields exerted by the nucleus or electrons.

Let E_1 be the energy of the neutron and due to the collision, the atom is scattered with energy T transferred from the neutron. There exists a threshold energy E_d for the atom to be displaced that is related to the binding energy of the atom. Thus, the struck atom will be displaced only if the transferred energy T is at least equal to or greater than this threshold value. As we have noted before, the displacement threshold energy is very small on the order of ~25 eV, while the transferred energy might be in keV or MeV range. Thus, there is a very high probability that the struck atom will get displaced and become a PKA. This PKA with reasonably high energy will thus interact with other atoms and displace them from their positions leading to secondary, tertiary, and other higher order knock-on atoms.

In the collision process between the incoming high-energy particle and the atom being displaced, if the knocked-on atom flies away with energy, T the incoming neutron will have energy $E_1 - T - E_d$, at a scattering angle of θ. In general, however, the energy consumed in the scattering process is assumed negligible and thus the scattered neutron will have energy equal to $E_1 - T$. Of course, the neutron

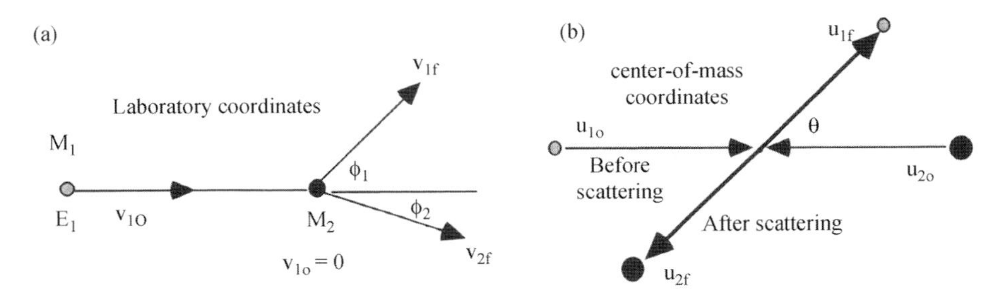

Figure 3.5 Various characteristics of an elastic collision event are shown in (a) the laboratory coordinates and (b) in the center-of-mass coordinates.

with energy $[E_1 - T]$ and the PKA with energy T produce other knock-on atoms. A simple calculation can show the likelihood of neutron–nucleus interaction. If we consider a neutron scattering cross section (σ) as 1 b (i.e., 10^{-24} cm) and the number density of atoms (N) in a material as $0.85 \times 10^{23}\,\text{cm}^{-3}$ (typical of alpha-Fe), the distance between collisions, that is, the mean free path of the neutron (calculated by $1/N\sigma$) will be several centimeters (in this case, ~11.8 cm).

Analytical derivation based on the assumption of the kinetic energy and momentum conservation for an *isotropic elastic scattering* event between a neutron and an atom nucleus gives the following relations:

$$v_{2f}^2 = \frac{2M_1^2}{(M_1 + M_2)^2} v_{10}^2 (1 - \cos\theta). \tag{3.6}$$

$$T = \frac{1}{2} M_2 v_{2f}^2 = \frac{2M_1 M_2}{(M_1 + M_2)^2} E_{10}(1 - \cos\theta). \tag{3.7}$$

$$\text{Hence, } T = \frac{1}{2}\Lambda E_n (1 - \cos\theta), \tag{3.8}$$

where $\Lambda = (4M_1M_2)/(M_1 + M_2)^2$ and E_n is a generalized form of average incident neutron energy and is essentially equal to E_{10}.

From Eq. (3.8), it is clear that the value of T depends on the scattering angle, the energy of the incident particle and the masses of the collided particles.

We can see that the maximum energy that can be transferred is given for the angle of scattering of 180°. In other words, the minimum possible value of $\cos(\theta)$, that is, -1, gives the transferred energy the highest value (T_{max}), which is then given by the following relation:

$$T_{max} = \Lambda E_n. \tag{3.9}$$

Of course, the minimum energy (T_{min}) that can be transferred would be zero given a θ value of 0°.

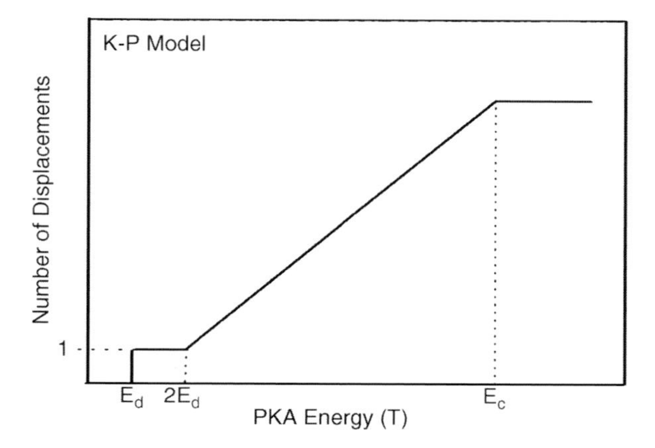

Figure 3.6 The number of atom displacements due to a PKA as a function of PKA energy according to the K–P model.

Table 3.2 A summary of maximum percentage energy transfer that takes place from a 2 MeV neutron to different nuclides in the event of an isotropic elastic scattering (mass of neutron = 1.008665 amu).

Nuclides	Nuclide mass M_2 (amu)	$\Lambda = \dfrac{4M_1M_2}{(M_1+M_2)^2}$	$T_{max} = \Lambda E_n$ (MeV)	%Energy approximately transferred
H^1	1.007825	1.0000	2.0000	100
H^2	2.014102	0.8894	1.7787	89
C^{12}	12.000000	0.2861	0.5722	28.6
Si^{12}	27.976927	0.1344	0.2687	13.4
Au^{197}	196.96654	0.0203	0.0406	2.0

However, the neutron–nucleus elastic scattering event could lead to glancing collisions. In that case, the use of average recoil energy (T_{avg}) is more appropriate:

$$T_{avg} = \frac{\Lambda E_n}{2}. \tag{3.10}$$

Table 3.2 lists the maximum percentage energy that can be transferred to different nuclides in the event of an elastic collision between a fast neutron (kinetic energy of 2 MeV) and the nuclides. One important observation from this table could be that the neutron transfers more energy to lighter nuclei. Thus, nuclides with low mass numbers generally act as good moderators. As we have discussed in brief, the materials often used are light water (H_2O), heavy water (D_2O), and graphite (C).

Even though earlier in the chapter we assumed that neutrons are scattered through every possible direction (Isotropic Scattering), in reality they tend to scatter preferentially in the forward directions, and hence the scattering behavior is anisotropic. That is why Eqs (3.9) and (3.10) may need to be corrected using a correction factor (f). The value of f varies with the nuclide type; for example, for Be it is 0.56, whereas for C and Cu they are 0.84 and 0.6, respectively.

Example 3.2

A beryllium (Be) reflector is exposed to a neutron of 1 MeV energy. Let us assume the masses of the neutron and the Be atom are 1 and 9.01 amu, respectively. (a) Calculate the maximum and average energies transferred from the neutron to the beryllium atom based on isotropic elastic scattering event. (b) Determine the average energy transferred if the anisotropic scattering occurs ($f = 0.56$).

Solution

From Eq. (3.8), we know that $T_{max} = \Lambda E_n$, where $\Lambda = (4M_1M_2)/(M_1+M_2)^2$.

a) In order to find out T_{max}, we need to first calculate Λ.
Given $M_1 = 1$ amu and $M_2 = 9.01$ amu,

$$\Lambda = \frac{(4)(1)(9.01)}{(1+9.01)^2} = \sim 0.36.$$

Hence, $T_{max} = (0.36) \times (1\ \text{MeV}) = 0.36$ MeV.

$$T_{avg} = \frac{\Lambda E_n}{2} = 0.18\ \text{MeV}.$$

b) For anisotropic elastic scattering event, $T_{aniso(avg)} = f(\Lambda E_n/2) = (0.56 \times 0.18\ \text{MeV}) = \sim 0.1$ MeV.

Now, let us try to get an expression of the displacement damage rate (R_d) defined as the number of displacements per unit volume (cm^3) per second. R_d is basically proportional to the number of target atoms per cm^3, N (i.e., number density of atoms), and the displacement cross section $\sigma_d(E_n)$ for neutrons with energy E_n, and the neutron flux $\phi(E_n)$, and can generally be written as

$$R_d = N \cdot \sigma_d(E_n) \cdot \phi(E_n). \tag{3.11}$$

The important unit of radiation damage, displacements per atom (or dpa), can now be defined as

$$\text{dpa} = \frac{R_d t}{N} = t \int_{E_d/\Lambda}^{\infty} \sigma_d(E_n) \cdot \phi(E_n) \mathrm{d}E_n \approx t \int_{0}^{\infty} \sigma_d(E_n) \cdot \phi(E_n) \mathrm{d}E_n \tag{3.12}$$

for neutrons with energies varying from 0 to ∞.

Note that the neutrons with $E_n < E_d/\Lambda$ do not generate any displaced atoms since T should be $\geq E_d$ (i.e., the maximum transferable energy given by $\Lambda E_n \geq E_d$ to produce a knock-on atom).

The displacement cross section is a function of the sum of the number of atomic displacements, $\nu(T)$, produced by PKAs with energies T from E_d (the minimum needed) to $T_{max}(= \Lambda E_n)$. The interaction probability is given by the differential energy transfer cross section, $\sigma_d(E_n, T)\mathrm{d}E$, for producing a PKA with energy $(T, \mathrm{d}T)$ due to the interaction with a neutron of energy E_n.

Therefore, we can write

$$\sigma_d(E_n) = \int_{E_d}^{\Lambda E_n} \sigma_n(E_n, T) \cdot \nu(T) \mathrm{d}T. \tag{3.13}$$

Note that the minimum energy of the PKA to produce displacements is E_d ($\Lambda = 1$) since scattering between like atoms and the maximum energy of PKA is $T_{max} = \Lambda E_n$.

The Kinchin–Pease model is the simplest of all radiation damage models. It gives an estimate of the number of displacements produced by a PKA. The model is based on certain assumptions:

1) The cascade is created due to a sequence of binary (two-body) elastic collisions and the potential is based on the hard sphere model approximation.

2) Atomic displacements occur only when $T > E_d$ (i.e., no quantum effects).
3) No energy is passed to the lattice during the collision phase.
4) Energy loss by electronic stopping is given by a cutoff energy known as electronic cutoff energy (E_c). If the PKA energy is more than E_c, no additional displacements occur until electron energy losses reduce the PKA energy to E_c. For all energies less than E_c, electronic stopping is ignored and only atomic collisions take place.
5) Atomic arrangement is random (i.e., the effect of crystal structure is neglected).
6) No annihilation of defects is assumed.

Special Note

As noted before, understanding interatomic potentials is very important to describe the radiation damage behavior of a material. The hard sphere approximation is assumed in the K–P model. This is the simplest of all the potential functions (note there are many other potential functions, such as Born–Mayer potential, simple Coulomb, screened Coulomb, Brinkman potential, inverse square potential, and so forth). The potential is described in this model as follows: If the separation distance between two atoms is greater than the atom radius, the interaction vanishes. So, it means that the situation is much like interaction between billiard balls. However, this may not be a very realistic situation as the electron shells do overlap in reality. But it suffices for use in the derivation of the simple K–P model.

For the derivation of the K–P model, readers are referred to Refs [4, 5]. Here, we present the end result of the K–P model derivation:

$$\nu(T) = 0, \quad \text{for} \quad T < E_d. \tag{3.14a}$$

$$\nu(T) = 1, \quad \text{for} \quad E_d < T < 2E_d. \tag{3.14b}$$

$$\nu(T) = \frac{T}{2E_d}, \quad \text{for} \quad 2E_d < T < E_c. \tag{3.14c}$$

$$\nu(T) = \frac{E_c}{2E_d}, \quad \text{for} \quad T \geq E_c. \tag{3.14d}$$

According to Eq. (3.14), the number of displaced atoms can be plotted as a function of PKA energy, as shown in Figure 3.6. The relation (3.14a) is self-explanatory from the very definition of E_d. That is, if T is less than E_d, we expect no atom displacement (i.e., $\nu(T) = 0$). However, for the relation (3.14b), when T is greater than E_d but less than $2E_d$, there could be two possibilities. One scenario could be that the struck atom does not get energy E_d from the PKA and stays in the same place. Second scenario could be that the struck atom gets more than E_d energy to be dislodged from the lattice site. But the PKA now with less than E_d energy falls back to the original site of the lattice atom. So, in either case, only one displacement takes place. When the PKA energy is greater than $2E_d$ but less than E_c, the number of displacements increases monotonically with PKA energy until it reaches the electronic cutoff energy (E_c). Beyond that, the number of displacements does not change. Even though in the simple K–P model it was assumed that all energy losses of the PKA go toward elastic collisions, with the increasing PKA energy, a greater fraction of energy will be lost into electronic excitation and ionization. In this case, we need to recognize that PKAs are basically fast ions. So, the number of displacements calculated at higher PKA energy would not be all consumed in creating the displacement damage. There are several modifications to the simple K–P model depending on the relaxation of the various assumptions.

Now getting back to the expression of R_d developed in Eq. (3.11) and combining with Eq. (3.13), we get the following expression:

$$R_{\mathrm{d}} = N \int_{E_{\mathrm{d}}}^{\infty} \phi(E_{\mathrm{n}}) \left(\int_{E_{\mathrm{d}}}^{\lambda E_{\mathrm{n}}} \nu(T) \sigma_{\mathrm{n}}(E_{\mathrm{n}}, T) \mathrm{d}T \mathrm{d}E_{\mathrm{n}} \right) \mathrm{d}E_{\mathrm{n}} \tag{3.15}$$

From the theory of isotropic elastic scattering and hard sphere approximation, we can get a relation between double differential energy transfer cross section and scattering cross section:

$$\sigma_{\mathrm{d}}(E_{\mathrm{n}}, T) = \frac{\sigma_{\mathrm{n}}^{\mathrm{el}}(E_{\mathrm{n}})}{\Lambda E_{\mathrm{n}}}, \tag{3.16}$$

where elastic collision cross section weakly depends on the neutron energy (E_n). Thus, it can be shown that

$$R_{\mathrm{d}} = \frac{N \cdot \Lambda \cdot \sigma^{\mathrm{el}}(\bar{E}_{\mathrm{n}})}{4E_{\mathrm{d}}} \bar{E}_{\mathrm{n}} \Phi \tag{3.17}$$

and

$$\mathrm{dpa} = \frac{R_{\mathrm{d}} t}{N} = \frac{\Lambda \cdot \sigma^{\mathrm{el}}(\bar{E}_{\mathrm{n}})}{4E_{\mathrm{d}}} \bar{E}_{\mathrm{n}} \Phi \mathrm{t}, \tag{3.18}$$

where Φ is the total integrated neutron flux given by

$$\Phi = \int_0^{\infty} \phi(E_{\mathrm{n}}) \mathrm{d}E_{\mathrm{n}}. \tag{3.19}$$

Note that the weighted average energy of neutrons is given by

$$\bar{E}_{\mathrm{n}} = \frac{\int_0^{\infty} E_{\mathrm{n}} \phi(E_{\mathrm{n}}) \mathrm{d}E_{\mathrm{n}}}{\int_0^{\infty} \phi(E_{\mathrm{n}}) \mathrm{d}E_{\mathrm{n}}}. \tag{3.20}$$

Example 3.3

Consider 0.5 MeV neutrons ($\sigma^{el} = 3$ b) with a flux of 5×10^{15} n cm^{-2} s^{-1} interacting with alpha-iron (atomic mass number 56, lattice constant 0.287 nm) target. Find R_d.

Solution

Given are $\bar{E}_n = 0.5$ MeV, $\sigma^{el} = 3$ b, $E_d = 24$ eV, $\Phi = 5 \times 10^{15}$ n cm^{-2} s^{-1}, and $a_0 = 0.287$ nm.

From Eq. (3.17), we have

$$R_d = \frac{N \cdot \Lambda \cdot \sigma^{el}(\bar{E}_n)}{4E_d} \bar{E}_n \Phi.$$

We have values of all the terms in the above equation except N and Λ.

For Fe, we can calculate the number density of atoms (N) in the target material. There are two ways to do it.

1) If the material is a crystalline solid and we know its crystal structure (here, alpha-Fe that is BCC), we first find out the volume (V_{cell}) of the unit cell of alpha-iron from its lattice parameter ($a_0 = 0.287$ nm). We know that the effective number of atoms (q) in a BCC unit cell is 2. So, $N = q/V_{cell} = 2/a_0{}^3 = 2/(0.287\ \text{nm})^3 = 0.85 \times 10^{23}$ atoms per cm^3.
2) Another way to calculate N can be applied to all state of matter. This has been explained in Chapter 1. This needs knowledge of density and atomic mass of the material.

For Λ, we use

$$\Lambda = \frac{4M_1M_2}{(M_1 + M_2)^2} = \frac{4.1.56}{(1 + 56)^2} = 0.069.$$

Hence,

$$R_d = \frac{(0.85 \times 10^{23}\ \text{cm}^{-3})(0.069)(3 \times 10^{-24}\ \text{cm}^2)}{4(24\ \text{eV})} (0.5 \times 10^6\ \text{eV})(5 \times 10^{15}\ \text{n cm}^{-2}\ \text{s}^{-1})$$
$$= \sim 4.5 \times 10^{17}\ \text{displacements cm}^{-3}\ \text{s}^{-1}.$$

Now if we divide R_d just by N, we obtain $\sim 5 \times 10^{-6}$ dpa s^{-1}.

We can also find the dpa value by multiplying the above number by the days of neutron exposure (say, 30 days), we get a dpa of ($\sim 5 \times 10^{-6}$ dpa s^{-1}). ($30 \times 24 \times 3600$s) = ~ 13.0.

Furthermore, if we wish to calculate the number of displaced atoms per neutron, we can use the following expression:

$$\frac{\Lambda \bar{E}_n}{4E_d} = \frac{(0.069)(0.5 \times 10^6\ \text{eV})}{4(24\ \text{eV})} = \sim 360\ \text{displaced atoms per neutron collision.}$$

Note

As pointed out earlier, K–P model is based on a number of assumptions. Relaxing any of these assumptions will lead to a smaller number of atomic displacements (or dpa), which is often expressed as

$$v(T) = \xi(T)\left(\frac{T}{2E_\mathrm{d}}\right). \tag{3.21}$$

Here $\xi(T)$ is the correction factor (also known as damage efficiency factor) generally less than 1. In a model proposed by Norgett, Robinson and Torrens (NRT model), this factor is taken as 0.8.

3.3 Summary

In this chapter, we focused our attention on understanding the concepts of primary radiation damage involving neutron–nucleus collision and subsequent PKA and lattice atoms interaction leading to the formation of displacement cascades. The displacement threshold concept is elucidated with a simple example. The value of displacement energy is generally about 25 eV. Then, the K–P model is introduced. This gives the number of atomic displacements produced by a PKA. Finally, the methods of calculating the damage displacement rate and dpa are described. The field of radiation damage is vast, but discussion on all these aspects is out of the scope of this chapter. Interested readers may refer to texts listed in Bibliography.

Problems

3.1 a) If a copper target (mass number 64) is exposed to monoenergetic neutrons of flux $2 \times 10^{15}\,\mathrm{n\,cm^{-2}\,s^{-1}}$ for 6 days continuously, show that the number of atomic displacements per atom (dpa) is 2.856 ($E_\mathrm{d} = 22\,\mathrm{eV}$, $\sigma^{\mathrm{el}} = 2\,\mathrm{b}$).

b) What will be the enhancement in self-diffusion in copper at 600 °C if only vacancy survived for million displacements (due to *in situ* annealing)? Given: Coordination number of copper = 12, lattice constant = ~0.22 nm, vacancy formation energy = 20 kcal mol^{-1}, and vacancy migration energy = 18 kcal mol^{-1}.

3.2 a) Using the simple Kinchin–Pease model, evaluate the number of atomic displacements per atom (dpa) of iron due to a monoenergetic neutron (2 MeV) flux of $3 \times 10^{15}\,\mathrm{n\,cm^{-2}\,s^{-1}}$ for 1 year (assume isotropic elastic scattering $\sigma^{\mathrm{el}} = 3$ b and $E_\mathrm{d} = 40\,\mathrm{eV}$. Iron is BCC with $A = 56$ and $\varrho = 7.9\,\mathrm{g\,cm^{-3}}$).

b) Calculate the dpa for neutrons of energy 0.25 MeV and compare with the above.

3.3 a) Calculate the average energy of primary knock-ons for Zr when struck with 2 MeV neutrons (Zr: (HCP) $A = 91$, $a = 3.23$ Å, $c = 5.147$ Å, $E_v = 30$ kcal mol^{-1}, $E_D = 59$ kcal mol^{-1}, $\sigma^{el} = 2$ b, $E_d = 25$ eV).

b) If in a zirconium target exposed to high-energy neutrons, PKAs were produced with 240 keV, calculate the number of atoms displaced due to a PKA (assume Kinchin–Pease model).

c) Where do the displaced atoms go to and what is the primary defect created by these atomic displacements?

d) Some of these zirconium specimens were exposed to high-energy neutrons and dpa was calculated to be 2.5, while only one vacancy survived for billion displacements. Calculate the vacancy concentration in the irradiated sample at 500 °C.

e) Evaluate the %increase in lattice diffusivity of the irradiated material.

Bibliography and Suggestions for Further Reading

1 Brinkman, J.A. (1956), Production of atomic displacements by high-energy particles, **24**, 246–267.

2 Seeger, A. (1962) *The Nature of Radiation Damage in Metals, Radiation Damage in Solids*, IAEA, p. 1.

3 Kinchin, G.H. and Pease, R.S. (1955) The displacement of atoms in solids by radiation. *Reports on Progress in Physics*, **17**, 1.

4 Was, G.S. (2007) *Fundamentals of Radiation Materials Science: Metals and Alloys*, Springer.

5 Olander, D.R. (1976) *Fundamental Aspects of Nuclear Reactor Fuel Elements*, Technical Information Center, ERDA TID-26711-P1.

Additional Reading

Zinkle, S.J. and Kinoshita, C. (1997) Defect production in ceramics. *Journal of Nuclear Materials*, **251**, 200–217.

Ma, B.M. (1983) *Nuclear Reactor Materials and Applications*, Van Nostrand Reinhold.

Smith, C.O. (1967) *Nuclear Reactor Materials*, Addison-Wesley Publishing Company.

Bush, S.H. (1965) *Irradiation Effects in Cladding and Structural Materials*, Rowman & Littlefield.

Norgett, M.J., Robinson, M.T., and Torrens, I. M. (1975) "A proposed method of calculating displacement dose rate," Nuclear Engineering and Design, **33**, 50–54.

4
Dislocation Theory

> "Science is facts; just as houses are made of stones, so is science made of facts; but a pile of stones is not a house and a collection of facts is not necessarily science."
>
> —*Henri Poincare*

The dislocation concept has already been introduced in Section 2.2 dealing with crystal defects. Now we need to develop the concept further. The importance of dislocations in plastic deformation (i.e., permanent deformation) is well documented. But the question arises as to why we should be concerned about them in a textbook on nuclear materials. We will see in a later chapter how dislocation loops can form from the primary radiation damage; the dislocation loops can either stay as loops or join the overall dislocation networks in the irradiated materials. Indeed, dislocations are the major microscopic defects that are created during irradiation. Hence, this chapter serves as a prelude to understanding these different aspects of dislocations and their significance.

4.1
Deformation by Slip in Single Crystals

We have already gained some basic idea about slip from the previous chapters. Slip is nothing but the movement of one crystal part over another causing plastic deformation. Slip occurs only when the shear stress on the slip plane along the slip direction attains a critical value (known as *critical resolved shear stress* (CRSS)). Generally, the slip planes are the crystallographic planes with the highest atomic density (closest-packed planes (CPPs)) in that particular crystal structure, and the slip directions are the closest--packed directions (CPDs) in the respective crystal structures.[1] A combination of slip plane and slip direction is called a *slip system*. Due to the slip, steps are formed on the prepolished surface of a material that has been plastically deformed. Due to the height variations in the different slip steps, they are observable on the sample surface as lines, and hence known as *slip lines*. Several slip lines banding together are

1) This is because CPPs are farthest apart and the atoms are closest along CPDs so that the force/stress required for slip to occur on CPPs along CPDs will be the lowest.

An Introduction to Nuclear Materials: Fundamentals and Applications, First Edition.
K. Linga Murty and Indrajit Charit.
© 2013 Wiley-VCH Verlag GmbH & Co. KGaA. Published 2013 by Wiley-VCH Verlag GmbH & Co. KGaA.

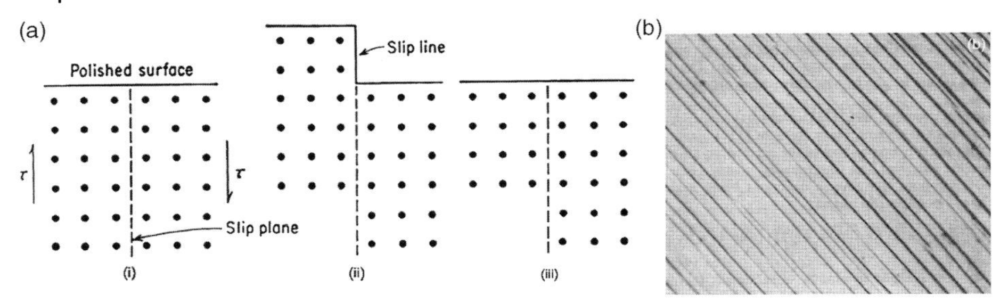

Figure 4.1 (a) Schematic of classical slip concept showing the origin of slip line in different stages – (i)–(iii). (b) Microscopic observation of straight slip lines in copper [1].

called *slip bands*. The slip bands can be seen by observing the prepolished surface of a deformed sample with an optical microscope or a scanning electron microscope. Note that slip is not just a surface phenomenon, the manifestation of slip can be tracked by observing the slip steps on the rightly conditioned surface. If the surface is later repolished, the slip lines will be removed as the slip steps showing the height variations will no longer be present. This is demonstrated in Figure 4.1a. For example, a micrograph of copper with slip lines is shown in Figure 4.1b.

Let us first discuss an example from an FCC metal. As seen in Chapter 1, the close-packed planes in the FCC crystal are $\{111\}$ with the close-packed directions being $\langle 110 \rangle$ (face diagonals). They are the slip planes and slip directions in FCC, respectively. Planes $\{111\}$ are called *octahedral planes* as they form the faces of an octahedron inside the FCC crystal. There are eight (effective number) such octahedral planes per FCC unit cell. However, one plane is parallel to the other plane, thus leaving four independent slip planes. Now each such $\{111\}$ plane contains three $\langle 110 \rangle$ directions (reverse directions are not taken into consideration as they are essentially the same directions), as shown in Figure 4.2. Thus, an FCC crystal would have 12 (4×3) slip systems. Generally, FCC metals show straighter slip lines as shown in Figure 4.1b. An easy way whether a crystallographic direction $[uvw]$ is

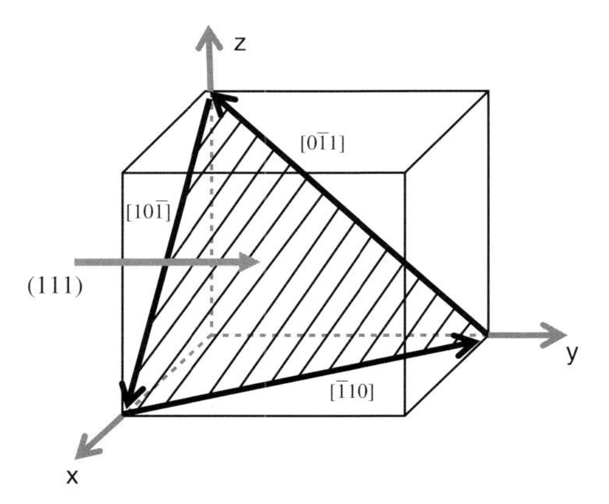

Figure 4.2 A schematic representation of a (111) slip plane with three slip directions in an FCC unit cell (atoms in the unit cell are not shown).

indeed a slip direction on a slip plane (*hkl*) is to satisfy the relation $h{\cdot}u + k{\cdot}v + l{\cdot}w = 0$ (dot product of the direction and the plane normal). For example, $[\bar{1}10]$ resides on the slip plane (111) as $(-1)(1) + (1)(1) + (0)(1) = 0$.

A BCC crystal does not have a close-packed plane. Its closest-packed plane is {110}, closely followed by {112} and {123} due to their relatively high atomic density. However, the BCC crystal has only one slip direction, the close-packed direction ⟨111⟩. There are 48 possible slip systems in a BCC crystal. As none of the slip planes is close packed in BCC crystals, higher shearing stresses are required to create slip. So, there are multiple slip planes, but slip always occurs in a close-packed direction. As (screw) dislocations in BCC crystals can move from one slip plane to another, the slip lines produced have irregular wavy appearance. By observing the slip in the ⟨111⟩ direction more or less independent of the slip plane, Taylor coined the term *pencil glide* for describing slip in BCC crystals. Screw dislocations can cross-slip readily from one plane to another, thus forming such slip bands.

In the HCP metals, the close-packed plane is basal plane {0002}. The *a*-axes are the close-packed directions having Miller index, $\langle 11\bar{2}0 \rangle$, serving as the slip direction. Slip along this direction regardless of the slip plane does not produce any strain parallel to *c*-axis. Only certain HCP metals like zinc, magnesium, cobalt, and cadmium show basal slip. There is only one type of close-packed plane and three slip directions in HCP crystal leading to the total number of available slip systems to be only three. That is why they exhibit limited ductility and extreme orientation dependence of properties. They all have one thing in common – their c/a ratios are close to the ideal (1.633). Interestingly, beryllium with a c/a ratio quite less than the ideal ratio primarily shows basal slip closely followed by prismatic slip. The stress required for pyramidal slip is much greater and can lead to fracture. Figure 4.3 shows the scenario in terms of stress–strain curves. On the other hand, alpha-Ti

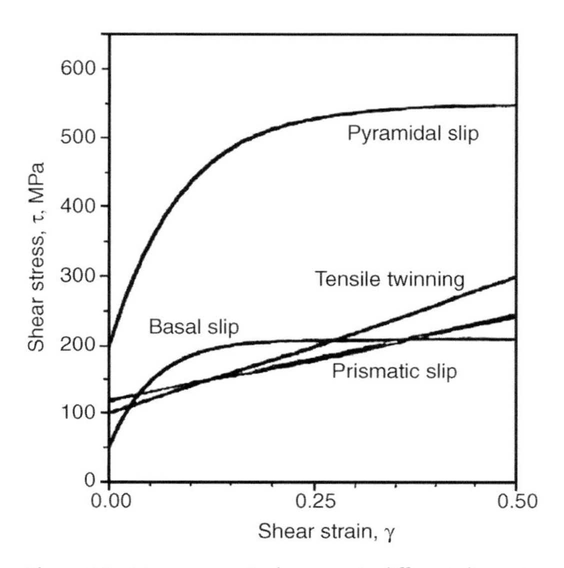

Figure 4.3 Stresses required to operate different slip systems in beryllium [2].

and alpha-Zr (c/a ratio less than the ideal ratio) under normal conditions undergo prismatic slip ($\{10\bar{1}0\}\langle 11\bar{2}0\rangle$). But Poirier downplayed the effect of c/a ratio on the slip behavior of HCP metals and attributed the diverse slip behaviors to the anisotropy of the HCP crystals. Twinning can produce small strains in the HCP crystals even along the c-axis, but the main role of twinning in HCP crystals is to help orient unfavorable slip systems favorably for slip to take place.

Additional slip systems can be activated depending on test temperature. For example, $\{110\}$ slip planes in aluminum start taking part in slip deformation at elevated temperatures even though the crystal structure remains FCC. Magnesium is known for its basal slip, but at a higher temperature (~225 °C), secondary slip systems involving $\{10\bar{1}1\}$ pyramidal planes get activated.

Example 4.1

Zr alloys with low c/a ratio exhibit prism slip ($\{10\bar{1}0\}\langle 11\bar{2}0\rangle$). Show on a single HCP crystal a slip system and the Miller indices of the plane and direction chosen.

Solution

In the following figure, the slip plane ($ABDF$) is $(10\bar{1}0)$ and the slip direction AB is $[\bar{1}2\bar{1}0]$ so that the slip system is $(10\bar{1}0)[\bar{1}2\bar{1}0]$.

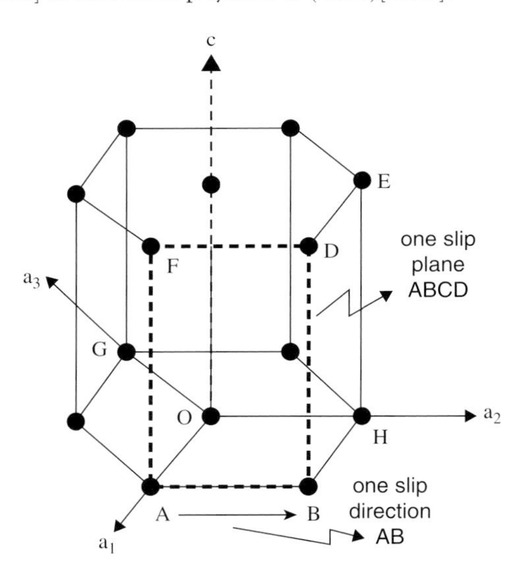

4.1.1
Critical Resolved Shear Stress

We already know that shear stress causes slip in crystals. The resolved shear component based on the magnitude of the external load, geometry of the crystal

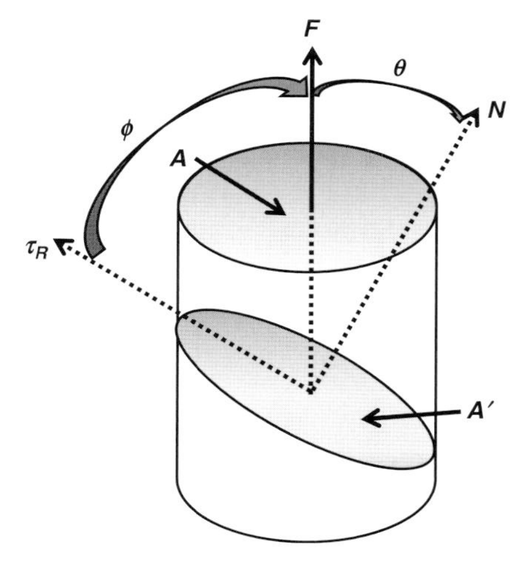

Figure 4.4 A schematic representation of a single-crystal cylinder used to calculate critical resolved shear stress.

structure, and orientation of the active slip systems needs to reach or exceed a critical value in order for the slip to occur. This critical shear stress is called critical resolved shear stress. The CRSS is the single crystal equivalent of yield stress of a polycrystalline material as obtained in a standard stress–strain curve. The CRSS value for a slip system in a crystal depends on purity and temperature. The following derivation of resolved shear stress makes clear how a single crystal can undergo slip.

Erich Schmid [15] was the first to analyze the problem, one of his contributions to understanding crystal plastic deformation. Let us take a cylindrical single crystal under tensile force F with a cross-sectional area A, as illustrated in Figure 4.4. We need to find out the resolved component of stress.

The normal stress (σ) is given by F/A. Now we can define the slip system with respect to the tensile axis and normal to the slip plane. Slip plane contains the slip direction along which the resolved shear stress (τ_R) would act. As shown in Figure 4.4, the included angle between the tensile axis (along F) and the slip plane normal (along N) is θ, and the angle between the tensile axis and the slip direction is ϕ. The slip plane is inclined at θ to the tensile axis, and thus its area (A') is given by $A/\cos\theta$. The normal force F can be resolved along the slip direction as $F{\cdot}\cos\phi$. Hence,

$$\tau_R = \frac{F\cdot\cos\phi}{A/\cos\theta} = \frac{F}{A}\cos\theta\cos\phi = \sigma(\cos\theta\cos\phi). \tag{4.1}$$

Equation (4.1) is also known as *Schmid's law*, while the term $\cos\theta\cos\phi$ is called *Schmid factor*. Schmid factor gives a measure of the slip system orientation. The resolved shear stress is the maximum when $\theta = \phi = 45^\circ$ so that $\tau_R = \sigma\cdot(1/\sqrt{2})\cdot(1/\sqrt{2}) = (1/2)\sigma$. In the extreme cases, where either θ or ϕ is 90°, there will be no resolved shear stress in the slip plane, and the material tend to fracture rather than deform. A numerical example on this topic is worked out in Example 4.2. Note that even though the derivation of resolved shear stress has been made in a uniaxial tensile situation, it is equally applicable to plastic deformation under compression.

Example 4.2

Calculate the critical resolved shear stress for a metallic single crystal (BCC) that starts deforming in the slip system $(\bar{1}10)[111]$ when the tensile stress along [010] is 100 MPa.

Solution

To solve this problem, let us first refer to Eq. (4.1).

Given the tensile stress $(\sigma) = 100$ MPa; we need to calculate $\cos\theta$ and $\cos\phi$. Refer to Figure 4.4 to locate the slip plane, slip direction, and tensile axis with their Miller indices. Note that the orientation of slip plane and slip direction can be taken as arbitrary with respect to the tensile axis. It is just a representation of the problem statement so that the steps below can be followed easily.

Before we start calculation of the cosines, let us turn our attention to Section 2.1. We know that in a cubic system the normal direction to a crystallographic plane bears the same digits of the Miller indices. So, the normal direction to the slip plane, $(\bar{1}10)$, is $[\bar{1}10]$.

Following Eq. (2.4), the cosine of the angle between the tensile axis [010] and slip plane normal $[\bar{1}10]$ is given by the dot product of the unit vectors along these two directions:

$$\cos\theta = \frac{u_1u_2 + v_1v_2 + w_1w_2}{\sqrt{u_1^2 + v_1^2 + w_1^2}\sqrt{u_2^2 + v_2^2 + w_2^2}}$$
$$= \frac{(0)(-1) + (1)(1) + (0)(0)}{\sqrt{0^2 + 1^2 + 0^2}\sqrt{(-1)^2 + 1^2 + 0^2}} = \frac{1}{\sqrt{2}}.$$

Furthermore, the cosine of the angle between the tensile axis [010] and the slip direction [111] is

$$\cos\phi = \frac{(0)(1) + (1)(1) + (0)(1)}{\sqrt{0^2 + 1^2 + 0^2}\sqrt{1^2 + 1^2 + 1^2}} = \frac{1}{\sqrt{3}}.$$

Hence, using Eq. (4.1), we obtain the resolved shear stress: $\tau_R = (100\ \text{MPa})(1/\sqrt{2})(1/\sqrt{3}) =\sim 40.8$ MPa.

The problem statement states that the single crystal starts deforming at the tensile stress of 100 MPa on the given slip system. So, the resolved shear stress must reach the corresponding critical value to accomplish that. Hence, it is clear that the resolved shear stress of 40.8 MPa calculated above is indeed the CRSS.

Table 4.1 A comparison of CRSS versus theoretical shear strength in some metals.

Metal	Purity (%)	Slip system	CRSS (MPa)	Theoretical shear yield stress (MPa)
Copper (FCC)	>99.9	(111)[110]	~1	~4800
Silver (FCC)	99.99	(111)[110]	0.48	~3000
Alpha-iron (BCC)	99.96	(110)[111]	27.5	~8199
Cadmium (HCP)	99.996	$(0002)\langle 11\bar{2}0\rangle$	0.58	~543

Special Note

Note that θ and ϕ are complementary angles (sum of the two angles 90°) only if a special condition is met, that is, when the slip direction happens to be in the same imaginary plane containing the stress axis and the slip plane normal. But that is clearly not the case above. The angles θ and ϕ are approximately 45.0° and 54.8°, respectively, and their sum is 99.8°, not 90°!

CRSS value in a single crystal is determined by the interactions between various dislocations and their interactions with other types of defects present. However, this value is quite smaller than the theoretical shear strength of the crystal. Table 4.1 lists comparative values of the CRSS for four different metals. The theoretical shear strength values for respective metals are calculated from $G/10$ (where G is the shear modulus of the metal), and CRSS values are all experimentally determined [1].

4.1.2 Peierls–Nabarro (P–N) Stress

A dislocation experiences an opposing force (the basic level of lattice friction) when it tries to move through an otherwise perfect crystal (i.e., without any other defect acting as obstacles). The corresponding stress needed to move a dislocation in a particular direction in the crystal is known as Peierls–Nabarro (P–N) stress. The P–N stress is a direct consequence of the periodic force field present in crystal lattice and is very sensitive to any changes in the individual atom positions. That is, it is a function of the dislocation core structure, and hence developing a single analytical expression is difficult. However, the analysis forwarded by Peierls (1940) and Nabarro (1947) still gives us some important qualitative understanding that is of definite value. The P–N stress ($\tau_{\text{P-N}}$) is given by the following relation:

$$\tau_{\text{P-N}} \approx \frac{2G}{1-\nu} \exp\left(-\frac{2\pi w}{b}\right), \tag{4.2}$$

where w is the dislocation core width, b is the distance between atoms in the slip direction, that is, the Burgers vector of the dislocation involved, G is the shear modulus, and ν is the Poisson's ratio of the material. It can be shown that for screw dislocations, w is close to the interplanar spacing between slip planes (d); whereas for edge dislocations, w is given by $(d/(1-\nu))$. Dislocation core widths generally seem to vary between b and $5b$ (sometimes on the order of $10b$ for ductile metals), depending again on the interatomic potential and crystal structure. The related energy barrier is called *Peierls energy*.

Despite its limitations, the concept of P–N stress explains some important qualitative aspects of plastic deformation in various crystalline materials. From Eq. (4.2), one can see that a wider dislocation core (i.e., with larger w) leads to lower value of P–N stress. This situation arises in ductile metals where the dislocation core is highly distorted and is not localized. However, ceramic materials (with covalent and/or ionic bonds) that show a very low or no ductility at lower temperatures have dislocations of narrower width. This also means that the P–N stress in ceramic materials is correspondingly high. Also, the presence of electrostatic forces in ceramic materials makes the dislocation movement difficult leading to lower plasticity. However, some ductility in ceramic materials can be obtained by increasing temperature that provides for thermal activation surmounting the relevant Peierls barrier. The relative higher P–N stress in BCC metals compared to FCC metals can also be explained using Eq. (4.2). In FCC metals, slip occurs on the close-packed planes that are greater distance apart, that is, d is higher). In the close-packed structure, the magnitude of the Burgers vector (b) is smaller. That means $d > b$. It makes the magnitude of $\tau_{\mathrm{P-N}}$ smaller. The opposite thing happens in BCC crystals that contain loosely packed slip planes with close separation from each other, that is, $d < b$. Again, from Eq. (4.2), we can see that $\tau_{\mathrm{P-N}}$ will be higher in BCC and like typical not-so close-packed crystal lattices.

4.1.3
Slip in Crystals: Accumulation of Plastic Strain

We have noted in the preceding chapters that the dislocation movement under applied stress creates plastic strain (permanent deformation) in crystals. This is different from the elastic deformation under applied stress where the atomic bonds are stretched without the help of a dislocation. It is not simple to relate plastic strain to applied stress as it depends on temperature, strain rate, and microstructural factors. However, Orowan developed a simple expression, known as *Orowan's equation*, relating the macroscopic strain rate to the microscopic parameters such as dislocation density, velocity, and Burgers vector. This derivation is based on the recognition of the fact that when a single dislocation moves, it creates a displacement of the magnitude equal to the Burgers vector (b).

Consider a crystal of volume HLD containing a number of straight edge dislocations (for the sake of simplicity, dislocation type is assumed to be edge), as shown in Figure 4.5. Now let us consider an applied shear stress causing a dislocation move by δ_i. We know that if the dislocation moves the distance D along the Burgers vector, it produces b displacement. Therefore, it can be said that the contribution of the dislocation moving by δ_i distance to the displacement is $(b/D) \times \delta_i$. The displacement generated by a single dislocation is pretty small. To produce plastic strain of engineering significance, there are a large number of dislocations which

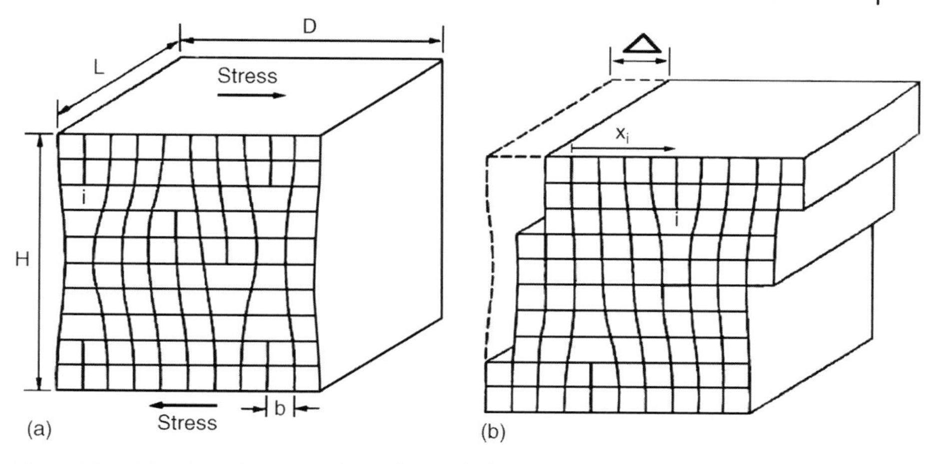

Figure 4.5 (a) A schematic crystal subjected to applied shear stress. (b) After deformation, a total plastic displacement of Δ is produced [3].

when move produce the total displacement Δ given by

$$\Delta = \frac{b}{D}\sum_{i=1}^{N} \delta_i, \tag{4.3}$$

where N is the total number of dislocations that have moved in the crystal volume.

The macroscopic plastic shear strain (γ) is given by

$$\gamma = \frac{\Delta}{H} = \frac{b}{HD}\sum_{i=1}^{N} \delta_i. \tag{4.4}$$

Equation (4.4) can be further simplified by defining the average distance $\overline{x}$ traveled by a dislocation as

$$\overline{x} = \frac{1}{N}\sum_{i=1}^{N} \delta_i. \tag{4.5}$$

Hence, by replacing Eq. (4.5) into Eq. (4.4), we obtain

$$\gamma = \frac{bN\overline{x}}{HD} = \left(\frac{NL}{HLD}\right) b\overline{x}. \tag{4.6}$$

The expression NL/HLD gives the total dislocation line length (NL) per unit volume (HLD), that is, mobile dislocation density. Let us denote the mobile dislocation density by ϱ_m. Note that the term ϱ_m is not the total dislocation density as the immobile dislocation does not contribute to the plastic strain. Therefore, Eq. (4.6) can be rewritten as

$$\gamma = \varrho_\mathrm{m} b\overline{x}. \tag{4.7}$$

Equation (4.7) can be further expressed in terms of shear strain rate by taking differential of both sides of the equation with respect to time t.

$$\dot{\gamma} = \frac{\mathrm{d}\gamma}{\mathrm{d}t} = \varrho_\mathrm{m} b \frac{\mathrm{d}\overline{x}}{\mathrm{d}t} = \varrho_\mathrm{m} b \overline{v}_\mathrm{d}, \tag{4.8}$$

where $\bar{v}_d$ is the average velocity of dislocations. The equation is universal in nature and also applicable for climb of edge dislocations. Furthermore, it can be universally applied to screw dislocations and mixed dislocations. Equation (4.8) is a nice example of relating macroscopic behavior of a material described by a set of microscopic parameters. The Burgers vector (b) is determined by crystal structure of the material. The other two quantities ϱ_m and $\bar{v}_d$ are the parameters that depend on several other factors such as stress, temperature, prior processing history, and so on.

4.1.4
Determination of Burgers Vector Magnitude

So far, we have come across a couple of equations that include the magnitude of Burgers vector in this chapter. The concept of Burgers vector was first introduced in Section 2.2. Now we need to understand how the magnitude of the Burgers vectors can be determined. Burgers vector is basically the shortest lattice vector joining one lattice point to another. This type of Burgers vector is associated with dislocations termed as *perfect* or *unit dislocation*. For example, in a BCC crystal, the shortest lattice vector is the distance between a corner atom and the body center. We can resolve the vector by $a_0/2$ length from the origin along X-, Y-, and Z-axes, where a_0 is the lattice constant. The standard notation for Burgers vector is then $[(a_0/2),(a_0/2),(a_0/2)]$ or $(a_0/2)[111]$. The magnitude (strength) of the Burgers vector of a perfect dislocation in a BCC crystal is

$$b=\left(\frac{a_0^2}{4}+\frac{a_0^2}{4}+\frac{a_0^2}{4}\right)^{1/2}=\frac{\sqrt{3}}{2}a_0 \quad \text{(i.e., half of the body diagonal).}$$

Similarly, in FCC metals, the Burgers vector is $(a_0/2)$ [110], that is, half of the face diagonal. The magnitude of the vector is $a_0/\sqrt{2}$. In a cubic crystal, we can follow the following procedure to find out the Burgers vector of a dislocation (does not need to be a perfect dislocation). If a dislocation has Burgers vector of $q[uvw]$ with x being a fraction or a whole number, the magnitude of the Burgers vector is $q\cdot(u^2+v^2+w^2)^{1/2}$. For instance, Burgers vector of $a_0/3$ [112] dislocation has a magnitude of $(a_0/3)(1^2+1^2+2^2)^{1/2}=(a_0/3)\sqrt{6}$.

Determination of the Burgers vector of a perfect dislocation in a HCP crystal is much simpler than the cubic metals. The strength of the Burgers vector is given by the edge of the base, that is, a_0.

Example 4.3

An FCC crystal (lattice constant $a_0=0.286$ nm) contains a total dislocation density of 10^9 m^{-2}, of which only half are mobile (i.e., glissile). If we assume that the Burgers vector of all the mobile dislocations is $a_0/2$ [110] and the shear strain rate is 10^{-1} s^{-1}, what is the average dislocation velocity?

Solution

To solve the above problem, we need to use Eq. 4.8.

Given are the shear strain rate $(\dot{\gamma}) = 10^{-1}\,\mathrm{s}^{-1}$, mobile dislocation density $(\varrho_m) = 0.5 \times 10^9\,\mathrm{cm}^{-2}$, and the magnitude of the Burgers vector $= a_0/\sqrt{2} = 0.202\,\mathrm{nm}$. Therefore, the average dislocation velocity is given by

$$\bar{v}_d = \frac{\dot{\gamma}}{\varrho_m b} = \frac{= 10^{-1}\,\mathrm{s}^{-1}}{(0.5^*\,\mathrm{m}^{-2})(0.202 \times 10^{-9}\,m)} =\sim 1\,\mathrm{m\,s}^{-1}.$$

4.1.5
Dislocation Velocity

Dislocation velocity depends on the purity of the crystal, applied shear stress, temperature, and dislocation type. Johnston and Gillman (1959) developed an expression for the dislocation velocity in freshly grown lithium fluoride crystals. It was found that the edge dislocations travel about 50 times faster than screw dislocations. Studies on the close-packed FCC and HCP metals have revealed that the dislocation velocity approaches $\sim 1\,\mathrm{m\,s}^{-1}$ at the critical resolved shear stress of the specific crystal. Dislocation velocities have been found to be a very strong function of applied shear stress as shown in Eq. (4.9):

$$v_d = A\tau^{m'}, \tag{4.9}$$

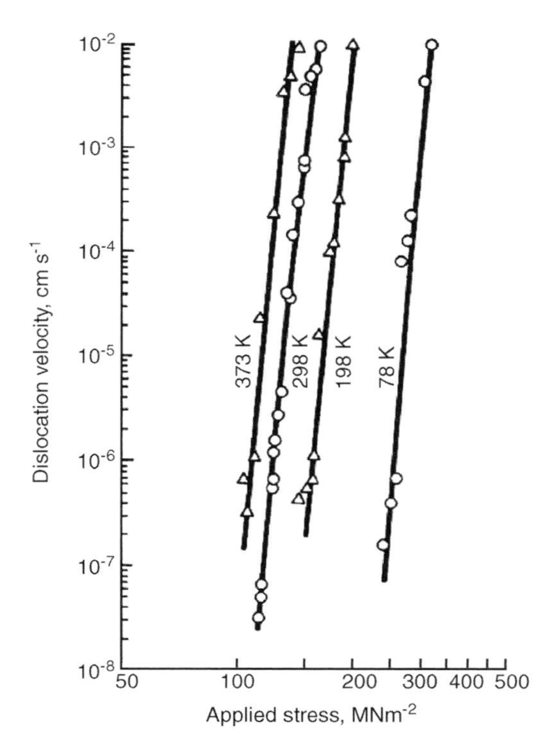

Figure 4.6 Dislocation velocity as a function applied stress in 3.25% Si containing iron from D.F. Stein and J.M. Low [28].

where m' is a material constant with values ranging from 1.5 to 40 for various types of materials and A is a constant. However, for pure crystals, m' generally remains $\sim$1 at 300 K and 4–12 at 77 K. At lower temperatures, dislocation velocity is higher compared to that at higher temperatures because phonons (lattice vibrations) are more at higher temperatures obstructing the dislocation velocity. The theoretical maximum velocity of dislocation in a crystalline solid is the velocity of the transverse shear wave propagation. However, damping forces (related to phonons) are enhanced as the dislocation velocity reaches 1/1000 of the theoretical limit. Figure 4.6 illustrates the variation of dislocation velocity as a function of applied stress in an iron alloy containing 3.5% Si.

Observing Dislocations

As mentioned previously, the concept of dislocations was first introduced by Taylor, Polanyi, and Orowan in 1930s. Following the discovery of dislocation, several theories were proposed supported by the *indirect* observation of dislocations. However, later with the advent of more sophisticated characterization techniques, the dislocations were directly observed and likewise related theories were rapidly developed. Almost all techniques used to visualize the dislocations utilize the strain field around the dislocations. Some methods are discussed very briefly.

Etch Pit Technique

This is one of the simplest chemical technique methods to observe dislocations indirectly. Dislocations intersecting the surface etch at a different rate than the surrounding matrix and the region appears as *pits*. An example of etch pits in a lithium fluoride crystal is shown in Figure 4.7. The relative position of the etch pits represents the location and number of dislocations. However, the etch pit method has serious limitation when the etch pits tend to overlap. That is why the etch pit technique is applicable only for a low dislocation density ($10^6\,cm^{-2}$).

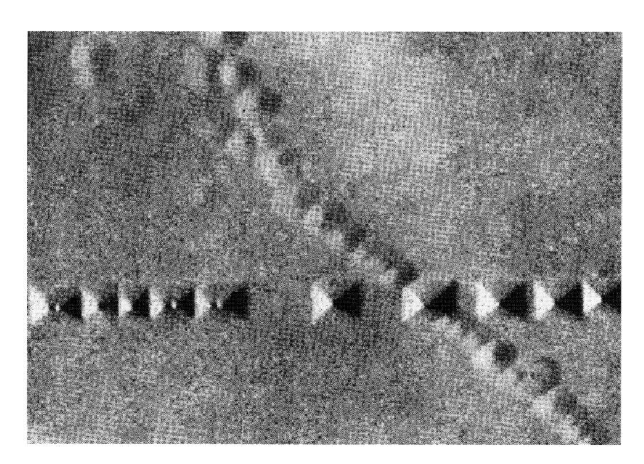

Figure 4.7 An optical micrograph of etch pits produced on a lithium fluoride crystal [4].

Decoration Technique

This is another type of chemical technique in which dopants are added to the crystal. By suitable heat treatment, these foreign atoms precipitate near the dislocation core and decorate the dislocations. Dislocations in KCl are revealed by adding AgCl to the melt prior to the crystal growth. An optical micrograph in Figure 4.8 shows silver particles decorating dislocations in KCl. This technique has been mainly used in ionic crystals.

Transmission Electron Microscopy

Transmission electron microscopy is the most powerful technique for direct visualization of dislocations. Transmission electron microscopes (TEM) use an electron beam in a high-vacuum environment to pass through the electron-transparent region (~100 nm or so) of a thin foil specimen. Very high resolutions on the order of few angstroms can be easily achieved in advanced TEMs. Generally, a semester-long, stand-alone graduate course on TEM is offered in most research universities. The topic in itself is complex enough to be covered in a single paragraph. TEM is a versatile tool that can be used to detect not only the dislocations but also a host of other defects ranging from stacking faults, twins, voids, and so forth. The key to imaging dislocations with TEM is the way the electrons interact with the dislocation strain field, as illustrated inFigure 4.9b. Bragg's law is the guiding principle behind the electron diffraction responsible for the dislocation contrast and specialized technique such as weak beam imaging is used to image dislocations with better clarity. A TEM micrograph of the titanium alloy samples with a number of dislocations is shown in Figure 4.9a. There are several limitations of the TEM technique, including the limited sample volume that can be examined.

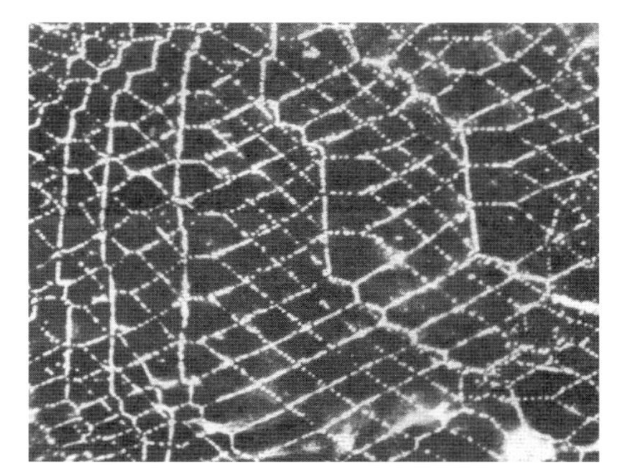

Figure 4.8 An optical micrograph of a KCl crystal with dislocations decorated by silver due to the addition of AgCl [5].

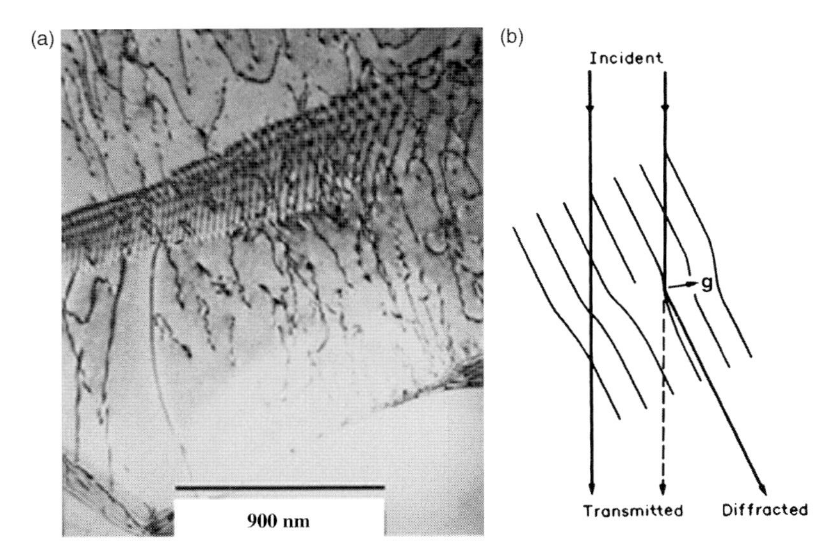

Figure 4.9 (a) Morphology of a dislocation pileup [6]. (b) The electron diffraction occurs differently near the plane of the edge dislocation compared to the dislocation-free crystal portion [3].

X-Ray Diffraction Topography

This is another technique of direct dislocation observation, albeit with much lower resolution. Figure 4.10 shows an X-ray diffraction topograph in a single crystal of silicon. Dislocation widths imaged are quite coarse (on the order of 1 μm). Hence, it is not possible to image dislocations of samples with dislocation density higher than $10^6\,\text{cm}^{-2}$. X-rays generally got higher penetration compared to the electrons. Thus, the specimen used is large single crystal oriented in such a way that strong reflections are obtained. The difference in the intensities of the diffracted X-rays when recorded as photomicrographs shows the dislocation structures, as depicted in Figure 4.10. Nowadays, this technique is not much used for characterizing dislocations.

4.2 Other Dislocation Characteristics

4.2.1 Types of Dislocation Loops

We have noted that pure edge and pure screw dislocations are rarely observed. Most dislocation lines are of mixed type. Many of them stay in the form of dislocation loops. Let us discuss two types of dislocation loops here.

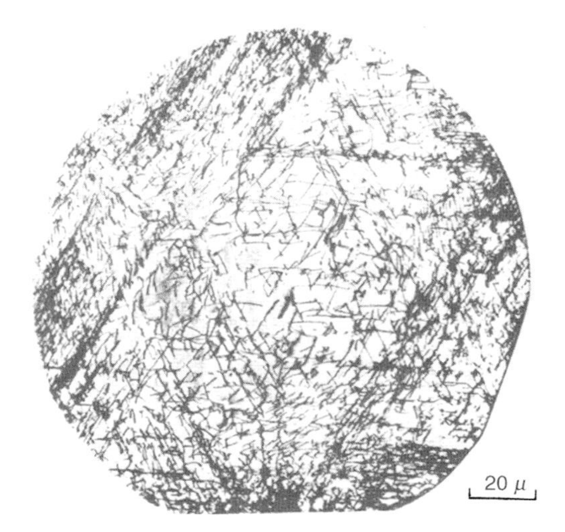

Figure 4.10 An X-ray diffraction topograph showing multiple dislocations in a single crystal of silicon [3].

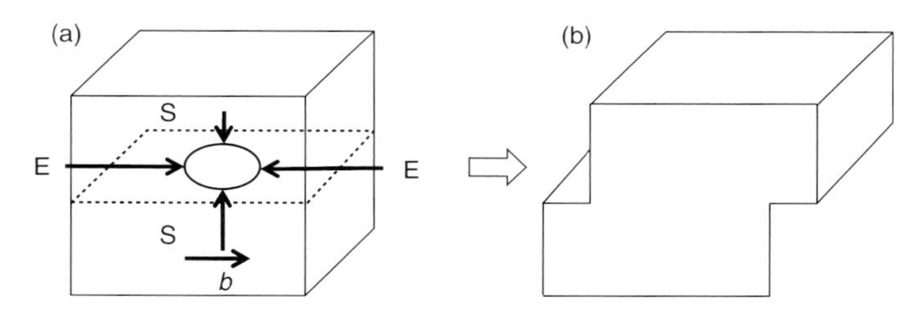

Figure 4.11 (a) A schematic of a glide loop. (b) Result of the glide loop expansion producing crystal slip.

4.2.1.1 **Glide Loop**

In this type of dislocation loops, the Burgers vector is on the same plane as the dislocation loop. Hence, the loop expands or contracts under an applied stress. The loop here has edge, screw, and mixed orientations. Figure 4.11 shows how the glide loop can produce plastic deformation.

4.2.1.2 **Prismatic Loop**

In this type of dislocation loops, the Burgers vector lies perpendicular to the plane of the dislocation loop, as shown in Figure 4.12a. As the Burgers vector is perpendicular to the dislocation loop, the loop is entirely of pure edge orientation. The glide plane is perpendicular to the loop. Thus, the loop cannot expand or contract on the plane of the loop conservatively unlike the glide loop. If the loop movement

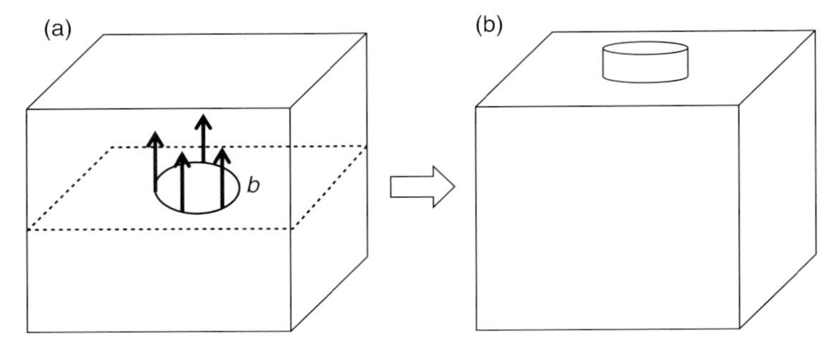

Figure 4.12 (a) A schematic of a prismatic loop. (b) Result of the prismatic loop movement producing crystal slip perpendicular to the original dislocation loop plane.

does take place, it would produce crystal slip, as shown in Figure 4.12b. In latter sections, we will again discuss about this type of dislocation loops.

4.2.2 Stress Field of Dislocations

The dislocations are associated with elastic stress fields around them. This stress field is crucial for the dislocation to interact with other dislocations and crystal defects. For a straight edge and screw dislocation in an isotropic medium, the stress/strain field analysis is quite straightforward. However, it becomes complicated with the elastic anisotropy of the crystal and other variations in dislocation characteristics.

4.2.2.1 Screw Dislocation

Assume an elastic distortion in a hollow cylinder, as shown in Figure 4.13a, creating a straight screw dislocation. This type of construction is known as *Volterra* dislocation, named after the Italian mathematician who first considered such distortions even though the concept of dislocations was not introduced. In the figure, MN appears as a screw dislocation due to the way radial slit (LMNO) is cut in the hollow cylinder and displaced by a distance of b, which is the magnitude of the Burgers vector of the screw dislocation in the z-direction. We can confirm it as a screw dislocation since the Burgers vector and the dislocation line are parallel to each other. Without getting into the details of the derivation, only the stress components of the stress field that are nonzero are found to be shear components. There is no dilatational (i.e., tensile or compressive) stress component associated with the screw dislocation. The stress field around a screw dislocation is given by

$$\tau_{\theta z} = \tau_{z\theta} = \frac{Gb}{2\pi r}. \tag{4.10a}$$

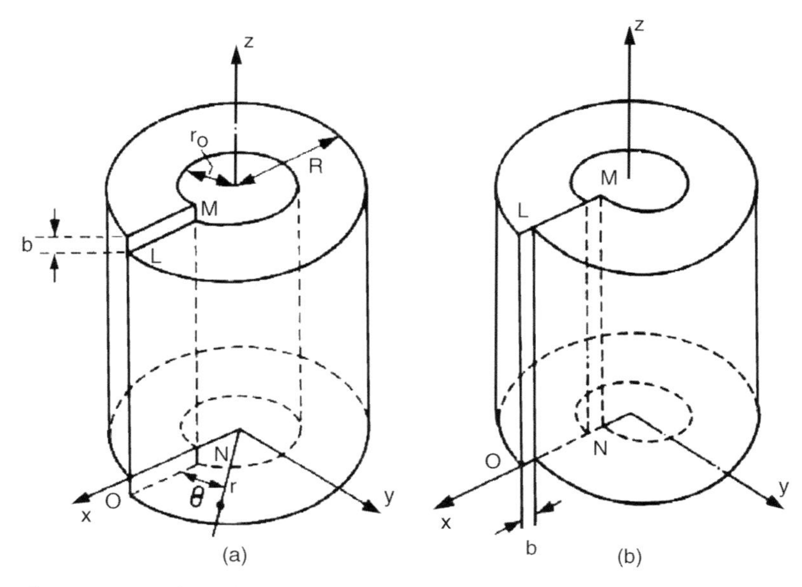

Figure 4.13 (a) Elastic distortion of a hollow cylinder simulating the presence of a screw dislocation. (b) Elastic distortion of a hollow cylinder simulating the presence of an edge dislocation.

The strain field is given by the following equation:

$$\gamma_{\theta z} = \gamma_{z\theta} = \frac{b}{2\pi r}. \tag{4.10b}$$

The above equations imply that the stress and strain fields consist of pure shear components. The stress field exhibits complete radial symmetry. As one can see, the stress varies inversely with the radius of the cylinder. The stress then becomes infinite when r tends to zero (i.e., approaching the center of the cylinder). The cylinder cannot physically sustain infinite stress. That is why a hollow cylinder (with a radius of r_0) assumption is justified. Also, the region in the hollow region does not follow the linear elastic theories. This region is akin to the dislocation core. In order to know the stress field in this region, one needs to employ nonlinear atomistic theories leading to complex derivations. In most cases, the dislocation core dimension remains approximately at <1 nm.

4.2.2.2 **Edge Dislocation**

The stress field of an edge dislocation is more complex than that of a screw dislocation. However, in a similar way, a Volterra-type edge dislocation can be created as done for screw dislocation. In Figure 4.13b, such a case is shown. But in this case, the cut has been made perpendicular to the z-axis and a displacement of b has been made. Thus, the magnitude of the simulated edge dislocation is b that is perpendicular to the dislocation line, and hence an edge dislocation. Here also the solution breaks down when x and y approach zero. So, even in the edge dislocation, the core region does not follow linear elastic solutions. Without getting into the details of derivation, the stress field of an edge dislocation is shown below. It contains both dilatational (σ_{xx}, σ_{yy}, and σ_{zz}) and shear components (τ_{xy} and τ_{yx}).

$$\sigma_{xx} = -Cy\frac{(3x^2 + y^2)}{(x^2 + y^2)^2}, \tag{4.11a}$$

$$\sigma_{yy} = Cy\frac{(x^2 - y^2)}{(x^2 + y^2)^2}, \tag{4.11b}$$

$$\tau_{xy} = \tau_{yx} = Cx\frac{(x^2 - y^2)}{(x^2 + y^2)^2}, \tag{4.11c}$$

$$\sigma_{zz} = \nu(\sigma_{xx} + \sigma_{yy}), \tag{4.11d}$$

where $C = Gb/(2\pi(1 - \nu))$.

Here, ν is the Poisson's ratio. At $y = 0$ where the slip plane lies, all but the shear stress components given in Eq. (4.11c) are zero and the maximum compressive stress acts just above the slip plane with maximum tensile stress acting immediately below the slip plane. The stress field of mixed dislocations could also be found out.

4.2.3
Strain Energy of a Dislocation

The distortion associated with a dislocation implies that it must have certain strain energy. The total strain energy (E_{total}) of a dislocation is composed of two parts – elastic strain energy (E_{el}) and core energy (E_{core}). The elastic strain energy is stored outside the core region, and hence can be estimated using the linear elastic theories. We have discussed the shear stress and strain components associated with a screw dislocation in Eqs. (4.10a) and (4.10b). Due to the complete radial symmetry of the stress/strain field associated with the screw dislocation, we derive its elastic strain energy using a simple calculation. Let us consider a volume element shell of the cylinder with an arbitrary radius r, thickness dr, and length l, and thus the elastic strain energy is given by

$$E_{\text{el}} = \frac{1}{2}\tau_{\theta z}\gamma_{\theta z} \cdot (\text{volume of the element}) = \frac{1}{2}\frac{Gb}{2\pi r}\frac{b}{2\pi r}(l \cdot 2\pi r \cdot \mathrm{d}r) = \frac{Gb^2 l}{4\pi r}\mathrm{d}r.$$

Note the above equation is applicable only for an isotropic continuous medium. The hollow cylinder of length l, inner radius r_1, and outer radius r_2 will have an elastic strain energy determined by the following integral:

$$E_{\text{el}} = \int_{r_1}^{r_2} \frac{Gb^2 l}{4\pi}\frac{\mathrm{d}r}{r} = \frac{Gb^2 l}{4\pi}\ln\left(\frac{r_2}{r_1}\right). \tag{4.12}$$

As noted earlier, the dislocation core can be *1b–5b* radius, and in most cases can be taken as ~1 nm radius. So, r_1 value can be assumed to be of 1 nm. Furthermore, dislocations tend to form networks where the long-range stress fields are superimposed and may get canceled. Thus, the effective energy of the dislocation can get reduced. That is why if one takes the value of r_2 as approximately half the average spacing between dislocations in a random arrangement, a realistic value could be 10^6 nm. Then the value of r_2/r_1 becomes 10^6. So, Eq. (4.12) is reduced to

$$E_{el} = (Gb^2 l)\frac{13.8}{4 \times 3.14} = \alpha G b^2 l,$$

where α is about ~1.1.

So, the elastic strain energy of a screw dislocation *per unit length* is

$$E_{el} = \alpha G b^2, \tag{4.13}$$

where $\alpha = \sim 1$ for screw dislocation.

An analysis can be performed for deriving the energy of an edge dislocation incorporating its complex stress field into the derivation. A similar expression can be obtained with a constant, $\alpha = \sim 1/(1-\nu) = \sim 1.5$ (for $\nu = 0.33$). So, in simple terms, the edge dislocation has greater energy than that of a screw dislocation of similar length. The elastic strain energy of a mixed dislocation can also be obtained, and is generally between the energies of a pure edge and a pure screw dislocation.

However, the value of the constant may undergo changes depending on the precise values of r_1 and r_2, which are difficult to determine. However, whatever their values are, the constant α would still remain close to unity.

The estimation of the dislocation core energy is only approximate as linear elastic theories are no longer valid in this region. Approximate calculations show that the core energy is estimated to be on the order of 0.5 eV per atom plane threaded by the dislocation, whereas the elastic strain energy of a dislocation is 5–10 eV per atom plane. Despite variation in the core energy during the dislocation movement, the core energy still remains a small fraction of the elastic strain energy. Thus, the dislocations possess large positive strain energy increasing the overall free energy of the crystal. But the system would like to attain a more stable state via lowering its free energy. That is why given the chance the dislocation density can be reduced through dislocation annihilation processes such as annealing. The same is not true for point defects (such as vacancies) as they attain different equilibrium concentrations at different temperatures. That is why dislocations are thermodynamically unstable defects, whereas vacancies are thermodynamically stable defects.

4.2.3.1 **Frank's Rule**

From the definition of a dislocation being the line of demarcation between slipped and unslipped regions, dislocations cannot end inside a crystal except at a *node*. The node is a point where two or more dislocations meet. The existence of dislocation nodes affirms that different types of *dislocation reactions* occur inside a crystal. However, a dislocation reaction is totally different from a chemical reaction. So, we should not confuse ourselves looking for similarities between the two. The dislocation reactions pertain to the addition or dissociation of initial dislocation(s) into new product dislocation(s). However, they cannot just happen arbitrarily. The dislocation reaction should follow certain geometrical and energetic rules. The energy criterion used to decide whether a dislocation addition or dissociation reaction is feasible or not is known as *Frank's rule*. The underlying basis of this rule is that the total energy needs to be minimized for the reaction to be energetically feasible. To elucidate the rule, let us take an example of a dislocation reaction where two dislocations with Burgers vectors b_1 and b_2 meet to form a third dislocation with a Burgers vector b_3. That is, the reaction is $b_1 + b_2 \rightarrow b_3$. According to Frank's rule, the reaction will be feasible when $b_1^2 + b_2^2 > b_3^2$. In case of a reaction $b_1 \rightarrow b_2 + b_3$, the energetic feasibility condition of the reaction would be $b_1^2 > b_2^2 + b_3^2$, according to the Frank's rule. This implies that the reaction should be vectorially correct and energetically favorable. With little introspection, we can find out that the Frank's rule actually stems from Eq. (4.13).

Example 4.4

Show that the following dislocation reaction is feasible:

$$\frac{a_0}{2}[10\bar{1}] \rightarrow \frac{a_0}{6}[2\bar{1}\bar{1}] + \frac{a_0}{6}[11\bar{2}].$$

Solution

To test the feasibility of the reaction, a two-step procedure needs to be followed. The first step is to ascertain whether the reaction is geometrically sound or not. This can be accomplished by verifying that the x, y, and z components are equal on both sides of the reaction:

$$x \text{ components: } \frac{a_0}{2} = \frac{2a_0}{6} + \frac{a_0}{6}$$

$$y \text{ components: } \frac{a_0}{2}(0) = 0 = -\frac{a_0}{6} + \frac{a_0}{6}$$

$$\text{z components: } \frac{a_0}{2}(-1) = -\frac{a_0}{2} = -\frac{a_0}{6} - \frac{2a_0}{6}$$

Therefore, the given dislocation is geometrically (or vectorially) possible. If any reaction that does not happen, there is no need to go to the second step for validation.

For the dislocation reaction to be energetically favorable, we need to show that $b_1^2 > b_2^2 + b_3^2$.

$$\text{Original dislocation: } b_1 = \frac{a_0}{2}[10\bar{1}].$$

The magnitude of the Burgers vector is

$$b_1 = \frac{a_0}{2}(1^2 + 0^2 + (-1)^2)^{1/2} = \frac{a_0}{\sqrt{2}}.$$

Hence,

$$b_1^2 = \frac{a_0^2}{2}. \tag{4.14a}$$

Product dislocation : $b_2 = \frac{a_0}{6}[2\bar{1}\bar{1}]$.

The magnitude of Burgers vector is

$$b_2 = \frac{a_0}{6}(2^2 + (-1)^2 + (-1)^2)^{1/2} = \frac{a_0}{6}\sqrt{6} = \frac{a_0}{\sqrt{6}}.$$

Hence,

$$b_2^2 = \frac{a_0^2}{6}.$$

Product dislocation: $b_3 = \frac{a_0}{6}[11\bar{2}]$.

The magnitude of the Burgers vector is

$$b_3 = \frac{a_0}{6}(1^2 + 1^2 + (-2)^2)^{1/2}.$$

Hence,

$$b_3^2 = \frac{a_0^2}{6}.$$

$$b_2^2 + b_3^2 = \frac{a_0^2}{6} + \frac{a_0^2}{6} = \frac{a_0^2}{3} < \frac{a_0^2}{2} = b_1^2. \tag{4.14b}$$

In other words,

$$b_1^2 > b_2^2 + b_3^2.$$

Hence, the dislocation reaction is feasible.

4.2.4 Force on a Dislocation

With a large enough applied stress to a crystal containing dislocations, some dislocations can move and create plastic deformation. Thus, work is done on the crystal by the dislocation as it moves. The dislocation also experiences an opposing force against its movement. One can then evaluate the force on a dislocation due to the applied stress using the principle of virtual work. The work done by the applied stress when the dislocation moves completely through the crystal is equated to the work done by the resisting force ($f = Fl$, where l is the dislocation length and F is the force per unit length), as shown in Figure 4.14. The resulting displacement due to the force ($\tau l_1 l_2$) is b and thus the work done by the applied stress (τ) is $\tau l_1 l_2 b$. The dislocation moves through the crystal a distance l_2 doing work against the resistance force f ($=Fl_1$) given by $Fl_1 l_2$. Thus, the force (F) per unit length on the dislocation is given by the following relation:

$$F = \tau b, \tag{4.15a}$$

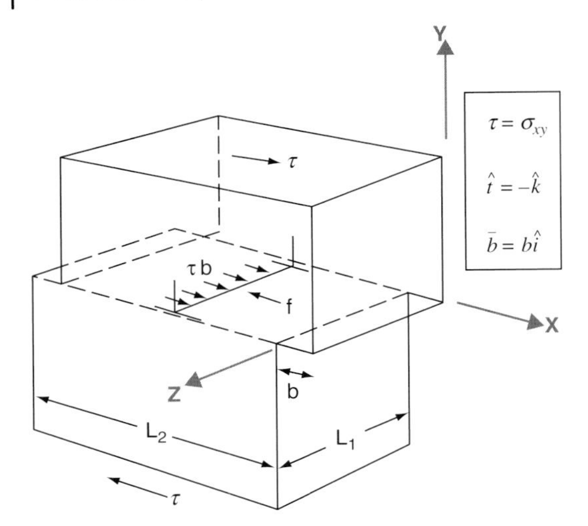

Figure 4.14 Force acting on a dislocation [7].

where τ is the shear stress resolved in the glide plane along the direction of the Burgers vector (b). Since the force is acting along x-direction, it is indeed $\overline{F} = \tau b\hat{i}$.

In the general case, in three- dimensions with a dislocation and Burgers vector ($\overline{b} = a[hkl]$) exposed to a stress tensor σ_{ij}, one can use the Peach–Koehler formula to calculate the force vector per unit length:

$$\overline{F} = \overline{G}x\hat{t} = (\sigma_{ij} \cdot \overline{b})x\hat{t}. \tag{4.15b}$$

Equation (4.15b) is very useful in evaluating the forces on dislocations due to complex stress states.

Example on Using Peach–Koehler Formula

Let us consider the example in Figure 4.14 as done earlier using virtual work. In this example, the Burgers vector and unit line vector of the dislocation are given by

$$\begin{aligned} \overline{b} &= b\hat{i} \\ \hat{t} &= -\hat{k} \end{aligned}$$

and the stress state is

$$\sigma_{ij} = \begin{pmatrix} 0 & \tau & 0 \\ \tau & 0 & 0 \\ 0 & 0 & 0 \end{pmatrix}$$

so that

$$\overline{G} = \sigma_{ij} \cdot \overline{b} = \begin{pmatrix} 0 & \tau & 0 \\ \tau & 0 & 0 \\ 0 & 0 & 0 \end{pmatrix} \begin{pmatrix} b \\ 0 \\ 0 \end{pmatrix} = \begin{pmatrix} 0 \\ \tau b \\ 0 \end{pmatrix}$$

Thus,

$$\overline{F} = \overline{G} x \hat{t} = \begin{vmatrix} i & j & k \\ 0 & \tau b & 0 \\ 0 & 0 & -1 \end{vmatrix} = \tau b \hat{i}$$

as per the same result as in Eq. (4.15a).

In addition to the force, a dislocation has a *line tension*. The concept of line tension is much akin to the surface tension prevalent in a soap bubble. The line tension of a dislocation arises because the length of a dislocation increases with the applied stress, and the dislocation can reduce its energy by going back to the earlier configuration, that is, with less length. The line tension has a unit of energy per unit length. We have already noted such unit in Eq. (4.13). Hence, the line tension (T) can be defined as the enhancement of energy in a dislocation line per unit increase in its length, and is given by the following relation:

$$T = \alpha G b^2. \tag{4.16}$$

Consider a curved dislocation as shown in Figure 4.15. The line tension (T) acting along the length of the dislocation line would try to straighten it. However, shear stress acting on the dislocation line would try to resist that effect. Let us now

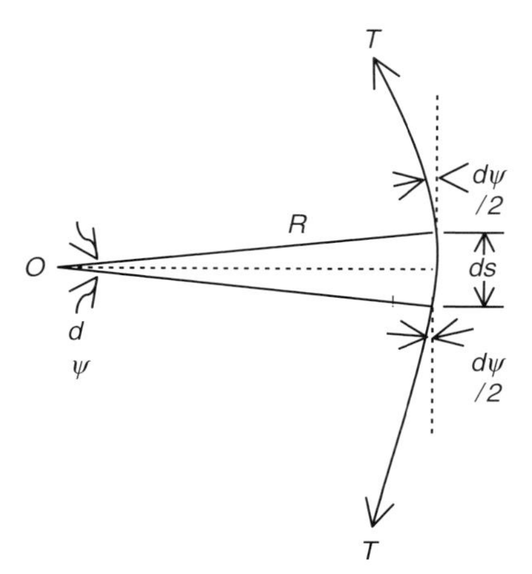

Figure 4.15 Curved dislocation line under the line tension T.

determine the shear stress (τ) needed by a dislocation line to maintain a radius of curvature R (with the center at O). The angle subtended by a very small arc segment of length ds is given by $\mathrm{d}\psi = \mathrm{d}s/R$, which is a very small value in radians. The outward force on the dislocation segment is $(\tau b)(\mathrm{d}s)$. On the other hand, the opposing line tension force is $2T\sin(\mathrm{d}\psi/2)$. For small values of $\mathrm{d}\psi$, this force becomes $T\mathrm{d}\psi$. The dislocation line will be in equilibrium if the outward force and the inward force are equal, that is,

$$T\mathrm{d}\psi = (\tau b)(\mathrm{d}s)$$

or

$$\tau = \frac{T\mathrm{d}\psi}{b\mathrm{d}s} = \frac{T}{bR}.$$

Using Eq. (4.16), we can write

$$\tau = \frac{\alpha G b^2}{bR} = \frac{\alpha G b}{R}. \tag{4.17a}$$

This is thus the stress required to keep the dislocation curved with a radius of curvature R. Alternatively, a dislocation pinned between two points tends to become curved with a radius R under the applied stress τ given by

$$R = \frac{\alpha G b}{\tau}. \tag{4.17b}$$

This implies that in the absence of any external stress, the dislocation wants to be as straight as possible with the minimum possible length.

Example 4.5

A stress

$$\sigma\,(\mathrm{MPa}) = \begin{pmatrix} 100 & 0 & 25 \\ 0 & 50 & 0 \\ 25 & 0 & 0 \end{pmatrix}$$

acts on an edge dislocation with b along $[1\bar{1}0]$ and line vector along $[11\bar{2}]$ in Ni (FCC) ($G = 110\,\mathrm{GPa}$, $a = 3.6\,\mathrm{Å}$).

a) What is the magnitude of the Burger's vector in millimeter and how is it represented?
b) What are the x, y, z components of the *unit* line vector?
c) Calculate the force per unit length ($\mathrm{N\,mm^{-1}}$) on the dislocation due to the applied stress state? (*Hint*: Use Peach–Koehler formula.)
d) What is the force (per unit length) along the Burger's vector (i.e., perpendicular to the dislocation line)?
e) If the dislocation is pinned between two points separated by 1000 Å, determine the radius of curvature.

Solution

a) Representation: $\bar{b} = (a/2)[1\bar{1}0]$ and magnitude:

$$|\bar{b}| = \frac{a}{\sqrt{2}} = \frac{3.6 \times 10^{-7}}{1.414} = 2.546 \times 10^{-7} \text{ mm}.$$

b) $\hat{t}_x = \frac{1}{\sqrt{6}}, \quad \hat{t}_y = \frac{1}{\sqrt{6}}, \quad \hat{t}_z = -\frac{2}{\sqrt{6}}.$

c) $\bar{F} = \bar{G} \times \hat{t}, \quad \text{where } \bar{G} = \sigma_{ij} \cdot \bar{b} = \begin{pmatrix} 100 & 0 & 25 \\ 0 & 50 & 0 \\ 25 & 0 & 0 \end{pmatrix} \cdot \begin{pmatrix} 1 \\ -1 \\ 0 \end{pmatrix} \frac{a}{2} = \frac{a}{2}\begin{pmatrix} 100 \\ -50 \\ 25 \end{pmatrix}$

so that

$$\bar{F} = \bar{G} \times \hat{t} = \frac{a}{2}\begin{pmatrix} 100 \\ -50 \\ 25 \end{pmatrix} \times \frac{1}{\sqrt{6}}\begin{pmatrix} 1 \\ 1 \\ -2 \end{pmatrix} = \frac{a}{2\sqrt{6}}\begin{vmatrix} i & j & k \\ 100 & -50 & 25 \\ 1 & 1 & -2 \end{vmatrix}$$
$$= \frac{a}{2\sqrt{6}}\{i(100 - 25) - j(-200 - 25) + k(100 + 50)\}$$

Thus,

$$\bar{F} = \frac{a}{2\sqrt{6}}\{75i + 225j + 150k\}.$$

d) Force along the Burgers vector: $F_{\bar{b}} = \bar{F} \cdot \hat{b}$, where

$$\hat{b} = \frac{1}{\sqrt{2}}\begin{pmatrix} 1 \\ -1 \\ 0 \end{pmatrix}$$

so that

$$F_{\bar{b}} = \frac{a}{2\sqrt{6}}\begin{pmatrix} 75 \\ 225 \\ 150 \end{pmatrix} \frac{1}{\sqrt{2}}(1 \quad -1 \quad 0) = \frac{a}{2\sqrt{12}}(75 - 225 + 0) = \frac{a}{2\sqrt{12}}(-150),$$

implying that the force acts along the $-b$ direction.

e) Recall:

$$R = \frac{\alpha G b^2}{F} = \frac{110 \times 10^3 \times (2.546\ \text{Å})^2}{(150 \times 3.6\ \text{Å})/2\sqrt{12}} = \frac{713 \times 10^3}{77.94} = 9.148\ \text{Å}.$$

4.2.5
Forces between Dislocations

Generally, two dislocations with the same sign on the same slip plane would repel each other, whereas dislocations with the opposite signs would attract each other.

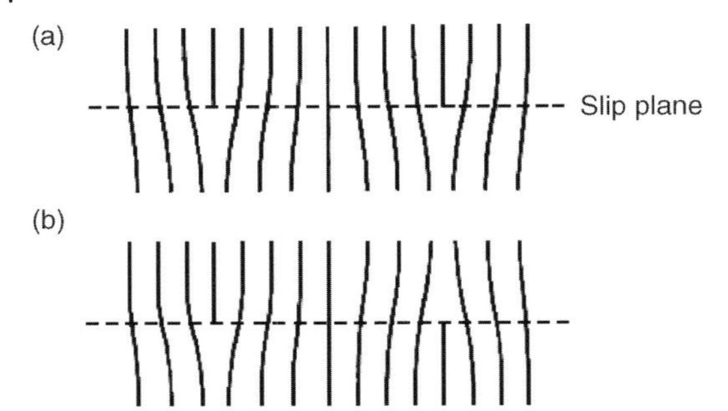

Figure 4.16 (a) Like edge dislocations and (b) unlike edge dislocations lying on the same slip plane.

Let us assume that two parallel edge dislocations of the same sign (both positive edge in the case) are on the same slip plane, as illustrated in Figure 4.16a. If the dislocations come very close together, the configuration can be assumed to be a dislocation with double the Burgers vector (i.e., $2b$) of the individual dislocations, and the elastic strain energy would be given by $\alpha G(2b)^2$ or $4\alpha Gb^2$. However, when they are separated by a large distance, they have a total energy of $2\alpha Gb^2$. The dislocation configuration with a smaller Burgers vector is more stable. Thus, the dislocations would repel each other. However, when two edge dislocations of the opposite signs lie on the same slip plane (one positive and another negative) as illustrated in Figure 4.16b, the effective Burgers vector is zero, and thus the elastic strain energy becomes zero. As in this way the dislocation configuration reduces their energy, the two dislocations will be attracted to each other and will get annihilated. Similarly, while the like screw dislocations on the slip plane will repel each other, the unlike dislocations will attract each other.

Now let us consider the case of the interaction between two parallel screw dislocations (not necessarily on the same slip plane). As the screw dislocation has a radially symmetrical stress field, the force between the two dislocations is given by the following equation and depends only on the distance of separation.

$$F_i = \tau_{\theta z} b = \frac{Gb^2}{2\pi r}. \tag{4.18}$$

This interaction force is repulsive between two like screw dislocations and attractive between two unlike dislocations. This can easily be shown by evaluating the force on a screw dislocation using Peach–Koehler formula (Eq. (4.17b)) due to the stress fields from the second dislocation using Eq. (4.10a) (or using forces in Cartesian coordinates).

The interaction between two edge dislocations is much more complex because of their more complicated stress fields. If we consider one of the edge dislocations to be lying parallel to the z-direction with Burgers vector parallel to the x-direction, the interaction would change from attractive to repulsive and vice versa depending on how the two dislocations are positioned with respect to each other. The various cases are shown in Figure 4.17 in terms of coordinates.

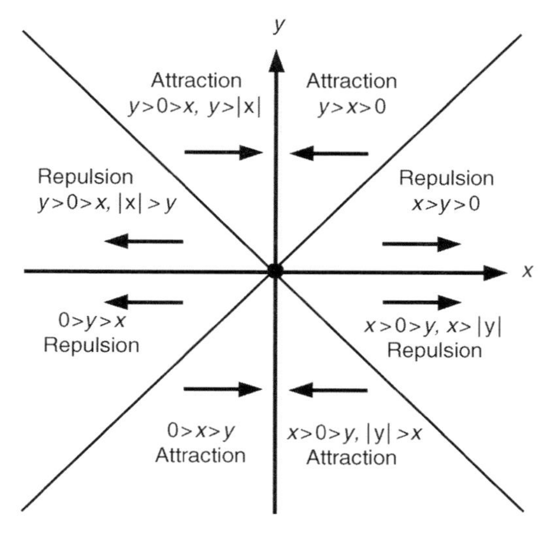

Figure 4.17 Type of interaction forces of an edge dislocation with attraction or repulsion for another parallel edge dislocation based on how they are positioned relative to each other.

The arrows in Figure 4.17 correspond to forces in the x-direction, while the forces along the y-direction are repulsive between two like edge dislocations although they cannot glide in the y-direction. The edge dislocations can move along the y-direction only by nonconservative climb that involves vacancies and/or atoms move away/to the dislocation core. From Figure 4.17, certain simple cases can be derived. Figure 4.18a shows an edge dislocation parallel to another like edge dislocation just vertically above its slip plane. This is a stable configuration. This type of configuration is found in the low angle tilt boundaries (see Figure 2.37). Two stable forms of dipole pairs are shown in Figure 4.18b where two edge dislocations of opposite signs glide past each other in parallel slip planes at low applied stresses.

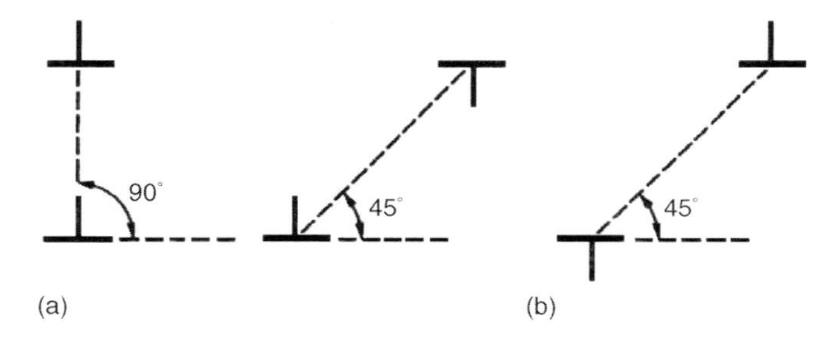

Figure 4.18 Stable positions for two edge dislocations of (a) the same sign and (b) the opposite sign.

Example 4.6

Derive the force between two like dislocations lying along the z-axis.

Solution

$\perp_1 : \overline{b}_1 = b_1\hat{k}, \hat{t}_1 = \hat{k}$ and $\perp_2 : \overline{b}_2 = b_2\hat{k}, \hat{t}_2 = \hat{k}$.
Force on dislocation 2 due to dislocation 1 : $\overline{F}_2 = (\underset{\approx}{\sigma}^{(1)} \cdot b_2)x\hat{t}_2$.

Recall stresses around screw dislocations : $\underset{\approx}{\sigma}^{(1)} = \begin{pmatrix} 0 & 0 & \sigma^1_{xz} \\ 0 & 0 & \sigma^1_{yz} \\ \sigma^1_{xz} & \sigma^1_{yz} & 0 \end{pmatrix}$.

$$\boxed{\sigma^1_{xz} = -\frac{Gb_1}{2\pi}\frac{y}{r^2} \text{ and } \sigma^1_{yz} = \frac{Gb_1}{2\pi}\frac{x}{r^2}} \quad \overline{G} = \underset{\approx}{\sigma}^{(1)} \cdot b_2 = \begin{pmatrix} 0 & 0 & \sigma^1_{xz} \\ 0 & 0 & \sigma^1_{yz} \\ \sigma^1_{xz} & \sigma^1_{yz} & 0 \end{pmatrix}\begin{pmatrix} 0 \\ 0 \\ b_2 \end{pmatrix} = \begin{pmatrix} \sigma^1_{xz}b_2 \\ \sigma^1_{yz}b_2 \\ 0 \end{pmatrix}.$$

$$\text{So, } \overline{F}_2 = (\underset{\approx}{\sigma}^{(1)} \cdot b_2) \times \hat{t}_2 = \begin{pmatrix} \sigma^1_{xz}b_2 \\ \sigma^1_{yz}b_2 \\ 0 \end{pmatrix} \times \begin{pmatrix} 0 \\ 0 \\ 1 \end{pmatrix} = \begin{vmatrix} i & j & k \\ \sigma^1_{xz}b_2 & \sigma^1_{yz}b_2 & 0 \\ 0 & 0 & 1 \end{vmatrix}.$$

$$\boxed{\text{Or, } \overline{F}_2 = \sigma^1_{yz}b_2 i - \sigma^1_{xz}b_2\, j = \frac{Gb_1b_2}{2\pi r^2}(xi + yj) = \frac{Gb_1b_2}{2\pi r}\hat{r}}.$$

$\overline{F}_1$ force on $\perp_1$ due to dislocation 2 is equal and opposite to $\overline{F}_2$ – repulsion.

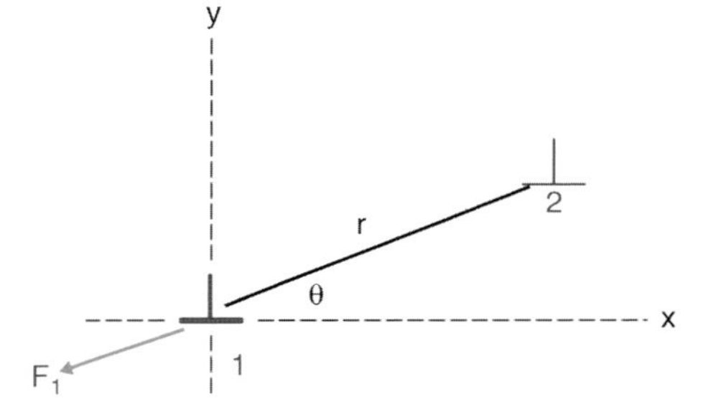

4.2.6 Intersection of Dislocations

A direct result of dislocation intersection is a contribution of strain hardening, that is, the increase in flow stress with increasing strain. In reality, dislocations need to intersect forest dislocations (dense existing dislocation networks) in order for the plastic deformation to continue. These dislocation intersection phenomena could

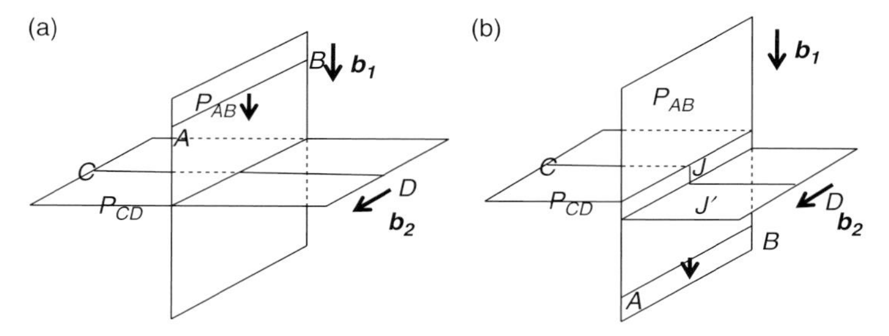

Figure 4.19 Intersection of two edge dislocations (*AB* and *CD*) with Burgers vectors perpendicular to each other. (a) An edge dislocation *AB* is moving toward an edge dislocation line *CD*. (b) A jog *JJ′* is produced on dislocation line *CD*.

be quite complex. Here, we discuss only the cases of intersection between straight edge and/or screw dislocations. The intersection of dislocations may create two types of sharp breaks, only a few atoms wide – (a) *Jog*: A jog is a sharp break on the dislocations moving it out of the slip plane. (b) *Kink*: It is a sharp break in the dislocation line but lies in the same slip plane.

To understand the dislocations intersection event better, let us consider first the case of two edge dislocations with their Burgers vectors perpendicular to each other, as shown in Figure 4.19a. An edge dislocation *AB* (with Burgers vector b_1) is gliding on the slip plane P_{AB}. The edge dislocation *CD* (Burgers vector b_2) lies on its slip plane P_{CD}. The dislocation *AB* cuts through the dislocation *CD* and creates a sharp break (*JJ′*) on dislocation *CD*. The sharp break produced on dislocation *CD* is called a *jog*, as depicted in Figure 4.19b. Hence, the jog has a Burgers vector of b_2, but with a length (or height) of b_1. Hence, the strain energy of the jog (E_j) would be $\alpha G b_2^2 b_1$. If $b_1 \approx b_2 = b$, we can write $E_j = \alpha G b_2^2 b_1 = \alpha G b^3$. However, the jog is a very small dislocation segment, no long-range elastic strain energy is possible. The jog energy mostly consists of the core energy. So, instead of $\alpha = 1$, α with a value of 0.1–0.2 is more appropriate.

Now another example of two orthogonal edge dislocations with parallel Burgers vectors is discussed (Figure 4.20a). Dislocation *XY* is moving toward dislocation *WV*.

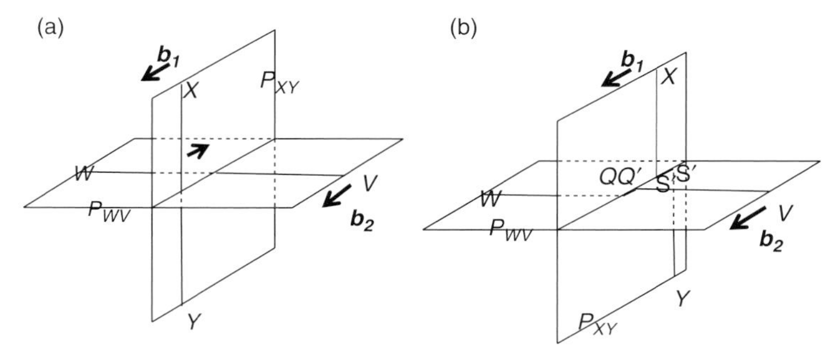

Figure 4.20 Intersection of two edge dislocations (*XY* and *WV*) with parallel Burgers vectors. (a) Dislocation *XY* is moving toward dislocation *WV*. (b) A jog *SS′* is produced on dislocation *XY* and *QQ′* on dislocation *WV*. These jogs are actually kinks.

In this case, jogs SS' and QQ' are created on dislocations XY and WV, respectively, as illustrated in Figure 4.20b. But these are not real jogs as they are on the same slip plane as the dislocation lines. They are called *kinks* as we have already defined. These kinks are generally not stable and disappear as the dislocations glide.

The intersection of a screw dislocation with an edge dislocation creates a jog with an edge orientation on the edge dislocation and a kink with an edge orientation on the screw dislocation (right-handed). Intersection between two like screw dislocations would produce jogs of edge orientation on both the dislocations involved. Here, all jogs have height in the order of atomic spacing. These are called *elementary* or *unit jogs*. There are many cases where jogs with height of more than one atom spacing have been found. They are called *superjogs*. However, the discussion on superjogs is outside the scope of this book.

Jogs produced by the intersection of edge dislocations are of edge orientations and they lie on the original slip planes of the dislocations. They can glide with the edge dislocations on the stepped surface instead of a single slip plane. Hence, jogs found on such edge dislocations do not hinder their motion. However, jogs produced by the intersection of screw dislocations are of edge orientation. As we know, edge dislocations can only glide in the plane that contains both the dislocation line and its Burgers vector. The screw dislocation can slip (i.e., move conservatively) with its jog if it glides on the same plane. However, if the screw dislocation tries to move to a different plane ($MNN'O$) as illustrated in Figure 4.21, it can take its jog with it only via nonconservative motion such as dislocation climb. As the dislocation climb process requires higher temperatures (i.e., thermally activated), the movement of jogged screw dislocations becomes temperature dependent. That is why at lower temperatures, the movement of screw dislocations is sluggish compared to edge dislocations as their motion is impeded by the existing jogs. At high stress regimes, the movement of jogs would leave behind a trail of vacancies or interstitials based on the dislocation sign and the direction of its movement. If the jog leaves behind vacancies, it is called *vacancy jog*, and if the jog goes in the opposite direction producing a trail of interstitials, it is called *interstitial jog*.

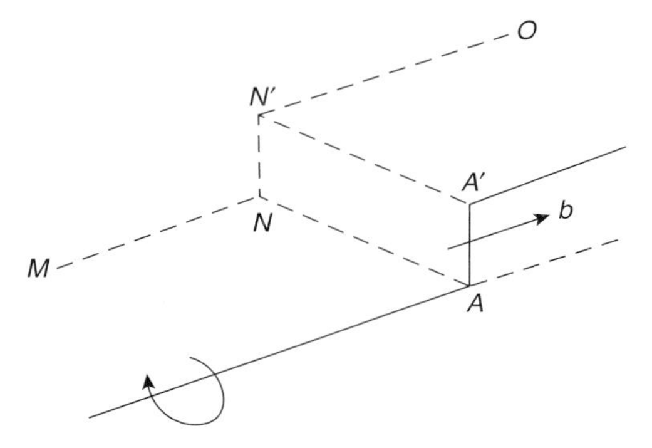

Figure 4.21 Movement of a jog on a screw dislocation [1].

4.2.7
Origin and Multiplication of Dislocations

All real crystals contain some dislocations except in the case of tiny, carefully prepared whiskers. Nevertheless, dislocations are not produced like intrinsic point defects such as vacancies or self-interstitials. The net free energy change due to the presence of dislocations is positive. Dislocations in freshly grown crystals are formed for various reasons: (a) Dislocations that may appear as preexisting in the seed crystal to grow the crystal. (b) "Accidental nucleation" – (i) heterogeneous nucleation of dislocations due to internal stresses generated by impurity, thermal contraction, and so on, (ii) impingement of different parts of the growing interface, and (iii) formation and subsequent movement of dislocation loops formed by the collapse of vacancy platelets. When a dislocation is created in a region of the crystal that is free from any defects, the nucleation is called "homogeneous." This occurs only under extreme conditions and requires rupturing of atomic bonds, which need very high stresses. Nucleation of dislocations at stress concentrators is important – generation of prismatic dislocations at precipitates/inclusions (Figure 4.22a), misfit dislocations during coherency loss, dislocation generation at other surface irregularities and cracks. As a matter of fact, grain boundary irregularities such as grain boundary ledges/steps (Figure 4.22b) are considered to be important sources of dislocations, especially in the early stages of deformation.

Regenerative multiplication of dislocations is key to sustaining large plastic strains. Frank–Read (F–R)-type sources and multiple cross-glide can participate as a multiplication mechanism of dislocations. It requires a preexisting dislocation (DD') pinned down at two ends (dislocation intersections or nodes, composite jogs, precipitates, etc.) with distance between pinning points being L, as illustrated in Figure 4.23. An applied resolved shear stress (τ) makes the dislocation bow out and the radius of curvature R depends on the stress according to Eq. (4.17). When R

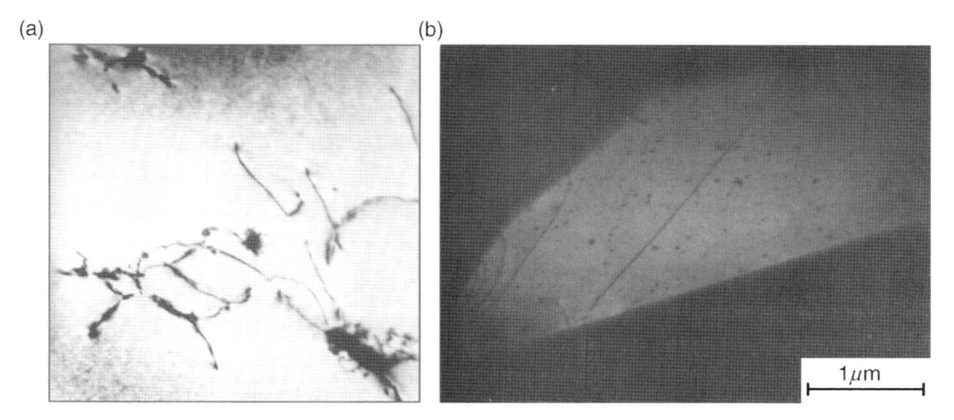

Figure 4.22 (a) TEM micrograph of irregular prismatic loops punched out at a carbide precipitate in iron. Precipitate formed during cooling [3]. (b) Dislocation sources at a grain boundary in a copper specimen strained to a plastic strain of 3×10^{-4} [8].

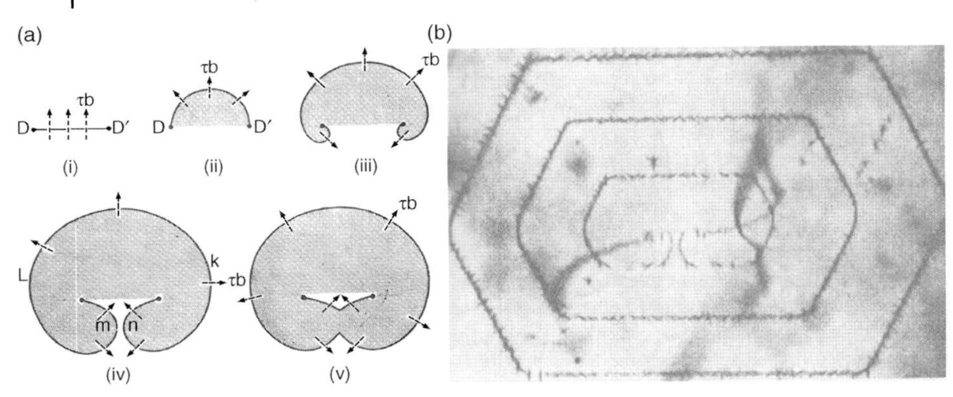

Figure 4.23 (a) Sequence of Frank–Read source operation. (b) Example of a Frank–Read source in silicon [3].

reaches a minimum value of $L/2$ and taking $\alpha = 0.5$, the stress required to produce the following configuration is given by

$$\tau = \frac{Gb}{L} = \tau_{\text{FR}}. \tag{4.19}$$

At this point in semicircular configuration, the segment becomes unstable and further increase in the dislocation loop does not require any more additional stress. As the loop expands, the dislocation at places such as m and n in Figure 4.23a(iv) attracts and annihilates leaving the loop expanding further while the original dislocation segment L is recovered. Under the applied stress (τ_{FR}), the dislocation segment expands again resulting in another loop, while the first loop keeps on expanding due to the applied stress. The process repeats itself until the lead dislocation loop gets stuck at some obstacle such as a grain boundary or another similar dislocation loop on a parallel glide plane. Since the dislocation loops generated are all alike, the second loop will be repelled by the first one, while the third loop will experience repulsive force arising from both the first and second loops, and so on, and finally the force due to all the dislocation loops in this "pileup" results in the original dislocation segment unable to produce any further dislocation loops until the lead dislocation climbs out of the glide plane. This is the Frank–Read source that is responsible for dislocation multiplication, and the stress (τ_{FR}) required for F–R source to operate is given by Eq. (4.19).

Total dislocation length can also be increased by climb: (a) the expansion of prismatic loops and (b) spiraling of a dislocation with a predominantly screw character. A regenerative multiplication known as the "Bardeen–Herring" source (Figure 4.24) can occur by climb in a way similar to the Frank–Read mechanism.

4.2.7.1 **Consequences of Dislocation Pileups**

Dislocations generated from a Frank–Read source pile up at barriers such as grain boundaries, second phases, and sessile dislocations or a lead dislocation from

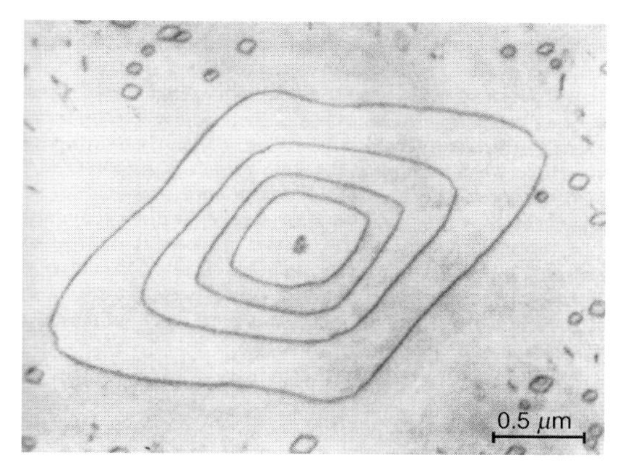

Figure 4.24 The operation of Bardeen–Herring mechanism during climb exemplified in an Al–3.5% Mg alloy quenched from 550 °C [3].

another such source on a parallel glide plane. The number of dislocations in the pileup would be directly proportional to the obstacle distance from the source (i.e., the pileup length L_{pileup}) and the applied stress. From the force balance between the dislocation–dislocation interactions and the external stress, one can show that

$$n = \frac{\pi(1-\nu)}{Gb}\tau_{xy}L_{\text{pileup}}, \tag{4.20}$$

where n is the number of dislocations in the pileup with pileup length L_{pileup} under an applied shear stress $\tau_{\xi\psi}$.

This leads to a high stress concentration ahead of the pileup that is usually relieved by plastic deformation. Just ahead of the pileup (i.e., r is small and $\theta \approx 0$ in Figure 4.25, which shows a dislocation pileup as a cross section of the loops coming out of the paper) the stress τ is enhanced by the number of dislocations in the pileup:

$$\tau = n\tau_{xy} = \frac{\pi(1-\nu)L_{\text{pileup}}}{Gb}\tau_{xy}^2. \tag{4.21}$$

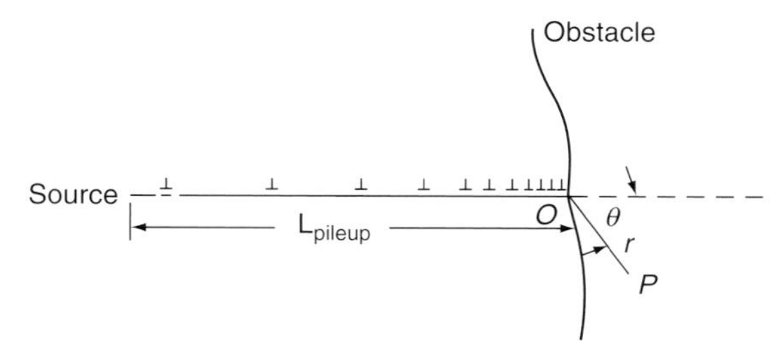

Figure 4.25 A dislocation pileup [1].

At large r but smaller than L (such as at point P in Figure 4.25), there exists a stress concentration due to the pileup:

$$\tau_{\text{at } P} \approx \sqrt{\frac{L_{\text{pileup}}}{r}}\tau_{xy} \quad (\text{true for } q < 70^\circ). \tag{4.22}$$

One can show from the above Eq. (4.22) the relation between the yield stress and grain size (namely, the Hall–Petch relation; refer to, for example, Ref. [1]):

$$\sigma_y = \sigma_0 + \frac{k_y}{\sqrt{D}}. \tag{4.23}$$

4.3
Dislocations in Different Crystal Structures

4.3.1
Dislocation Reactions in FCC Lattices

4.3.1.1 Shockley Partials

Let us now turn our attention to a situation in Figure 4.26a. Atomic arrangement (ABCABC . . .) on the {111} planes can show that slip via the motion of dislocations with Burgers vector $(b_1\ a_0/2)\langle 110\rangle$ dislocation may not occur easily. The vector $b_1 = (a_0/2)\langle 10\bar{1}\rangle$ defines one of such slip directions. On the other hand, if the same shear displacement can be accomplished by a two-step path following b_2 and b_3, it will be more energetically favorable. That is why the perfect dislocation with b_1

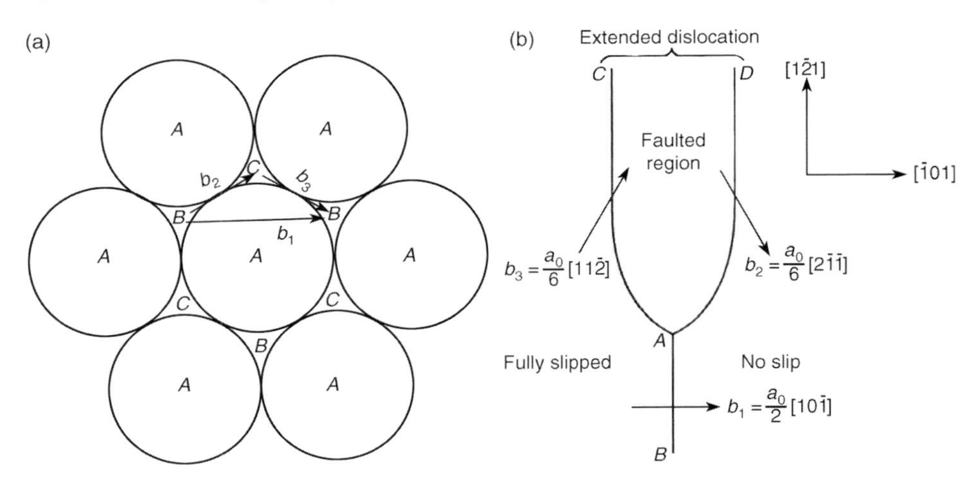

Figure 4.26 (a) A schematic showing slip in a close-packed (111) plane. (b) The dislocation dissociation reaction of a perfect dislocation into two Shockley partials is shown [1]. (c) Another view of the Shockley partials and stacking fault between them. (d) A dislocation is made by shifting atoms on cut surface (i) and a perfect dislocation (ii) and an imperfect dislocation (note the stacking mismatch in (iii)) [9].

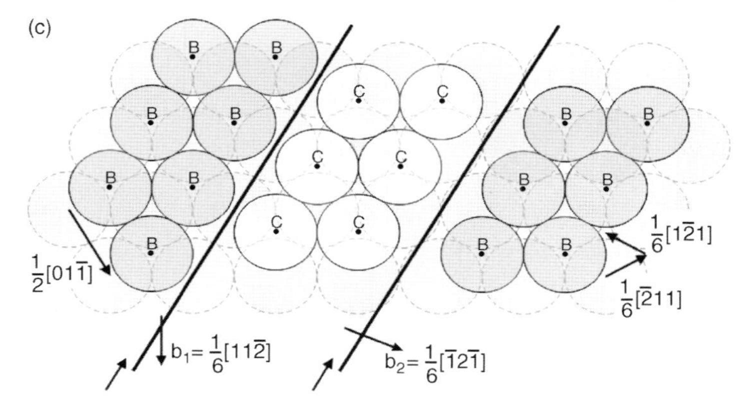

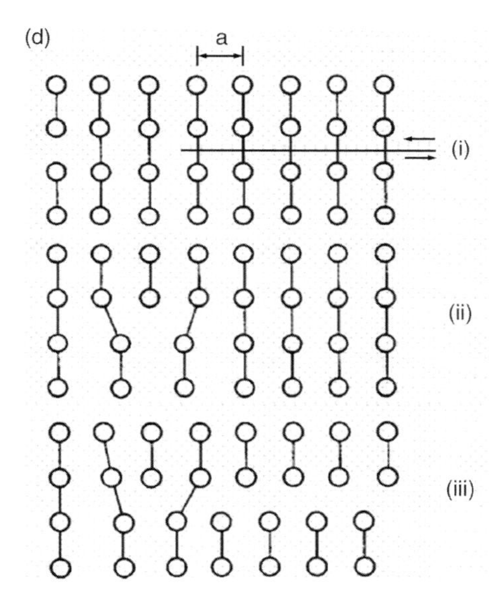

Figure 4.26 *(Continued)*

Burgers vector dissociates into two partial dislocations of b_2 and b_3 according to the following dislocation dissociation reaction:

$$\frac{a_0}{2}[10\bar{1}] \rightarrow \frac{a_0}{6}[2\bar{1}\bar{1}] + \frac{a_0}{6}[11\bar{2}]. \tag{4.24}$$

We have already seen in Example 4.3 how this reaction is feasible (recall Frank's rule). There could also be similar other reactions maintaining the reaction feasibility criterion intact. Slip via this two-stage process creates a stacking fault, *ABCAC|ABC*. Note the appearance of HCP-like stacking sequence (*CACA*) in the otherwise FCC stacking sequence. The product dislocations (*AD* and *AC* with Burgers vectors b_2 and b_3) of this dissociation reaction are known as *Shockley partials* (first described by Heidenreich and Shockley in 1948) as shown in Figure 4.26b and c. These are called partial dislocations as these dislocations are imperfect dislocations since they do not create complete lattice translations. The combination of the two partial dislocations is called *extended dislocation*. The region between the two partials is the stacking fault that is regarded as a region in a state intermediate between *full slip* and *no slip* (for more information on stacking faults, refer to Section 2.2). Dislocations such as $b_1 = (a_0/2)\langle 10\bar{1}\rangle$ are known as total or perfect dislocations and the stacking difference between the total (or perfect) and partial (or imperfect) dislocations is depicted in Figure 4.26d.

Noting that the Shockley partial dislocations lie on close-packed planes with Burgers vectors in the CPPs, these Shockley partials are considered *glissile* dislocations and they can move in the plane of the fault. So, the problem arises when the extended screw dislocation needs to cross-slip. It cannot glide only in the plane of the fault. The extended dislocation moves keeping an equilibrium separation between the two partials. The equilibrium separation (stacking fault width) is brought about by the balance of the repulsive force between two partials and the surface tension of the stacking fault. So, the partials must recombine into a perfect dissociation before it cross-slips. In the recombination process, the wider the stacking fault (or the lower the stacking fault energy), the more difficult the formation of constriction and vice versa. The stacking fault energy (E_{SF}) and width (w) are inversely related given by Eq. (4.25):

$$E_{\text{SF}} = \frac{G(b_2 \cdot b_3)}{4\pi w}. \tag{4.25}$$

Hence, materials (like aluminum) with higher stacking fault energy will have a narrow stacking fault width, and thus partial dislocations and stacking faults do not occur. However, for materials (like brass) with lower stacking fault energy, the recombination becomes difficult due to wider stacking fault width. Stacking fault energy is thus an important factor influencing the plastic deformation in different ways through strain hardening more rapidly, twinning easily on annealing, and showing different temperature dependence of flow stress compared to the metals with higher SF energy. Thus, appropriate alloying resulting in decreased stacking fault energy can result in higher strength as seen, for example, in Cu–Al alloys.

4.3.1.2 **Frank Partials**

Frank (1949) described another type of partial dislocation of $(a_0/3)\langle 111\rangle$ type (Figure 4.27). This type of dislocation is called *Frank partial dislocation* and its Burgers vector is perpendicular to the plane of the fault, and thus Frank partial cannot move by glide in the plane of the loop (i.e., a prismatic or a *sessile* dislocation). It can move only by climb. But as it requires higher homologous temperatures, sessile dislocations generally provide obstacles to the movement of other dislocations, and may aid in the strain hardening effect. A method by which a missing row of atoms can be created in the $\{111\}$ plane is by the collapse of vacancy clusters on that plane. These dislocation loops are frequently observed in irradiated materials. Also refer to the discussion on prismatic loops in Section 4.2.1.

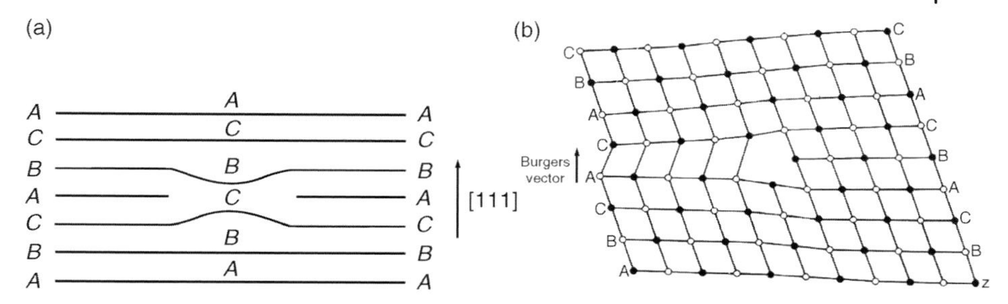

Figure 4.27 (a) The configuration of stacking sequence in the vicinity of a Frank partial. (b) Another view of a Frank partial dislocation [3].

4.3.1.3 Lomer–Cottrell Barriers

Sessile dislocations (such as *Lomer–Cottrell barrier*) that obstruct the normal dislocation movement are created in an FCC crystal by the glide of dislocations on intersecting $\{111\}$ planes during duplex slip. Two perfect dislocations, $(a_0/2)[101]$ and $(a_0/2)[\bar{1}10]$, lying on the intersecting slip planes $\{111\}$ attract each other and move toward their line of intersection, as illustrated in Figure 4.28a. Lomer (1951) suggested that the following reaction can happen to create a new dislocation of reduced energy:

$$\frac{a_0}{2}[101] + \frac{a_0}{2}[\bar{1}10] \rightarrow \frac{a_0}{2}[011]. \tag{4.26}$$

Try applying Frank's rule to show that the above dislocation reaction is energetically favorable. The new dislocation (called *Lomer lock*) with a Burgers vector $(a_0/2)[011]$ lies parallel to the line of the intersection but on a plane (100). Its Burgers vector lying in the (100) plane is perpendicular to the dislocation line (so, it is a pure edge dislocation), but as (100) slip plane is not a close-packed plane in FCC, this dislocation does not glide freely and is thus known as Lomer Lock.

Example 4.4 Show That the Lomer Lock is a Sessile Edge Dislocation?

Referring to Figure 4.28a, we note the following:

$$\bar{b}_1 = \frac{1}{2}[\bar{1}10] \text{ lies on } (111) \text{ plane}$$

$$\bar{b}_2 = \frac{1}{2}[101] \text{ lies on } (\bar{1}11) \text{ plane}$$

Once these two dislocations glide toward the line of intersection of the two planes, they combine to form a total dislocation $\bar{b}_3 = (1/2)[011]$ as per Eq. (4.26). The line vector of this product dislocation is the line of intersection of the two planes given by the cross-product of normals to the planes,

$$\bar{t}_3 = [111] \times [\bar{1}11] = \begin{vmatrix} i & j & k \\ 1 & 1 & 1 \\ -1 & 1 & 1 \end{vmatrix} = (1-1)i - (1+1)j + (1+1)k = [0\bar{2}2] \quad \text{or} \quad [0\bar{1}1].$$

This dislocation is an edge dislocation since b_3 is perpendicular to t_3 (since $[011] \cdot [0\bar{1}1] = 0$).

To see if this dislocation is a sessile or glissile dislocation, we need to find the glide plane of this dislocation that contains both the Burgers vector and the line vector (i.e., normal to the plane given by the cross-product of b_3 and t_3):

$$[011] \times [0\bar{1}1] = \begin{vmatrix} i & j & k \\ 0 & 1 & 1 \\ 0 & -1 & 1 \end{vmatrix} = (1+1)i - (0)j + (0)k = [200] \quad \text{or} \quad [100].$$

Or the glide plane is (100), which is not a close-packed plane and thus it is a sessile dislocation.

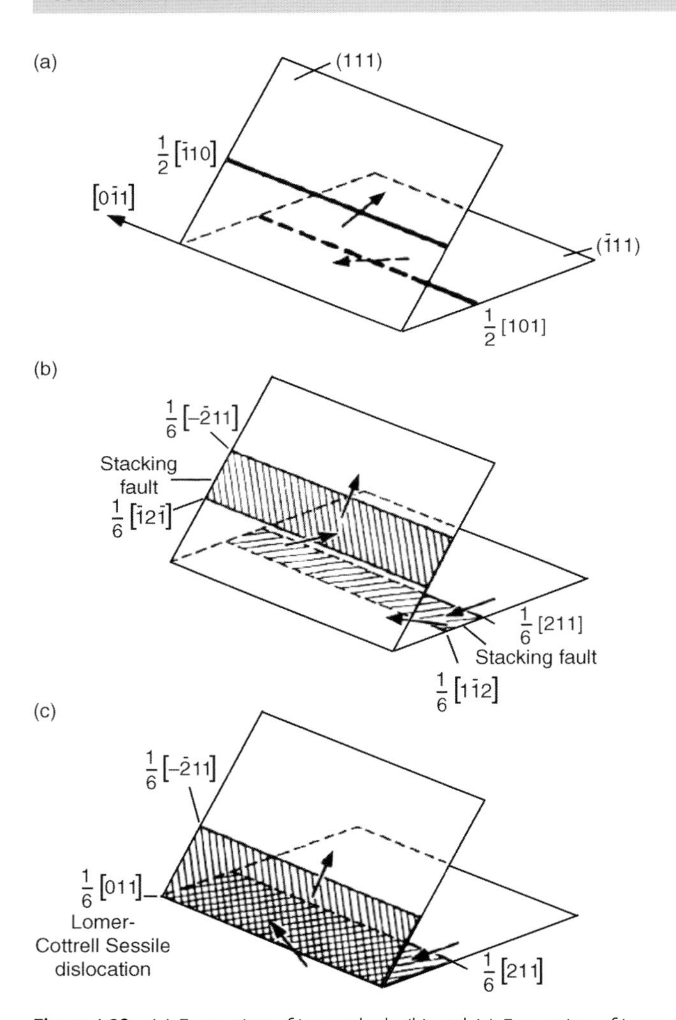

Figure 4.28 (a) Formation of Lomer lock. (b) and (c) Formation of Lomer–Cottrell lock [3].

Cottrell (1952) reexamined Lomer's reaction and suggested that it can be made strictly immobile if the perfect dislocations are considered to have dissociated into respective partial dislocations. Only the leading partials would interact with each other following the reaction shown in Eq. (4.24) to produce a truly imperfect, sessile dislocation with Burgers vector, $(a_0/6)[011]$ (Eq. (4.27)). The formation of this dislocation, called Lomer–Cottrell lock, is shown in Figure 4.28b and c. This new dislocation lies parallel to the line of intersection of the slip plane and has a pure edge character in the (100) plane. The dislocation is sessile because the product dislocation does not lie on either of the planes of its stacking faults. In analogy with the carpet on stair steps, this sessile dislocation is called a *stair-rod dislocation*. This type of sessile dislocation offers resistance to the dislocation movement and contributes to strain hardening, but it is, by no means, the major contributor. It is to be noted that the Lomer–Cottrell barriers can be surmounted at high-enough stresses and/or temperatures.

$$\frac{a_0}{6}\left[\bar{1}2\bar{1}\right] + \frac{a_0}{6}\left[1\bar{1}2\right] \rightarrow \frac{a_0}{6}[011]. \qquad (4.27)$$

4.3.2 Dislocation Reactions in BCC Lattices

In BCC crystals, slip occurs in the direction of $\langle 111 \rangle$. The shortest lattice vector extending along the body diagonal is $(a_0/2)[111]$, which is the Burgers vector of a perfect dislocation in a BCC lattice. While the slip occurs normally in $\{110\}$ planes, it is important to note that three $\{110\}$, three $\{112\}$, and six $\{123\}$ can intersect along the same $\langle 111 \rangle$ direction. Hence, screw dislocations may move at random on these slip planes under the action of high resolved shear stresses. This is a reason for observing wavy slip lines in BCC crystals. Extended dislocation formation is not common in BCC metals as they are in FCC and HCP metals due to their relatively high stacking fault energy. Even though theoretical research indicates such formation, it has not been substantiated well through experiments.

Cottrell (1958) proposed a dislocation reaction through which immobile dislocations can be produced in BCC lattice. It is a mechanism through which $a[001]$-type dislocation networks are produced in the BCC lattice (experimentally observed). This type of dislocation is produced when dislocation 1 with Burgers vector $(a_0/2)\left[\bar{1}\bar{1}1\right]$ glides on (101) plane, while dislocation 2 with Burgers vector $(a_0/2)[111]$ glides on the slip plane $(10\bar{1})$. Figure 4.29a illustrates the dislocation reaction involved following the dislocation reaction given below:

$$\frac{a_0}{2}[\bar{1}\bar{1}1] + \frac{a_0}{2}[111] \rightarrow a_0[001]. \qquad (4.28)$$

The two dislocations react and form a pure edge dislocation with a Burgers vector of $a_0[001]$, which resides on $\{100\}$-type planes that happen to be the cleavage plane, and thus do not take part in plastic deformation. It plays a role in the crack nucleation and consequent brittle fracture.

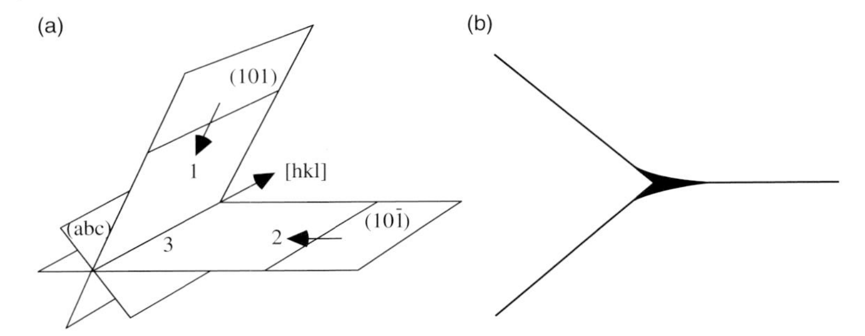

Figure 4.29 (a) Dislocation reaction in BCC crystal as proposed by Cottrell. (b) The nucleation of a crack due to the dislocation reaction.

4.3.3 Dislocation Reactions in HCP Lattices

Dislocation reactions in the HCP lattices are bit more complicated than those in the FCC and BCC ones. Some metals (such as magnesium and cobalt) with HCP lattices slip in their basal planes under normal deformation conditions. This is natural to expect given that the close-packed plane is (0002). The slip direction is $[11\bar{2}0]$, and the Burgers vector is given by $(a_0/3)[11\bar{2}0]$. The magnitude of the Burgers vector is a_0. This perfect dislocation can dissociate into two Shockley partials (each has a Burgers vector magnitude of $a_0/\sqrt{3}$ as shown in Eq. (4.29). A stacking fault with an FCC crystal structure is produced in between the partials.

$$\frac{a_0}{3}[11\bar{2}0] \rightarrow \frac{a_0}{3}[10\bar{1}0] + \frac{a_0}{3}[01\bar{1}0]. \tag{4.29}$$

Slip in some HCP lattices (alpha-Zr and alpha-Ti) with prismatic slip systems occurs along the close-packed direction $\langle 11\bar{2}0\rangle$ and dislocations involved have Burgers vector of $(a_0/3)[11\bar{2}0]$. Even though it is not very clear, the following types of dislocation reactions have been proposed that form stacking faults during prismatic slip:

$$\frac{a_0}{3}[11\bar{2}0] \rightarrow \frac{a_0}{18}[42\bar{6}\,3] + \frac{a_0}{18}[24\bar{6}3]. \tag{4.30}$$

$$\frac{a_0}{3}[11\bar{2}0] \rightarrow \frac{a_0}{9}[11\bar{2}0] + \frac{2a_0}{9}[11\bar{2}0]. \tag{4.31}$$

Slip has also been observed with Burgers vector $(a_0/3)[11\bar{2}3]$ with a magnitude of $(c^2+a^2)^{1/2}$, but only under exceptional conditions. The glide planes are generally pyramidal planes $\{10\bar{1}1\}$ and $\{10\bar{2}2\}$. Dislocation reactions are also possible, but they are outside the scope of this chapter.

4.3.4 Dislocation Reactions in Ionic Crystals

Dislocations in ionic crystals have unique electric charge effects that are not observed in the dislocations of simple metals or alloys. It stems from the very

Table 4.2 Slip planes and directions in some ionic crystals.

Structure	Slip plane	Slip direction
NaCl (rock salt)	{110}	⟨110⟩
CsCl	{100}	⟨001⟩
CaF_2 (fluorite)	{001}, {110}, {111}	⟨110⟩

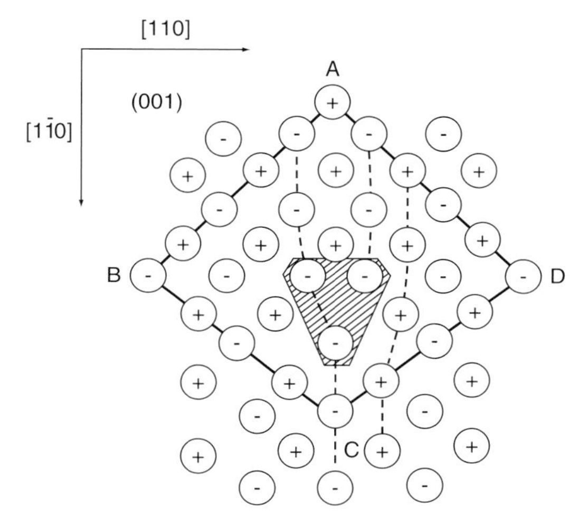

Figure 4.30 An edge dislocation configuration in a sodium chloride structure [3].

nature of the ionic crystal structure. Table 4.2 summarizes the slip planes and directions in three ionic lattices. Figure 4.30 shows a pure edge dislocation with Burgers vector of $(a_0/2)[110]$ lying on $(1\bar{1}0)$ slip plane.

4.4 Strengthening (Hardening) Mechanisms

Strength is an important property of a material and is very sensitive to the microstructure of the material. Therefore, understanding the ways in which strength of a material can be improved is of great technological significance. To be specific, we will discuss the strengthening mechanisms in terms of their effect on the yield stress of the material (defined as the stress at which gross plastic deformation begins). However, in order to accomplish that objective, we need to understand the science of the strengthening mechanisms. Dislocations are a common factor in almost all the important strengthening mechanisms we will discuss here. If dislocations are able to move in a crystal with relative ease, it means the material does not intrinsically offer any resistance to the dislocation movement and thus would be less strong. But if the microstructure is laden with various obstacles, dislocation movement will be effectively impeded and this movement obstruction will be translated into the increase of the strength of the materials. This principle is applicable to a wide array of materials with crystalline structures. The obstacle to dislocation movement could be dislocations themselves (strain hardening), grain boundaries (Hall–Petch or grain size strengthening), solute atoms (solid solution strengthening), precipitates (precipitate strengthening), and dispersions of fine stable particles (dispersion strengthening). There are a few other strengthening mechanisms like texture strengthening, composite strengthening, and so on, which are not directly but indirectly related to dislocations. However, they will not be discussed here for brevity.

4.4.1 Strain Hardening

Strain hardening is possible both in single crystals and polycrystalline materials. When metallic materials are cold worked, their strength increases. Generally, an annealed crystal contains a dislocation density of about $10^8\,m^{-2}$. However, moderately cold worked materials may contain 10^{10}–$10^{12}\,m^{-2}$ and heavily cold worked materials 10^{14}–$10^{16}\,m^{-2}$. Plastic deformation carried out in a temperature regime and over a time interval such that the strain hardening is not relieved is called cold working. At the first sight, one might think that with increase in the dislocation density, the material would be more ductile. But that is not the case. As the dislocation density increases, the movement of dislocation becomes increasingly difficult due to the *interfering effect* of the stress fields of other dislocations. In polycrystalline metals/alloys, multiple slips occur more readily due to the mutual interference of adjacent grains, leading to significant strain hardening. In the early stages of plastic deformation, slip is generally limited on primary glide planes and the dislocations tend to form coplanar arrays. However, as the deformation proceeds, cross-slip takes place and dislocation multiplication processes start to operate. The cold worked structure then forms high dislocation density regions or tangles that soon develop into *tangled networks*. Thus, the characteristic structure of the cold worked state is a *cellular substructure*. The cell structure is schematically shown in Figure 4.31.

With increasing strain, the cell size decreases at initial deformation regime. However, the cell size tends to reach a fixed size implying that as the plastic deformation proceeds, the dislocations sweep across the cells and join the tangles into the cell walls. The exact nature of the cold worked structure depends on the material, strain, strain rate, and deformation temperature. The development of cell structure, however, is less pronounced for low temperature and high strain rate deformation and in materials with low stacking fault energy.

Strain hardening is an important process that is used in metals/alloys that do not respond to heat treatment easily. Generally, the rate of strain hardening is lower in HCP metals compared to FCC metals and decreases with increasing temperature.

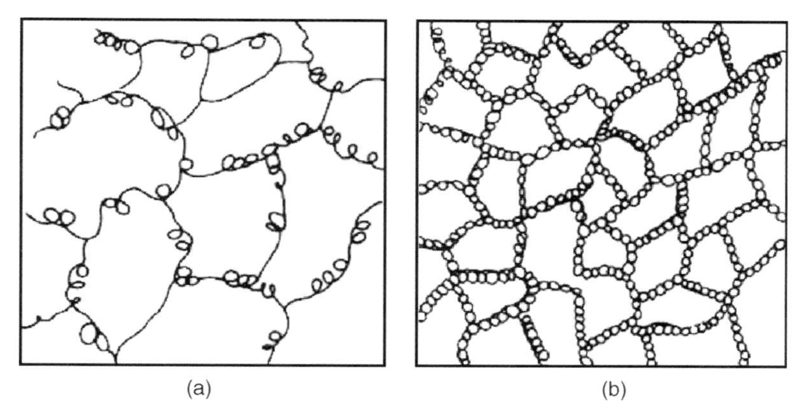

Figure 4.31 Schematics of dislocation microstructure (a) at early stage of deformation (~10% strain) – start of cell formation with dislocation tangles, and (b) deformed to 50% strain – equilibrium cell size with cell walls containing high dislocation density [1].

Furthermore, the final strength of the cold worked solid solution alloy is always greater than that of the pure metal cold worked to the same extent. Cold working may slightly decrease density and electrical conductivity, whereas thermal expansion coefficient and chemical reactivity increase (i.e., corrosion resistance decreases). Figure 4.32 shows the variation of tensile parameters (strength parameters of tensile strength, yield strength, and ductility parameters of reduction in area and elongation to be discussed in the Chapter 5) as a function of percentage cold

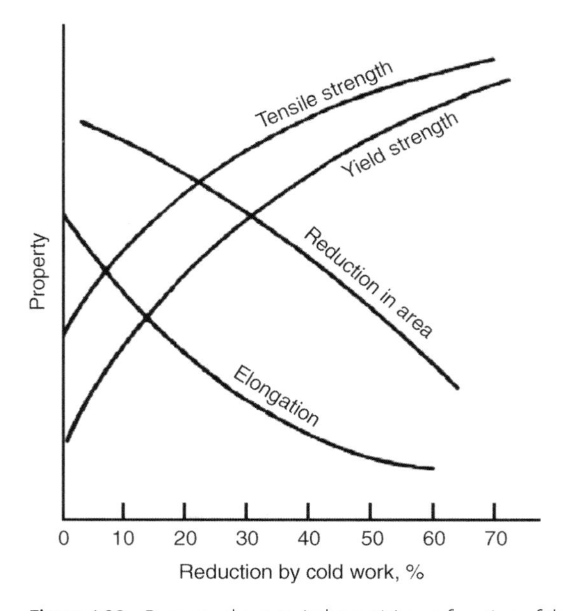

Figure 4.32 Property changes (schematic) as a function of the percentage cold reduction [1].

reduction. Torsteels are used as reinforcing bars for concrete is cold twisted to increase the strength. However, it acts upon the effective ductility of material.

A high rate of strain hardening implies mutual obstruction of dislocations gliding on intersecting slip systems. This can arise from three sources: (a) through interaction of the stress fields of dislocations, (b) through interactions that produce sessile locks, and (c) through the interpenetration of one slip system by another (like cutting trees in a forest) that results in the formation of jogs. The basic equation relating flow stress to structure is

$$\tau = \alpha \frac{Gb}{\ell}, \tag{4.32}$$

where ℓ is the obstacle–obstacle spacing with α having a value between 0.5 and 1. If dislocations are acting as obstacles like trees in dislocation forest, $\ell = 1/\sqrt{\varrho}$, so strengthening due to dislocations is given by

$$\tau = \tau_0 + \alpha G b \varrho^{1/2}, \tag{4.33}$$

where τ is the stress needed to move a dislocation in a matrix of dislocation density ϱ, τ_0 is the stress to move the dislocation in the same matrix with no dislocation density, G is the shear modulus, and b is the magnitude of Burgers vector. The constant α may change value, but may be considered as 0.5.

4.4.2
Grain Size Strengthening

Plastic deformation in single crystals is guided by relatively simple laws. However, in polycrystalline materials, it becomes complicated. Strength is inversely related to the dislocation mobility. That is, the dislocation movement can be obstructed by various factors that can, in turn, strengthen (or harden) the material. Grain boundaries can act as such an obstacle to the dislocation motion (refer to Section 2.2 for more information on the grain boundaries). The grain boundary ledges are a common feature – these steps can be produced by external dislocation joining the grain boundary or packs of dislocations in the grain boundary. Thus, the density of ledges increases with increasing misorientation angle. Grain boundary ledges act as an effective source of dislocations. Hence, before we introduce the topic of grain size strengthening, we need to know some fundamentals of the grain size effect on the plastic deformation in polycrystals. Grain size strengthening is not possible in a single crystal (i.e., single grain) for the obvious reasons.

In a polycrystal unlike a single crystal, during plastic deformation the continuity must be maintained between the deforming crystals. If not maintained, it will readily form flaws leading to fast failure. Although it is true that each of the grains attempts to deform as homogeneously as possible in concert with the overall deformation of the material, the geometrical constraints imposed by the continuity requirement cause significant differences between neighboring grains and even in parts of each grain. As the grain size decreases, the deformation becomes more homogeneous. That is why decreasing grain size in conventional materials generally improves ductility. Due to the constraints imposed by the grain boundaries, slip occurs on several systems even at low strains. As the grain size is reduced, more of the effects of grain boundaries will be felt at the grain center. Thus, the strain hardening of a fine grain size material will be greater than that of a coarse-grained polycrystalline material. Theoretically, it has been shown that the strain hardening rate $(\mathrm{d}\sigma/\mathrm{d}\varepsilon)$ for an FCC polycrystalline material is ~9.6 times that of $(\mathrm{d}\tau/\mathrm{d}\gamma)$ for a single crystal as noted in Eq. (4.34):

$$\frac{\mathrm{d}\sigma}{\mathrm{d}\varepsilon} = \overline{M}^2 \frac{\mathrm{d}\tau}{\mathrm{d}\gamma}, \tag{4.34}$$

where $\overline{M}$ is called Taylor's factor (for FCC lattice, it is taken as 3.1). This has also been verified by experiments.

Von Mises [27] showed that for a crystal to undergo a general change of shape by slip requires the operation of *five independent slip systems*. Crystals having less than five slip systems are never ductile in polycrystalline form even though small elongations can be obtained through twinning or a favorable preferred orientation. Generally, cubic metals easily satisfy this requirement (higher ductility), whereas HCP and other low symmetry metals do not (lower ductility). At elevated temperatures, other slip systems can get activated and thus increase the number of slip systems to at least five.

At temperatures above $\sim 0.5T_\mathrm{m}$ (T_m is the melting point), deformation can occur by sliding along grain boundaries. Grain boundary sliding becomes more prominent with increased temperature and decreasing strain rate. A general way to finding out when grain boundary sliding may start operating is the use of *equicohesive temperature* concept. Above this temperature, the grain boundary is weaker than grain interior and strength increases with increasing grain size. Below this temperature, the grain boundary region is stronger than the grain interior and strength increases with decreasing grain size. The grain size strengthening mechanism to be discussed below is thus applicable only to lower homologous temperature regimes.

A mathematical description of the grain size strengthening can be expressed in terms of a general relationship between yield stress and grain size, first proposed by Hall (1951) and later extended by Petch (1953):

$$\sigma_y = \sigma_\mathrm{i} + \frac{k_y}{\sqrt{d}}, \tag{4.35}$$

where σ_y is the yield stress, σ_i is the friction stress (also can be defined as yield strength at an infinite grain size), k_y is the unlocking parameter (that measures relative hardening contribution of the grain boundaries), and d is the grain diameter. The Hall–Petch relation has also been found to be applicable for a wide variety of situations, such as the variation of brittle fracture stress on grain size, grain size-dependent fatigue strength, and even to the boundaries that are not such grain boundaries as found in pearlite, mechanical twins, and martensite plates.

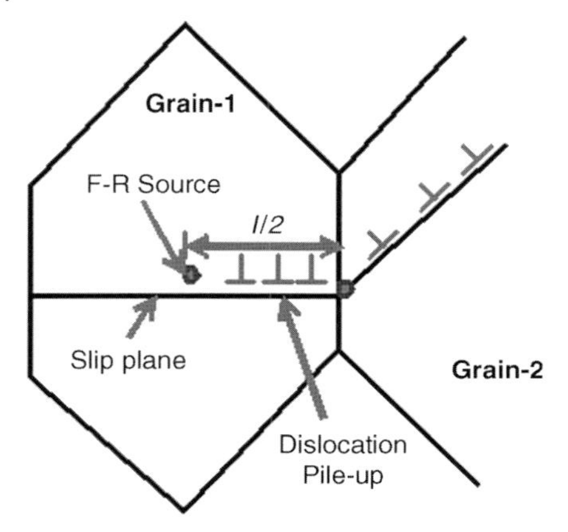

Figure 4.33 A schematic of a dislocation pileup is shown to form by the dislocations originating from a Frank–Read source in the center of Grain-1. The leading edge of the dislocation pileup produces enough stress concentration to create dislocations in Grain-2.

The Hall–Petch relation was developed based on the concept that grain boundaries act as barriers to the dislocation movement (as noted earlier during discussion of dislocation pileups). As the grain orientation changes at the grain boundary, the slip planes in a grain get disrupted at the grain boundary. This implies that slip planes are not continuous in a polycrystalline material from one grain to another. Thus, the dislocations gliding on a slip plane cannot burst through the grain boundary. Instead, they get piled up against it. In a larger grain, the number of dislocations in a pileup is higher. The magnitude of the stress concentration at the leading edge of the pileup (as shown in Figure 4.33) varies as the square root of the number of dislocations in the pileup. Hence, a greater stress concentration would be created at a larger grain. However, this event can trigger dislocation sources in the next grain quite readily. The situation is schematically shown in Figure 4.33. Now if the grain size is smaller, the stress concentration in Grain-1 will not be enough to produce slip in the next grain readily, and thus fine grain sized material will exhibit higher yield strength.

4.4.3
Solid Solution Strengthening

We know about solid solutions (substitutional and interstitial types) and the requirements for forming one or the other. Now let us get into little more details about solid solution strengthening. This mode of strengthening is present in both single and polycrystalline materials. Here, dislocations encounter barriers on their path from the solutes present in the host lattice.

Solute atoms can interact with dislocations through the following mechanisms based upon various situations: (a) elastic interaction, (b) modulus interaction,

(c) long-range order interaction, (d) stacking fault interaction, (e) electrical interaction, and (f) short-range order interaction. The first three interactions are long-range barriers, and they are relatively insensitive to temperature and continue to act upto $0.6T_m$. However, the last three are short-range barriers that contribute strongly to the flow stress only at low temperatures.

4.4.3.1 **Elastic Interaction**

This interaction takes place due to the mutual interaction of elastic stress fields surrounding misfitting solute atoms and the dislocation cores. The hardening effect related to the elastic interaction is directly proportional to the solute–lattice misfit. Substitutional atoms only obstruct the movement of edge dislocations. On the other hand, interstitial solutes impede both screw and edge dislocations.

4.4.3.2 **Modulus Interaction**

This occurs if the presence of a solute atom locally alters the modulus of the crystal. If the solute has a smaller shear modulus than the matrix, the energy of the strain field of the dislocation will be reduced and there will be an attraction between the solute and the matrix. The modulus interaction is quite similar to the elastic interaction, however, since it is accompanied by a change in the bulk modulus, both edge and screw dislocations will be subject to this interaction.

4.4.3.3 **Long-Range Order Interaction**

This interaction is produced only in superlattices (ordered alloys), not in conventional disordered alloys. In a superlattice, there is a long-range periodic arrangement of dissimilar atoms, such as in Cu_3Au. The movement of dislocations through a superlattice causes regions of disorder called *antiphase boundary* (APB). The dislocation dissociates into two ordinary dislocations separated by an APB. As the slip proceeds, more APBs are created. Ordered alloys with a fine domain size are stronger than the disordered state.

4.4.3.4 **Stacking Fault Interactions**

Stacking fault interactions are important as the solute atoms preferentially segregate to the stacking faults (contained in extended dislocation). This effect is also known as *Suzuki effect* or chemical interaction. The stacking fault energy gets reduced due to the increasing concentration of solutes in the SF, and thus the separation between the partial dislocations increases making it increasingly harder for the partial dislocations to move.

4.4.3.5 **Electrical Interactions**

Electrical interaction arises from the fact that solute atoms always have some charge on it due to the dissimilar valences and the charge remains localized around the solute atoms. The solute atoms can then interact with dislocations with electrical dipoles. This interaction contributes negligibly compared to the elastic and modulus interaction effects. It becomes significant only when there is large difference in valence between the solute and matrix and the elastic misfit is small.

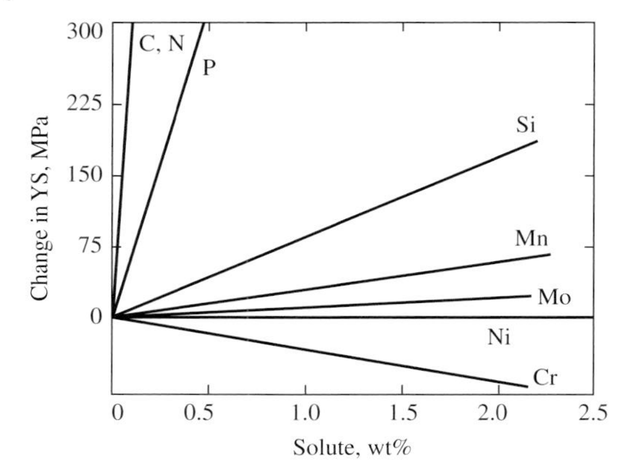

Figure 4.34 The variation of yield stress as a function of solute content in iron [10].

There were different theories proposed and debated over several decades. Some have found that the strength increment varies as the square root of the solute atom fraction ($c^{1/2}$) or $c^{1/3}$, or even c. One of the examples of solid solution strengthening is the martensite strengthening. Martensite phase in steels has a body-centered tetragonal (BCT) lattice structure and is considered as a supersaturated solid solution in the host lattice of iron. Carbon atoms present in the lattice strain the crystal creating nonspherical (tetragonal) distortion, which affects the dislocation movement severely. This is one of the reasons why the martensite phases are so hard and brittle. Figure 4.34 shows the variation in the yield strength of iron as a function of different solute concentrations. Note that the solid solution strengthening imparted by the interstitial elements like carbon and nitrogen is much higher than that imparted by substitutional solutes.

4.4.4 Strengthening from Fine Particles

Dispersed particles can impart significant obstacles to the dislocation motion, thus increasing the strength of the material. Particle hardening is a much stronger mechanism than the solid solution hardening. This type of strengthening could be effective even at higher temperatures based on the stability of the precipitates.

There are two types of fine particle strengthening mechanism: *precipitation hardening* and *dispersion strengthening*.

- For precipitation hardening (or age hardening) to happen, the second phase needs to be soluble at elevated temperatures and should have decreasing solubility with decreasing temperatures. However, in dispersion hardening, the second phases have very little solubility in the matrix (even at higher temperatures).
- Generally, there is atomic matching (i.e., coherency) between the precipitates and the matrix. However, in dispersion hardening system, there is no coherency between the second-phase particles and the matrix.

- The number of precipitation hardening systems is limited, whereas it is theoretically possible to create an almost infinite number of dispersion hardening systems.

As the particles are central to these kinds of strengthening mechanisms, the following particle characteristics are important: (a) particle volume fraction, (b) particle shape, (c) particle size, (d) nature of the particle–matrix interface, and (e) particle structure. Note that particle volume fraction can be related to the particle size through the interparticle spacing (λ):

$$\lambda = \frac{4(1-f)r}{3f}, \tag{4.36}$$

where f is the volume fraction of spherical particles of radius r. There are other similar relations also, depending on the assumptions of derivation.

4.4.4.1 Precipitation Strengthening

There are two basic mechanisms of precipitation strengthening: (a) particle shearing (particle cutting mechanism), and (b) dislocation bowing or bypassing mechanism (Orowan's theory).

Particle Shearing

In Section 2.2, we have already discussed the various types of precipitates (coherent, semicoherent, and incoherent). The understanding of these precipitates will prove to be helpful in the next discussion. When particles are small (typically less than 5 nm, however dependent on the particular alloy system) and/or soft and coherent, dislocations can cut and deform the particles. In this type of situation, six particle features become important to find out how easily they can be sheared: (a) coherency strains, (b) stacking fault energy, (c) ordered structure, (d) modulus effect, (e) interfacial energy and morphology, and (f) lattice friction stress.

As you can imagine, there are many aspects to the particle shearing mechanism, which is outside of the scope of this book. When the dislocation enters the particle, it produces a step at the interface, and when it leaves the particle, it produces another step at the opposite interface, as shown in Figure 4.35. The step size is generally on the order of the Burgers vector of the dislocation.

In the models of coherency strain and ordered structure, the strength increment due to the particle shearing mechanism varies directly with the square root of the product of the particle radius and volume fraction (i.e., $\propto \sqrt{fr}$ since $\ell \sim (1/\sqrt{fr})$. Note that deformation occurring through the shearing of particles does not entail a lot of strain hardening – rather it produces little strain hardening. The slip bands produced are planar and coarse. The steps produced may be detrimental to the corrosion properties of the materials involved.

Orowan Bypassing

With the increase in size (10 nm or so depending on the particular alloy system) in coherent precipitate, the coherency is eventually lost. This happens during

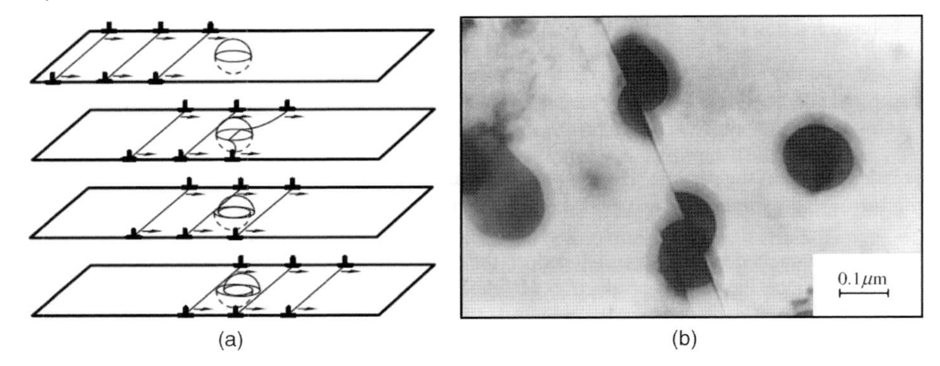

Figure 4.35 (a) The particle cutting in stages (Roesler). (b) A TEM view picture of particles cut by dislocations (Argon) [8].

overaging of precipitation hardenable alloys (such as Al–4.5 wt% Cu). The dislocation line is repelled due to the incoherency of the particles. Orowan (1947) proposed that the stress ($\Delta\tau$) required to bow a dislocation line between two particles separated by a distance λ is

$$\Delta\tau = \frac{Gb}{\lambda}, \tag{4.37}$$

where G is the shear modulus of the matrix material and b is the Burgers vector of dislocation.

Every dislocation gliding over the slip plane adds one dislocation loop around the particle (Figure 4.36) and these loops exert back stress on the dislocation sources that must be overcome to cause further slip. A real example of Orowan looping is shown in Figure 4.36b. Generally, Orowan bypassing leads to fine and wavy slip compared to the coarse and planar slip due to the particle cutting mechanism.

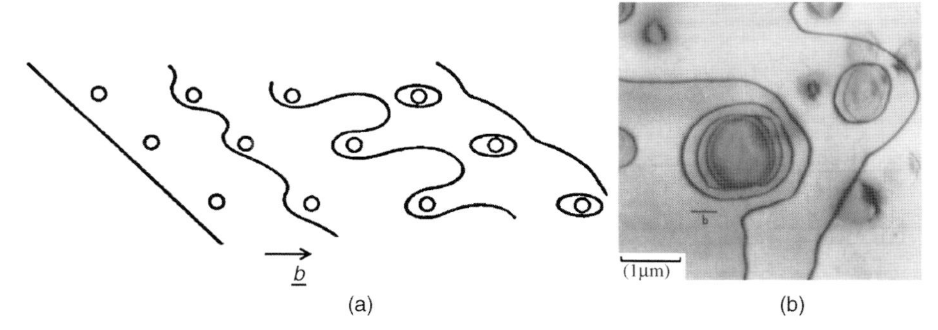

Figure 4.36 (a) A schematic illustration of dislocation line and particle interaction during the Orowan bypassing mechanism [14]. (b) A TEM image showing dislocation looping around incoherent particles [11].

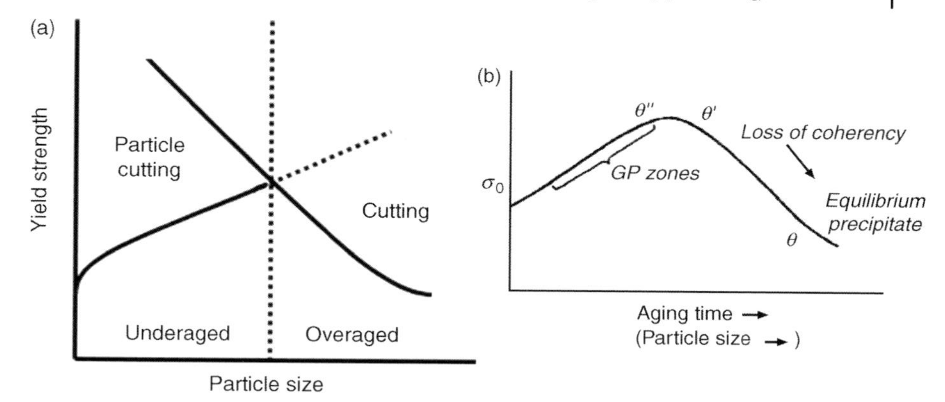

Figure 4.37 (a) The prediction of the strength contribution due to particle strengthening as a function of precipitate size. (b) A schematic illustration of the strength variation as a function of aging time (particle size) in an Al–Cu alloy (GP zones are the primitive particles formed during aging followed by transitional precipitates θ'' and θ' and finally equilibrium precipitate θ.

If we now take the appropriate models of the particle shearing and bypassing and try to see how the strength increments due to these mechanisms contribute to the overall particle strengthening, an interesting observation can be made. Figure 4.37a shows such a prediction. Particle cutting (underaged material) increases strengthening as the precipitate size increases, whereas Orowan bypassing increases with decreasing particle size (in the overaging regime). The crossover (the peak aging regime) gives the highest strength. Figure 4.37b shows why the aging experiments on aluminum alloys lead to such yield strength profiles as a function of aging time. With the aging time, the precipitate size increases and it follows first the particle cutting mechanism when the particles remain coherent, but when particles lose their coherency, the Orowan bypassing mechanism becomes important. Thus, the trend can be explained.

4.4.4.2 **Dispersion Strengthening**

In the beginning of the discussion on the fine particle strengthening, we have distinguished between the precipitation and dispersion strengthening. If the particles are fine, stable, and incoherent, dispersion strengthening is applicable. Dispersion strengthening is a significant strengthening mechanism utilized for high-temperature alloys. Orowan bypassing mechanism at lower temperature regime is applicable here also. One example of dispersion strengthening system is thoria-dispersed nickel (T-D Ni). This tends to have much greater high temperature deformation resistance compared to the nickel matrix itself.

Example 4.7

A Li {Li (bcc): $\mu = 32$ GPa; $a = 3$ Å; $\nu = 0.29$; stacking fault energy (Γ) = 1250 mJ m^{-2}; $T_M = 181$ °C} sample exhibited an yield strength of 10 MPa.

a) Estimate the dislocation density assuming that the strengthening is all due to dislocations (strain hardening)?
Solution

$$\tau = \alpha G b \sqrt{\varrho}$$
$$\text{for BCC} : b = \frac{\sqrt{3}a}{2} = \frac{\sqrt{3}(3 \times 10^{-10})}{2} = 2.598 \times 10^{-10}$$
$$\text{assume } \alpha = 1,\ G_{\text{Li}} = 32\ \text{GPa} = 32\,000\ \text{MPa}$$
$$\tau = 10 = (1)(32\,000)(2.598 \times 10^{-10})\sqrt{\varrho}$$
$$\text{So, } \sqrt{\varrho} = \frac{10}{(32\,000)(2.598 \times 10^{-10})}, \quad \sqrt{\varrho} = 1.203 \times 10^{6} \quad \text{and}$$
$$\varrho = 1.447 \times 10^{12}\ \text{m}^{-2}.$$

b) Estimate the strengthening if the dislocation density increases by a factor of 100 following cold working (or exposure to radiation).
Solution

$$\tau = \alpha G b \sqrt{\varrho} = (1)(32\,000)(2.598 \times 10^{-10})\sqrt{1.447 \times 10^{14}}$$
$$= (8.314 \times 10^{-6})(1.203 \times 10^{7}).$$
$$\text{So, } \tau = 100\ \text{MPa}.$$

4.5 Summary

This chapter introduced the dislocation concept in a greater detail. Various aspects of dislocation theory, including dislocation energy, line tension, velocity, and so on, are discussed. Also, dislocation reactions in various lattices are discussed. The chapter concludes with the introduction of various strengthening (hardening) mechanisms. This understanding will lead us to better understand the effect of increased dislocation density on the properties of the irradiated materials in the subsequent chapters.

Problems

4.1 We considered a cylindrical Ni single crystal (FCC) with a diameter of 0.1 mm pulled in tension with a stress of 1000 MPa (see Figure 4.4). The loading direction is along $[\bar{1}01]$, while the slip system is $(\bar{1}\bar{1}1)[\bar{1}10]$. We calculated the resolved shear stress along the slip direction in the slip plane.

a) What is the resolved shear stress along the slip direction in the slip plane?
b) If τ_{CRSS} for Ni is 550 MPa, would the crystal deform?

An edge dislocation (AB) is situated perpendicular to the slip direction of the slip system and pinned at two points (A and B) separated by 1000 Å.

c) If the Burgers vector of the dislocation is along the slip direction, what will be its direction and magnitude?
d) Determine the radius of curvature of the dislocation in part (c) above due to the applied stress?

e) Compute the minimum applied load (normal to the specimen cross section) required for the pinned dislocation (AB) to operate as a Frank–Read source.
f) If the dislocation line in the figure is a screw dislocation, what will be its Burgers vector (magnitude and representation)?

4.2 By increasing the temperature, the concentration of vacancies can be increased. Will this also increase the density of dislocations (be quantitative)? If not, why not and how may the dislocations be multiplied?

4.3 A prismatic loop has Burgers vector perpendicular to the plane of the loop and thus can glide in that plane under an applied shear stress. True or false?

4.4 In the figure, the dislocation AB lies in the plane (111) with b along $[1\bar{1}0]$ (perpendicular to the line AB).
a) Is the dislocation AB an edge or a screw or a mixed type?
b) What is the line vector of this dislocation?
c) If AB were a screw dislocation, what will be its Burger's vector?

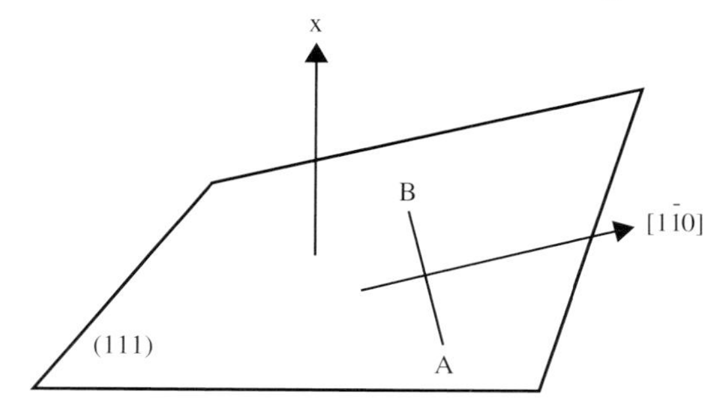

4.5 Show that a perfect dislocation, $(a/2)[110]$, in an FCC lattice splits into two Shockley partials with Burgers vectors of $(a/6)\langle 11\bar{2}\rangle$ type. (Write down the equation.)
a) Are the Shockley partials glissile or sessile? Why?
b) A Shockley partial $(a/6)[11\bar{2}]$ reacts with a Frank partial $(a/3)[111]$ to yield a perfect dislocation. What is the Burger's vector of the product dislocation?
c) Show that the reaction in problem 4.3 is valid.

(This reaction occurs in irradiated stainless steel when heated. Similar reaction could also occur in quenched Al–Mg alloy from 550 °C to −20 °C followed by heating (see Figure 5.12 in Ref. [3]). These reactions lead to unfaulting of faulted loops.)

4.6 a) Evaluate the force (magnitude and direction) acting on an edge dislocation ($b = b\hat{i}, \hat{t} = \hat{k}$) due to an external stress σxx.
b) How this dislocation may move due to this force?
c) Evaluate the force on the dislocation due to hydrostatic pressure p and comment on how the dislocation may move due to this force.

4.7 Evaluate the hardening (τ_c) of a Li (BCC, $G = 32$ GPa, $a = 3$ Å) alloy with a microstructure consisting of a dislocation density of $3 \times 10^{13}\,\text{m}^{-2}$ ($\alpha\perp = 0.5$) and 3 at %solutes ($\alpha_c = 0.005$).

Bibliography

1 Dieter, G.E. (1988) *Mechanical Metallurgy*, McGraw-Hill.
2 Hosford, W.F. (2010) *Mechanical Behavior of Materials*, Cambridge University Press.
3 Hull, D. and Bacon, D.J. (1984) *Introduction to Dislocations*, 3rd edn, Butterworth-Heinemann.
4 Gilman, J.J., Johnston, W.G., and Sears, G.W. (1958) *Journal of Applied Physics*, **29**, 747–754.
5 Amelinckx, S. (1958) *Acta Metallurgica*, **6**, 34
6 Gollapudi S. *et al.* (2010) "Creep Mechanisms in Ti-3Al-2.5V Tubing Deformed under Closed-end Internal Gas Pressurization," Acta Materialia, 56 (2008) 2406–2419.
7 Ashby M.F. and Jones D.R.H. Jones, Engineering Materials 1 - An Introduction to their Properties and Applications, International Series on Materials Science and Technology, Pergamon Press (1980).
8 Malis, T. and Tangri, K. (1979) *Acta Metallurgica* **27**, 25–32.
9 Weertman, J. and Weertman, J.R. (1992) Elementary dislocation theories, Oxford University Press, New York, USA.
10 Raghavan, V. (1995) *Physical Metallurgy: Principles and Practice*, Prentice-Hall, New Delhi, India.
11 Humphreys, F.J. and Hatherly, M. (2004) Recrystallization and related annealing phenomena, Pergamon, Oxford, UK.
12 Roesler, J., Harders, H., and Baeker, M. (2010) Mechanical behavior of engineering materials - metals, ceramics, polymers and composites, Springer, Berlin, Germany.
13 Argon, A.S. (2008) Strengthening mechanisms in crystal plasticity, Oxford University Press, New York, USA.
14 Barrett, C.R., Nix, W.D., and Tetelman A.S., (1973) The Principles of Engineering Materials, Prentice Hall, Englewood Cliffs, NJ, USA.
15 Schmid, E. (1935) Kristallplastizitat mit besonderer Berucksichtigung der Metalle (in German), Springer-Verlag, Berlin, Germany.
16 Nabarro, F.R.N. (1947) Dislocations in a simple cubic lattice, *Proc. Phys. Soc.*, **59**, 256.
17 Peierls, R.E. (1940) The size of a dislocation, *Proc. Phys. Soc.*, **52**, 34.
18 Johnston, W.G. and Gilman, J.H. (1959) Dislocation velocities, dislocation densities and plastic flow in lithium fluoride crystals, *J. Appl. Phys.*, **30**, 129.
19 Heidenreich, R.D. and Shockley, W. (1948) Report on Strength of Solids, Physical Society, London, UK.
20 Frank, F.C. (1949) Sessile Dislocations, *Proc. Phys. Soc.*, **62A**, 202.
21 Lomer, W.M. (1951) Philosophical Magazine, **42**, 1327.
22 Cottrell, A.H. (1952) Philosophical Magazine, **43**, 645.
23 Cottrell, A.H. (1958) Transactions of Metallurgical Society (AIME), **212**, 192.
24 Petch, N.J. (1953) Cleavage strength of polycrystals, J. Iron Steel Institute, 173, 25.
25 Hall, E.O. (1951) Deformation and ageing of mild steel, *Proc. Phys. Soc.*, **64B**, 742.
26 Orowan, E. (1947) Discussion on Internal Stresses, Institute of Metals, London.
27 Von Mises (1928) Z. Angew. *Math. Mech.*, **8**, 161.
28 Stein D.F. and Low J.M. (1960) Mobility of edge dislocations in silicon iron crystals, *J. Appl. Phys.*, **31**, 362.

Additional Reading

Meyers, M. and Chawla, K. (2009) *Mechanical Behavior of Materials*, 2nd edn, Cambridge University Press.

Reed-Hill, R.E. and Abbaschian, R. (1994) *Physical Metallurgy Principles*, 3rd edn, PWS Publishing, Boston.

5
Properties of Materials

"Inventing is a combination of brains and materials. The more brains you use, the less material you need."

—*Charles F. Kettering*

In the previous chapters, we did refer to materials properties without discussing them in details. It is difficult to cover all the aspects of materials properties in a single chapter. Hence, the readers are encouraged to refer to texts enlisted in Bibliography. Here, three broad categories of properties – mechanical, physical, and chemical are identified. From the viewpoint of nuclear reactor applications, all these properties are important. Hence, in this chapter, we will learn some basics of these properties, the test techniques used for their evaluation, and relevant test data analysis techniques.

5.1
Mechanical Properties

Load-bearing structures in engineering applications need some attributes in response to some particular forms of mechanical loading conditions. There are mechanical properties like strength and ductility that can be evaluated from simple uniaxial tensile testing, whereas dynamic mechanical properties like fatigue could only be evaluated from cyclic loading testing. Here, we will assume that the readers have been familiar with a basic engineering mechanics course. In this chapter, tensile, impact, fracture toughness, and fatigue and creep properties will be briefly discussed. The American Society for Testing of Materials (ASTM) has developed specific standards for each test procedure. These will be referred to here from time to time. It is important to understand them before we start deliberations on the effects of radiation on these properties in Chapter 6.

Before we go into the details of the various sections of the mechanical properties, we first describe the modes and types of deformation. There are, in general, four ways that a load may be applied: tension, compression, shear, and torsion, as depicted in Figure 5.1; we will not be concerned here of torsion. Force applied per unit area is known as stress (σ or τ) and the sample displacement per unit length is the strain (ε or γ); tension is regarded as positive while compression negative.

An Introduction to Nuclear Materials: Fundamentals and Applications, First Edition.
K. Linga Murty and Indrajit Charit.
© 2013 Wiley-VCH Verlag GmbH & Co. KGaA. Published 2013 by Wiley-VCH Verlag GmbH & Co. KGaA.

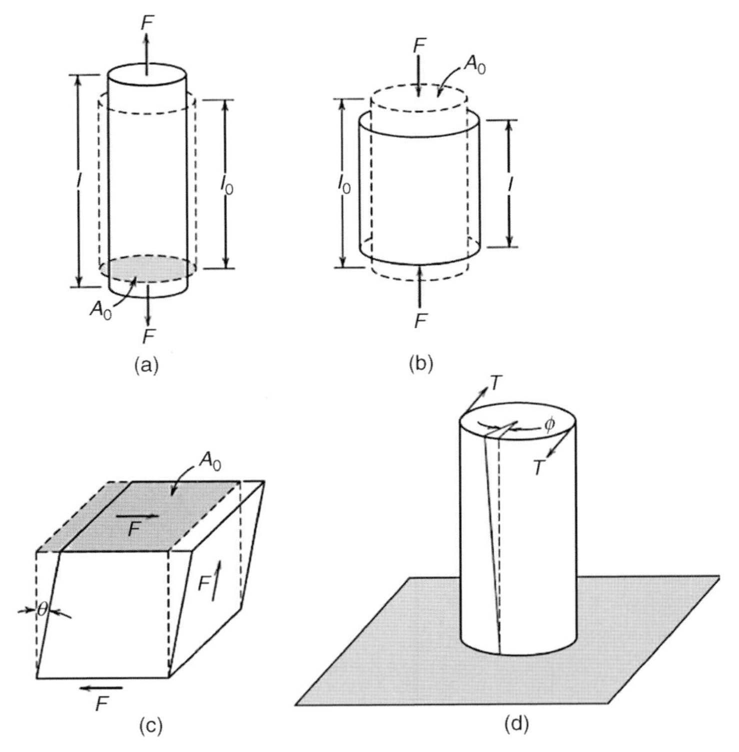

Figure 5.1 Schematic illustration of modes of deformation. (a) Tensile. (b) Compression. (c) Shear. (d) Torsion. From Ref. [1].

Thus, from Figure 5.1, we note the following relations:

$$\sigma = \frac{F}{A} \quad \text{and} \quad \varepsilon = \frac{\Delta l}{l} \quad \text{for tension and compression.} \tag{5.1a}$$

$$\tau = \frac{F}{A} \quad \text{and} \quad \gamma = \tan\theta \quad \text{for shear.} \tag{5.1b}$$

Note that the stress has units of $N\,m^{-2}$ known as Pascal (Pa), and we generally talk about MPa ($=10^6\,Pa = N\,mm^{-2}$), while the strain has no units and often referred to as %.

There are three types of deformation: elastic, anelastic, and plastic. Elastic deformation is instantaneous completely recoverable deformation where the body returns to the original shape once the external forces are removed. Anelastic deformation is time-dependent completely recoverable deformation, meaning that once the forces are removed, the body returns to the original shape in a time-dependent fashion. Plastic deformation is permanent deformation; i.e., it does not recover neither instantaneously nor as a function of time. During elastic deformation, force is directly proportional to the resulting strain as in Hooke's law (force proportional to displacement), and during uniaxial tension or compression, the proportionality constant is the Young's modulus (E) also referred to as elastic modulus:

$$E = \frac{\sigma}{\varepsilon}. \tag{5.2a}$$

Similarly, for shear loading,

$$G = \frac{\tau}{\gamma}. \tag{5.2b}$$

Here, for shear loading, G in Eq. (5.2b) is known as shear modulus or modulus of rigidity. In Section 5.1.1, we describe tensile properties, while shear characteristics will be mentioned as appropriate.

We note that when tensile stress is applied (Figure 5.1a), the diameter of the specimen decreases and the ratio of transverse contraction to longitudinal elongation is given by the Poisson's ratio (ν), which is positive and has a value close to 0.3. This implies that during elastic deformation, volume increases, which is known as elastic dilatation. For rectangular cross section in a tensile bar of an isotropic material with loading along the z-direction, the Poisson's ratio is given by

$$\nu = \frac{-\varepsilon_x}{\varepsilon_z} = \frac{-\varepsilon_y}{\varepsilon_z}. \tag{5.2c}$$

Some materials such as Zr and Ti alloys in general exhibit anisotropy in which case the ratios of contractile width to axial elongation and contractile thickness to axial elongation may not be the same.

The volume change is given by the sum of the three orthogonal strains,

$$\frac{\Delta V}{V} = \varepsilon_x + \varepsilon_y + \varepsilon_z. \tag{5.3a}$$

Under the application of hydrostatic pressure (P), the volume change is related to the bulk modulus (κ), the inverse of which is known as compressibility (β):

$$\frac{\Delta V}{V} = \frac{P}{\kappa} = \beta P. \tag{5.3b}$$

It turns out that E, G, and κ are related through the Poisson's ratio (ν):

$$G = \frac{E}{2(1+\nu)} \quad \text{and} \quad \kappa = \frac{E}{3(1-2\nu)} = \frac{2G(1+\nu)}{3(1-2\nu)}. \tag{5.4}$$

The above relationships are usually derived in mechanical property courses and it is sufficient here to know that such interrelationships exist.

Example Problem

A cylindrical steel specimen (10 mm in diameter and 40 mm in length) is subjected to a stress of 100 MPa resulting in the length and diameter of the elastically deformed specimen of 40.019 and 9.9986 mm, respectively. Evaluate (i) elastic modulus, (ii) Poisson's ratio, (iii) shear modulus, and (iv) change in volume.

Solution

We first need to calculate the longitudinal (ε_l) and diametral (ε_D) strains:

$$\varepsilon_l = \frac{\Delta l}{l} = \frac{0.019}{40} = 0.000475 = 0.0475\% \quad \text{and} \quad \varepsilon_D = \frac{\Delta D}{D} = \frac{-0.0014}{10}$$
$$= -0.00014 = -0.014\%.$$

i) $E = \dfrac{\sigma}{\varepsilon} = \dfrac{100}{0.000475} = 210\,526 = 210.5\ \text{GPa}.$

ii) $\nu = \dfrac{-\varepsilon_D}{\varepsilon_l} = \dfrac{0.014}{0.0475} = 0.295.$

iii) $G = \dfrac{E}{2(1+\nu)} = \dfrac{210.5}{2(1.295)} = 81.3\ \text{GPa}.$

iv) Change in volume can be found either by finding the final volume ($\pi D^2 l/4$) and subtracting from the initial volume or by adding the strains ($\varepsilon_l + 2\varepsilon_D$) to be 1.953×10^{-4} or 0.0195%, an increase.

5.1.1 Tensile Properties

Tension (or tensile) test is a popular method of studying the *short-term* mechanical behavior (strength and ductility) of a material under quasi-static uniaxial tensile state of stress. This test provides important design data for load-bearing components. Tension test involves stretching of an appropriately designed tensile specimen under monotonic uniaxial tensile loading condition. Generally, a constant displacement rate is used during the test. The load and sample elongation are measured simultaneously using load cells and strain gage/LVDT (linear variable differential transformer), respectively. Figure 5.2 shows a schematic of a basic tensile tester.

The test generally uses a standard specimen size shown in Figure 5.3. But for irradiated materials (considering the induced radioactivity among other factors), the use of the subsize tensile specimens is quite common. One important thing to observe is that the transition region between the gauge length and the shoulders is smooth so that stress concentration effect does not set in creating flaws. It is generally recognized that in order to compare elongation measurements with a good approximation in different sized specimens, the specimens must be geometrically similar. In this regard, Barba's law (1880), which states that $L_0/\sqrt{A_0}$ (L_0 is the gauge length and A_0 is the cross-sectional area) needs to be a constant, is useful. In the United States, $L_0/\sqrt{A_0}$ is taken as ~4.5 for round specimens. The relevant ASTM standard for tension testing is E8 (standard test method for tension testing of metallic materials).

5.1.1.1 Stress–Strain Curves

Any material would deform under an applied load. If all the deformation may recover upon unloading, this type of deformation is known as elastic deformation

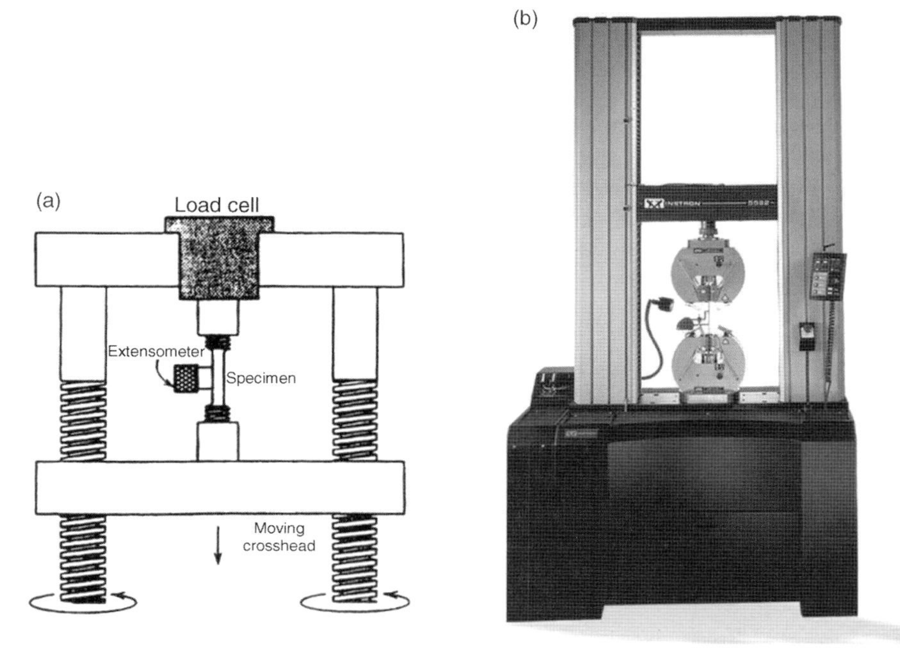

Figure 5.2 (a) A schematic of a tensile tester. Adapted from Ref. [1]. (b) An *Instron* universal tester that can perform tensile test among many other tests. *Courtesy:* Instron.

as we defined earlier. However, if the load is large enough, plastic deformation sets in so that the material undergoes a permanent deformation, that is, upon unloading only the elastic portion of the deformation recovers but the rest remains.

Even though the raw data obtained from the tension test are load and elongation, they need to be converted into stress and strain, respectively, to have meaningful use in engineering considerations. First, let us talk about the engineering stress–strain curve that is of prime importance in many load-bearing applications. Engineering stress and strain are determined based on the original dimensions of the specimens. Engineering stress (σ_e), also known as conventional or nominal stress,

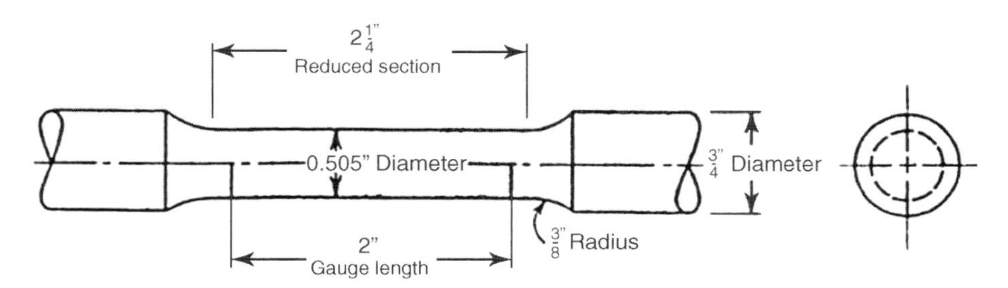

Figure 5.3 Design of an ASTM standard round tensile specimen.

thus is given by

$$\sigma_e = \frac{P}{A_0}, \tag{5.5}$$

where P is the instantaneous load and A_0 is the initial diameter of the gauge section. Engineering strain is defined as the ratio of the elongation of the gauge length ($\delta = L - L_0$) to the original gauge length (L_0) as shown in the equation below:

$$e = \frac{\delta}{L_0} = \frac{L - L_0}{L_0}, \tag{5.6}$$

where L is the instantaneous gauge length. Engineering strain is generally expressed in terms of percent elongation. As the engineering stress–strain curves are obtained based on the original dimensions (constant), the engineering stress–strain curve and load–elongation curve have the similar shape. A schematic engineering stress–strain curve is shown in Figure 5.4.

Let us talk about the general shape of the engineering stress–strain curve.

a) In the initial linear portion of the curve, stress is proportional to strain. This is the elastic deformation (instantaneously recoverable) regime where Hooke's law is applicable. The modulus of elasticity or Young's modulus (Eq. (5.3)) can be determined from the slope of the straight portion on the stress–strain curve. Young's modulus is determined by the interatomic forces that are difficult to change significantly without changing the basic nature of the materials. Hence, this is the most structure-insensitive mechanical property. However, any significant change in the crystal structure (such as material undergoing polymorphic transformation) would also change the elastic modulus. It is only affected to a small extent by alloying, cold working, or heat treatment. Most aluminum alloys have Young's moduli close to 72 GPa. Modulus of elasticity does decrease with increasing temperature as the interatomic forces become weaker.

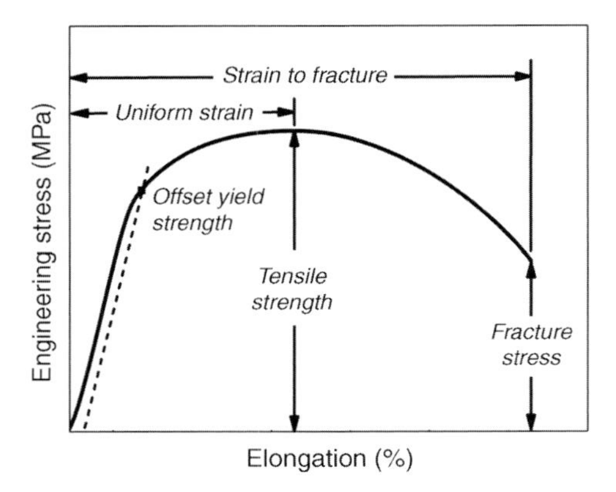

Figure 5.4 A typical engineering stress–strain curve.

b) When the deformation proceeds past a point, it becomes nonlinear, and the point at which this linearity ends is known as the proportional limit and the stress at which plastic deformation or yielding starts is known as yield stress or yield strength that depends on the sensitivity of the strain measurement. A majority of materials show a gradual transition from the elastic to the plastic regime and it becomes difficult to determine exactly what the yield stress actually is. That is why yield strength is generally taken at an offset strain of 0.2%, as shown in Figure 5.4. In cases, where there is no straight portion in the stress–strain curve (such as for soft copper and gray cast iron), the yield strength is defined as the stress that generates a total strain of 0.5%.

Another shape of stress–strain curve that is commonly observed in some specific materials represents discontinuous yielding. Figure 5.5 shows such an engineering stress–strain curve. Some metals/alloys (in particular, low carbon steels) exhibit a localized, heterogeneous type of transition from elastic to plastic deformation. As shown in the figure, there is a sharp stress drop as the stress decreases almost immediately after the elastic regime. This maximum point on the linear stress–strain curve is known as "upper yield point." Following this, the stress remains more or less constant, and this is called "lower yield point." Yield point phenomenon occurs because the dislocation movement gets impeded by interstitial atoms such as carbon, nitrogen, and so on forming solute atmospheres around the dislocations. However, at higher stress, the dislocations break away from the solutes and thus requires less stress to move. That is why a sharp drop in load or stress is observed. The elongation that occurs at constant load or stress at the lower yield point is called "yield point elongation." During yield point elongation, a type of deformation bands known as Lüders bands (sometimes known as Hartmann lines or stretcher strains) are formed across this regime. This particular phenomenon is called Piobert effect. After Lüders bands cover the whole gauge length of the specimen, the usual strain hardening regime sets in, as shown in Figure 5.5. However, temperature of testing could change the behavior drastically.

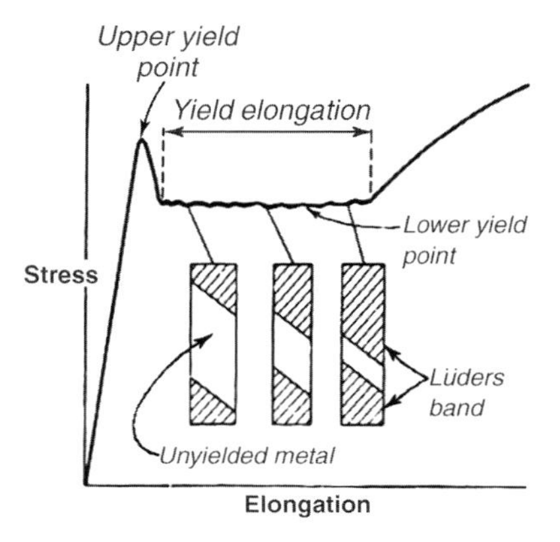

Figure 5.5 Typical discontinuous stress–strain curve with distinct yield point phenomenon. From Ref. [2].

c) After proportional limit or yielding, the material enters a strain hardening regime where stress increases with increase in strain until it reaches a stress where nonuniform plastic deformation (or necking starts). This stress is known as tensile strength or ultimate tensile stress (UTS) corresponding to the maximum load in the load–elongation curve. So, basically UTS is given by the maximum load divided by the original cross-sectional area of the specimen. UTS in itself is not a property of fundamental significance. But this has long been used in the design of materials with a suitable safety factor (~2). Nowadays yield strength rather than UTS is used for the purpose of designing. But still it can serve as a good quality control indicator and in specifications of the products. However, since UTS is easy to determine and is quite a reproducible property, it is still used in practice. For brittle materials, UTS is considered a valid design criterion.

d) Ductility as determined by the tension testing is a subjective property; however, it does have great significance: (i) Metal deformation processing needs a material to be ductile without fracturing prematurely. (ii) A designer is interested to know whether a material will fail in service in a catastrophic manner or not. That information may come from ductility. (iii) If a material is impure or has undergone faulty processing, ductility can serve as a reliable indicator even though there is no direct relationship between the ductility measurement and service performance.

Generally, two measures of ductility as obtained from a tension test are used – total fracture strain (e_f) and the reduction of area at fracture (q). These properties are obtained after fracture during tension test using the following equations:

$$e_f = \frac{L_f - L_0}{L_0}, \tag{5.7}$$

$$q = \frac{A_0 - A_f}{A_0}, \tag{5.8}$$

where L_f and A_f are the final gauge length and cross-sectional area, respectively. Generally total fracture strain is composed of two plastic strain components (prenecking deformation that is *uniform* in nature, and postnecking deformation that is nonuniform in nature). Both these properties are expressed in percentage. We note from Eq. (5.7) that the elongation to fracture depends on the gauge length (L_0). That is why it is customary to mention the gauge length of the tension specimen while reporting the elongation values; 2 in. gauge lengths are generally used. On the other hand, percentage reduction in area is the most structure-sensitive property that can be measured in a tension test even though it is difficult to measure very accurately or *in situ* during testing.

e) Fracture stress or breaking stress (Eq. (5.9)) are often used to define the engineering stress at which the specimen fractures. However, the parameter does not have much significance as necking complicates its real value.

$$\sigma_f = \frac{P_f}{A_0}, \tag{5.9}$$

where P_f is the load at fracture.

f) We discuss two more properties that are of importance – resilience and toughness. From dictionary meaning, they would be considered synonyms. But in the context of material properties, they are bit different. *Resilience* is the ability of a material to absorb energy when deformed elastically. Modulus of resilience (U_0) is used as its measure and is given by the area under the stress–strain curve up to yielding:

$$U_0 = \frac{\sigma_{e0}^2}{2E}, \tag{5.10}$$

where σ_{e0} is the yield strength and E is the modulus of elasticity.

Toughness is the ability of a material to absorb energy in the plastic range. As we will see later, there are two more types of toughness that are often used: fracture toughness and impact toughness. Hence, the toughness one obtains from stress–strain curves is known as *tensile toughness*. The area under the stress–strain curve generally indicates the amount of work done per unit volume on the material without causing its rupture. Toughness is generally described as a parameter that takes into account both strength and ductility. There are empirical relations that express toughness; however, as they are based on original dimensions, they do not represent the true behavior in the plastic regime.

Note

Recall Poisson's ratio – the ratio of lateral contraction to longitudinal elongation and given by the following relations:

$$\nu = -\frac{\varepsilon_x}{\varepsilon_z} = -\frac{\varepsilon_y}{\varepsilon_z}, \tag{5.11}$$

where ε_x and ε_y represent the lateral strains and ε_z is the longitudinal strain. The "minus" sign appears in the above relation due to the opposing sense of the lateral and longitudinal strains and Poisson's ratio is positive since contraction occurs in the lateral directions. Recalling that the Poisson's ratio has a value close to 0.3, volume increase (elastic dilatation) is noted in elastic deformation. However, volume is *conserved* during plastic deformation in which case the equivalent ratio will have a value of 0.5.

Even though we use the parameters obtained from engineering stress–strain curves for engineering designs, they do not represent the fundamental deformation behavior of the material as it is entirely based on the original dimensions of the tensile specimen. In the engineering stress–strain curve, the stress falls off after the maximum load due to the gradual load drop and calculation based on the original cross-sectional area of the specimen. But in reality, the stress does not fall off after maximum load. Actually, the strain hardening effect (the stress in fact increases) remains in effect until the fracture, as shown in Figure 5.6, but the cross-sectional area of the specimen decreases more compared to the load drop, thus increasing the stress. This happens because the true stress (σ_t) is based on the

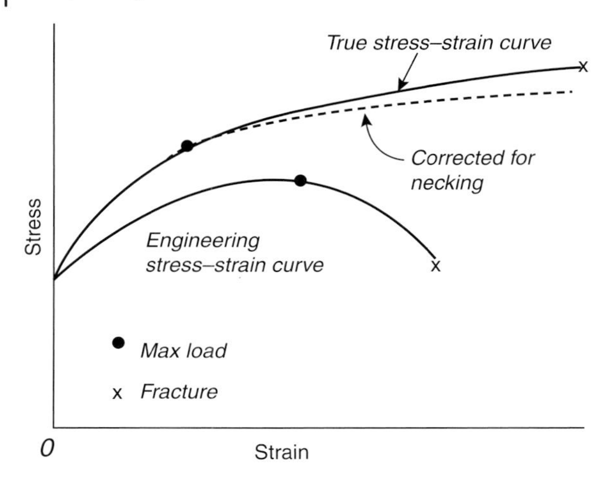

Figure 5.6 A comparison between an engineering stress–strain curve and a true stress–true strain curve.

instantaneous cross-sectional area (A) and is expressed as

$$\sigma_t = \frac{P}{A} = \sigma_e(e + 1). \tag{5.12}$$

Before necking, it is better to calculate the true stress from the engineering stress and engineering strain based on the constancy of volume and a homogeneous distribution of strain across the gauge length. However, after necking, these assumptions are hardly valid and the true stress should then be calculated by using the actual load and cross-sectional area in the postnecking regime. True strain, on the other hand, is generally calculated from the expression as given below:

$$\varepsilon = \ln\left(\frac{L}{L_0}\right) = \ln(e + 1). \tag{5.13}$$

However, the second equality of the equation does not hold valid after necking. That is why beyond the maximum load, true strain should be calculated based on actual area or diameter of the specimen measurement following the relation given below:

$$\varepsilon = \ln\left(\frac{A_0}{A}\right) = 2\ln\left(\frac{D_0}{D}\right). \tag{5.14}$$

For the true stress–strain curves, some parameters such as true stress at maximum load, true fracture stress, true fracture strain, true uniform strain, and so on can be calculated using appropriate relations.

One important parameter that can be obtained from the true stress–true strain curve is the strain hardening exponent (n). This is generally described by Hollomon's equation:

$$\sigma_t = K\varepsilon^n, \tag{5.15}$$

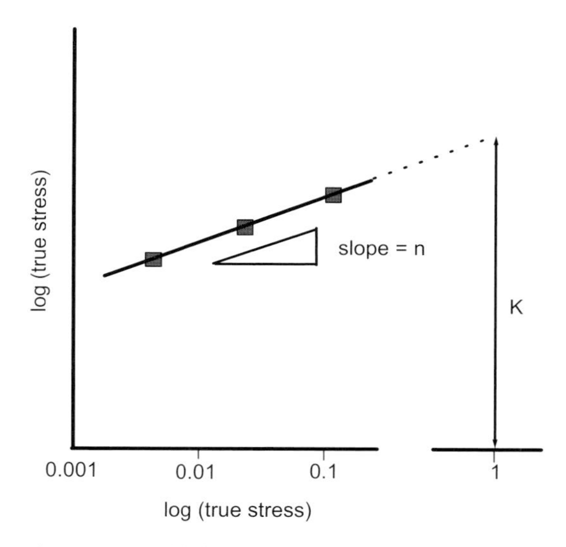

Figure 5.7 Double logarithmic plot illustrating the method of determining strain hardening exponent and strength coefficient.

where K is the strength coefficient and n is known as strain hardening (or work hardening) exponent or parameter or coefficient. As the above equation is in the form of a power law, the true stress and true strain data when plotted on a double logarithmic scale and fitted to a straight line will yield a slope that is equal to strain hardening exponent, and the strength coefficient (i.e., the true stress at $\varepsilon = 1.0$) can be calculated from the extrapolated line, as shown in Figure 5.7. Note that the data used for the calculation of n should not be taken beyond the maximum load (i.e., at or after maximum load). Theoretically, n value can range between 0 (elastic solid, follows Hooke's law) and 1 (perfectly plastic solid). For most metals/alloys, n values are found to be 0.1–0.5. Note that these aspects are valid for stresses and strains beyond the elasticity (i.e., in the plastic deformation regime). Strain hardening is sensitive to the microstructure, which in turn depends on processing. The rate of strain hardening is obtained by differentiating Eq. (5.15) with respect to ε and then replacing $K\varepsilon^n$ by σ_t, the following equation is obtained:

$$\frac{d\sigma_t}{d\varepsilon} = n\frac{\sigma_t}{\varepsilon}. \tag{5.16}$$

Note that often the data at low strains and high strains tend to deviate from Hollomon's equation. That is why many other techniques, such as Ludwik's equation, Ludwigson's equation, Ramberg–Osgood relation, and so on have been proposed over the past years to obtain a better estimate of strain hardening exponents.

One important aspect of strain hardening exponent is that it represents the true uniform strain, and is thus related to the ductility of a material. In order to derive the relation, we need to first discuss about the instability effect that occurs in tension. In tension, necking (i.e., localized deformation) occurs at the maximum load (generally in a ductile material since brittle materials fracture well before reaching that point). The load-carrying capacity of the material increases as the strain increases; however, as noted before, the cross-sectional area decreases as an opposing effect. However, at the onset of necking, the increase in stress due to decrease in cross-sectional area becomes higher than the concomitant increase in the load-carrying ability of the material because of strain hardening. The condition of instability can be written in a differential form as $\mathrm{d}P=0$ where P is a constant (the maximum load). Now replacing P with $(\sigma_t \cdot A)$ in this instability equation, we obtain

$$\mathrm{d}P = \mathrm{d}(\sigma_t \cdot A) = A \cdot \mathrm{d}\sigma_t + \sigma_t \cdot \mathrm{d}A = 0$$

or

$$-\frac{\mathrm{d}A}{A} = \frac{\mathrm{d}\sigma_t}{\sigma_t}. \tag{5.17}$$

Due to the constancy of volume (V) during plastic deformation, we can also write

$$\mathrm{d}V = 0, \quad V = AL; \quad \text{hence} \quad \mathrm{d}V = \mathrm{d}(AL) = A \cdot \mathrm{d}L + L \cdot \mathrm{d}A = 0$$

or

$$\frac{\mathrm{d}L}{L} = -\frac{\mathrm{d}A}{A} = \mathrm{d}\varepsilon. \tag{5.18}$$

By comparing Eqs (5.17) and (5.18), we can write

$$\frac{\mathrm{d}\sigma_t}{\mathrm{d}\varepsilon} = \sigma_t, \tag{5.19}$$

which is valid at the condition of tensile instability (or at maximum load).

Now comparing Eqs (5.16) and (5.19), we obtain

$$n = \varepsilon_u, \tag{5.20}$$

where ε_u is the true uniform strain, that is, true strain at the maximum load. Hence, it can be noted that the higher the strain hardening exponent, the greater the uniform elongation, which in turn helps in improving the overall ductility. Note that Considere's criterion ($\mathrm{d}\sigma/\mathrm{d}\varepsilon = \sigma$ or $\mathrm{d}\sigma/\mathrm{d}e = \sigma/(1+e)$) can be used to estimate the strain hardening exponent from a graphical plot, this is however outside the scope of this book. For further details on these relations, readers are referred to the suggested texts enlisted in Bibliography.

5.1.1.2 **Effect of Strain Rate on Tensile Properties**

The rate at which strain is imposed on a tensile specimen is called strain rate ($\dot{\varepsilon} = \mathrm{d}\varepsilon/\mathrm{d}t$). The unit is generally expressed in s^{-1}. It is instructive to know the ranges of generic strain rates used in different types of mechanical testing and deformation processing. Generally, quasi-static tension testing involves strain rates in the range of 10^{-5}–$10^{-1}\,\mathrm{s}^{-1}$. Generally, tension testing is done by placing the tension specimen in the cross-head fixture and running the test at a constant cross-head speed. However, it is important to know what the strain rate is in the specimen. Following Nadai's analysis, the nominal strain rate is expressed in terms of cross-head speed (v) and original gauge length (L_0):

$$\dot{e} = \frac{v}{L_0}. \tag{5.21}$$

However, the true strain rate changes as the gauge length changes:

$$\dot{\varepsilon} = \frac{v}{L}. \tag{5.22}$$

Therefore, it must be noted that most tensile tests are not conducted at constant true strain rates. However, specific electronic feedback (open loop before necking and closed loop after necking) system can be set up with the tension tester where the cross-head speed is continuously increased as the test progresses in order to maintain the same true strain rate throughout the test. But the test becomes more complicated without much benefit.

The flow stress increases with increasing strain rate. The effect of strain rate becomes more important at elevated temperatures. The following equation shows the relation between the flow stress and true strain rate at a constant strain and temperature:

$$\sigma_t = C\dot{\varepsilon}^m \tag{5.23}$$

where C is a constant and m is the strain rate sensitivity (SRS). The exponent m can be found out from the slope of the double logarithmic plot of true stress versus true strain rate. The value of strain rate sensitivity is quite low (<0.1) at room temperature, but it increases as the temperature becomes higher with a maximum value of 1 when the deformation is known as *viscous flow*. Figure 5.8 shows flow stress (at 0.2% strain) versus strain rate on a double logarithmic plot for an annealed 6063 Al–Mg–Si alloy. In superplastic materials, the strain rate sensitivity is higher (0.4–0.6). But these materials require finer grain diameter ($<10\,\mu m$) and temperatures at or above half the melting temperature (in K). Superplastic materials exhibit higher than normal ductility (as rule of thumb more than 200%) and utilize strain rate hardening instead of strain hardening.

5.1.1.3 **Effect of Temperature on Tensile Properties**

Temperature strongly affects stress–strain curves. Generally, strength decreases and ductility increases. However, this trend does change according to the microstructural evolution such as precipitation, strain aging, or recrystallization that may take place during testing. Thermally activated processes help in deformation process and result in reducing the strength. Figure 5.9 shows the stress–strain curves of a mild steel at three different temperatures.

Understanding of thermally activated deformation is important for structural materials serving at elevated homologous temperatures. The flow stress (shown as shear component) of a pure metal is composed of two parts:

$$\tau = \tau^* + \tau_G, \tag{5.24}$$

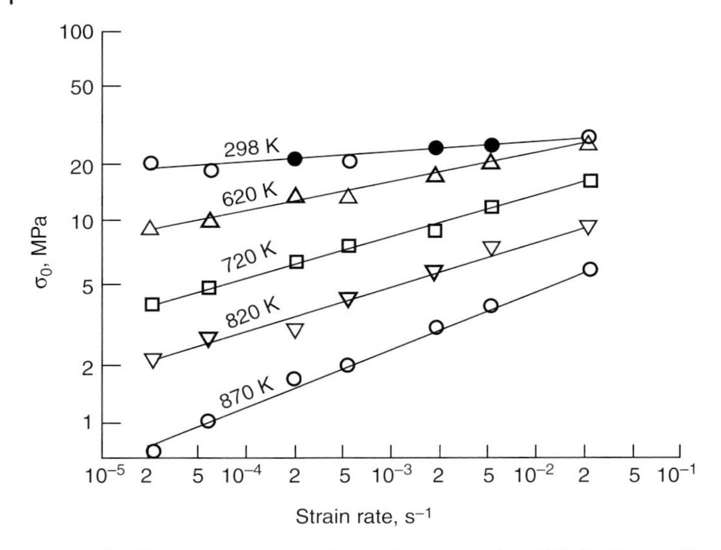

Figure 5.8 Flow stress versus strain rate for an annealed 6063 Al alloy at different temperatures. Note that the strain rate sensitivity increases with increasing temperature (the slope of the fitted lines at each temperature increases as the temperature increases).

where τ^* and τ_G are the thermally activated stress and athermal (temperature-independent) stress components, respectively. There are two types of obstacles in a material: long-range obstacles and short-range obstacles. The influence of long-range obstacles occurs over several atom distances and is difficult to surmount through pure thermal fluctuations. Hence, the athermal stress component comes from the long-range obstacles. The long-range stress field is not generally affected by temperature or strain rate except for the change in modulus due to temperature change (which is purely due to reduction in interatomic bonding forces). On the other hand, the short-range obstacles (less than 10 atom diameters) for which dislocations can surmount

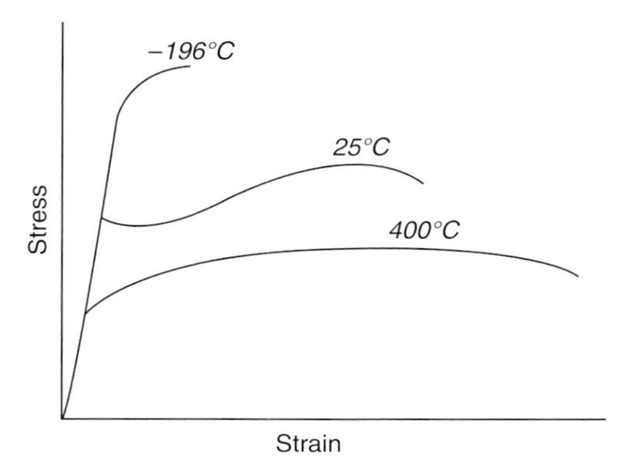

Figure 5.9 The effect of temperature on the engineering stress–strain curves of a mild steel.

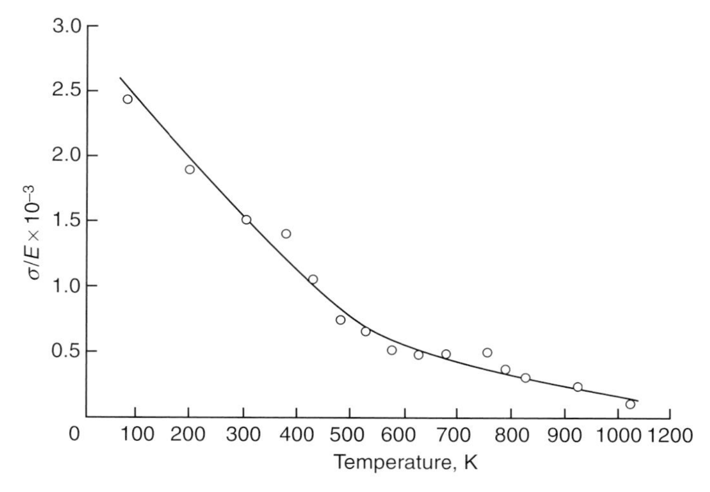

Figure 5.10 The variation of normalized flow stress as a function of temperature in a high-purity titanium.

these barriers with thermal fluctuations result in temperature-sensitive strengthening. These short-range obstacles are also known as thermal barriers and their influence on flow stress strongly depends on temperature and strain rate. Figure 5.10 shows the variation of normalized flow stress (tested at a strain rate of $3 \times 10^{-4}\,s^{-1}$) as a function of temperature in a high-purity titanium.

5.1.1.4 **Anisotropy in Tensile Properties**

When properties depend on orientation, the property is termed as anisotropic. Crystallographic anisotropy arises because of the crystallographic texture (preferred orientation of grains in a material). As we know, a single crystal is the most anisotropic. However, in polycrystalline materials, each grain tends to be oriented in different ways, thus reducing the texture and thus anisotropy. However, polycrystalline materials can also become textured depending on the processing it undergoes. Yield strength may then be strongly dependent on crystalline texture. Polycrystalline materials with anisotropic crystal structure like HCP tend to have strong texture and thus mechanical anisotropy. One such example would be zircaloy fuel cladding.

Example Problem

A tensile test of a metal was run in the laboratory, but only the maximum load (59 kN) was measured during the test. All other data were taken before and after the test:

Initial diameter = 13.0 mm; minimum diameter after test (at point of fracture) = 10.0 mm, maximum diameter after test (away from fracture) = 11.6 mm.

Assuming that the material follows the standard work hardening rule all the way to fracture, evaluate the (i) strain hardening exponent (n), (ii) strength coefficient (K), (iii) nominal yield strength of the material, (iv) load on the specimen at the instant just prior to fracture, and (v) toughness (plastic strain energy to fracture).

Solution

According to the standard work hardening rule, $\sigma = K\varepsilon^n$, where σ and ε are true stress and strain. We also note that $\varepsilon_u = n$.

i) Find

$$\varepsilon_u = \ln\left(\frac{l_0}{l_u}\right) = \ln\left(\frac{A_0}{A_u}\right) = 2\ln\left(\frac{D_0}{D_u}\right) = 2\ln\left(\frac{13}{11.6}\right) = 0.228 = n.$$

ii) Using the given maximum load, we can find

$$\sigma_{UTS} = \frac{P_{max}}{A_u} = \frac{59000N}{\pi(D_u^2/4)} = \frac{59000N}{\pi(11.6^2/4)\,mm^2} = 558\text{ MPa}; \quad \text{knowing } \varepsilon_u$$
$$= 0.228, \quad \text{find } K = 558/(0.228^{0.228}) = 781.7\text{ MPa}.$$

iii) First find true yield stress using the work hardening law: $\sigma_y = K(\ln(1.002))^{0.228} = 189.5$ MPa. Now find engineering yield strength $(S_y) = \sigma_y A/A_0 = \sigma_y/(1+0.002) = 189.5/1.002 = 189.1$ MPa.

iv) Assuming that the work hardening rule is valid all the way to fracture, first find the true fracture strain and fracture stress. $\varepsilon_f = \ln\left(\frac{A_0}{A_f}\right) = 2\ln\left(\frac{D_0}{D_f}\right) = 2\ln\left(\frac{13}{10}\right) = 0.525$ so that $\sigma_f = K\varepsilon_f^n = 781.7 \times 0.525^{0.228} = 674.9$ MPa. Thus, the load at fracture $P_f = \sigma_f A_f = 674.9(\pi/4)(10)^2 = 53.03$ kN.

v) Toughness $J = K(\varepsilon_f^n/1+n) = 781.7(0.525^{0.228}/1.228) = 549.6\text{ MPa} = 549.6\text{ MJ m}^{-3}$ recalling that J is energy per unit volume.

5.1.2 Hardness Properties

Hardness is defined as resistance to indentation. Hardness testing is simple and is a useful technique to characterize mechanical properties of materials. It provides a rapid and economical way of determining the resistance of materials to deformation. It generally does not involve a total destruction of the sample as needed in tension testing, and sometimes considered as a semi-nondestructive technique, and generally needs small test volume of materials. Thus, the hardness values are generally proportional to the strength values as obtained from conventional tension or compression tests. Materials scientists (metallurgists) are mostly interested in the hardness that is defined by the resistance and of a material against indentation. So, in all hardness tests, we use a hard indenter placed vertically against the sample surface and load the indenter with a specific load for a specified time into the sample, and measure the depth or lateral dimension of the indentation. Thus, relatively larger indentations are noted in softer materials.

There are two major kinds of hardness testing: macrohardness and microhardness. In the group of macrohardness test techniques, Brinell hardness test, Rockwell test, and Macro–Vickers test are important. Generally, larger loads are used giving rise to larger size indentation on the samples. In the group of microhardness test techniques, tests are generally done in two ways – micro-Vickers and Knoop indentation.

5.1.2.1 Macrohardness Testing

Brinell Hardness Test

In 1900, J.A. Brinell of Sweden invented ways to determine hardness by measuring the impression (or indentation) made by a steel ball forced into the metal under static loads. The definite advantage of this technique is that a single linear scale can be used to determine hardness. The standard Brinell test requires a 3000 kg load applied through a 10 mm diameter hardened steel or tungsten carbide ball. Brinell Hardness Number (BHN) is given by

$$\mathrm{BHN} = \frac{2P}{\pi D(D - \sqrt{D^2 - d^2})}, \tag{5.25}$$

where P is the applied load in kgf and D and d are the diameters of the ball and the impression, respectively, in mm.

Vickers Hardness Test

This test uses a square-based pyramid indenter (as shown in Figure 5.11) whose angle between the opposite faces is 136°. The Vickers Hardness Number (VHN) is

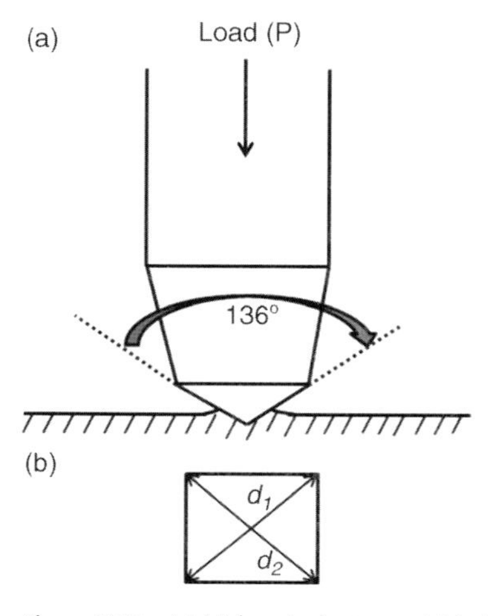

Figure 5.11 (a) Vickers indenter and (b) the corresponding impression created by the Vickers indenter.

determined by the load divided by the surface area of the indentation. The area is calculated from the average length of the diagonals (d_1 and d_2) of the impression (Figure 5.11). The following relationship is generally used:

$$\text{VHN} = \frac{2P\sin(\theta/2)}{d^2} = \frac{1.854P}{d^2}, \tag{5.26}$$

where P is the applied load in kgf, d is the average length (i.e., $(d_1 + d_2)/2$) of the diagonals in mm, and θ is the angle between the opposite faces of the indenter (136°). So, VHN is in the unit of kgf mm^{-2}. Vickers test is extremely useful as it can range from a scale of 5–1500. This can be used as a microhardness technique if the loads used are small (as discussed later). The disadvantages associated with Vickers hardness technique are the requirement of good surface preparation and errors in the determination of the diagonal length.

Rockwell Hardness Test

In 1908, Professor Ludwig of Vienna, Austria, first described a method by which a hardness of a material can be measured by a *differential depth measurement* technique. This mode of hardness test technique consisted of measuring the increment of depth of a cone-shaped diamond indenter (known as Brale) or spherical steel indenters of various diameters forced into the material by a minor load and a major load. The test first uses a minor load of 10 kg to set the indenter onto the material surface, and then a major load (as determined by the particular scale chosen) is applied and the penetration depth is shown instantly on the scale with 100 divisions (usually each division corresponds to 0.00008 in.). Although the required surface preparation is minimum, depending on the hardness of the material, there are several Rockwell scales that represent a particular combination of the indenters and major loads used; some of them are included in Table 5.1. Note that there is a *superficial Rockwell* test mode that uses lower load compared to that used in the standard Rockwell tests and is used for thinner material or probing surface-hardened materials. The martensite formed in eutectoid (0.77 wt% C) plain carbon steel can be very high (Rockwell hardness of up to 65 R_c).

5.1.2.2 Microhardness Testing

Microhardness technique is generally employed for measuring indentation hardness of very small objects, thin sheet materials, surface hardened materials, electroplated materials, structural phases in multiphase alloys, and so forth. There are two

Table 5.1 Some standard Rockwell scales.

Rockwell hardness scale	Indenter	Major load (kg)
A	120° diamond cone (Brale)	60
B	1/16 in. (~1.6 mm) diameter steel ball	100
C	Brale	150
D	Brale	100

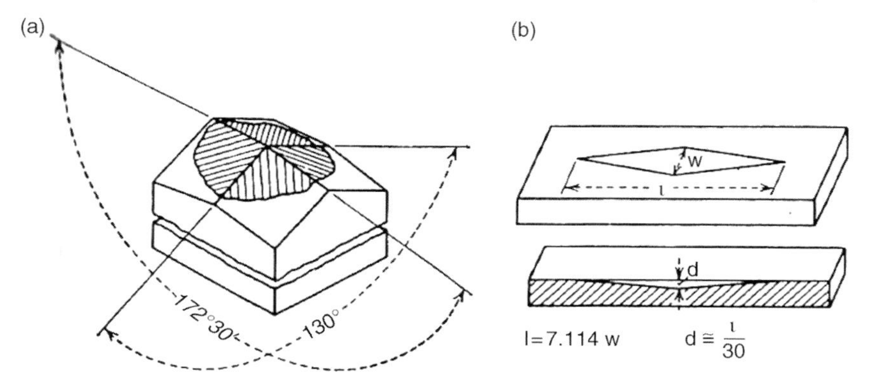

Figure 5.12 The configuration of the Knoop indenter (a) and the impression made by it (b). From Ref. [3].

standard indenters with which microhardness tests are done. One is Vickers indenter, the same one used in macro-Vickers test (square-based diamond pyramid with 136° apex angle), already illustrated in Figure 5.11. Knoop indenter is also a pyramid-shaped indenter (an included transverse angle of 130° and a longitudinal angle of 172°30′ such that it produces an indentation with a long to short diagonal ratio of ~7 : 1) (Figure 5.12). The Vickers indenter is pressed into the smooth, polished surface of the specimen. The load is applied for a predetermined time and removed. After a specified load and time, precision objectives are used to view the indentation and using scales attached on the eyepiece, one can measure the size of the two diagonal lengths of the square-shaped impression. Then, this square length is used to measure VHN from Eq. (5.1). There are also standard charts available that relate diagonals to the VHNs. Generally, indentation should be made away from the edge of the sample as much as possible. The thickness of the specimen should be at least 1.5 times of the diagonal length. There should be adequate spacing between the indentations so that their deformation zones do not affect the deformation zone of other indentation. Generally, one takes several indents from a region and measure hardness and take the mean with measurement error.

In all indentation hardness tests, when the applied load is removed, there always occurs some elastic recovery. The amount of recovery and the distorted shape depend on the size and precise shape of the indenter. Due to the unique shape of the Knoop indenter, elastic recovery of the projected impression occurs in a transverse direction, that is, shorter diagonal length, rather than long diagonal. Therefore, the measured longer diagonal length will give a hardness value close to what is given by the uncovered impression.

The Knoop hardness number (KHN) is given by the load divided by the unrecovered projected area of the impression. It should be noted that the area referred to here is the projected area and not the surface area of the indentation as in the Vickers and Brinell hardness techniques. Hence, KHN is given by

$$\mathrm{KHN} = \frac{P}{A_\mathrm{p}} = \frac{P}{Cl^2}, \qquad (5.27)$$

where P is the applied load in kgf, A_p is the unrecovered projected area of indentation (mm^2), l is the longer diagonal length (in mm), and C is the Knoop indenter constant that relates the longer diagonal length to the unrecovered projected area (generally 0.07028). Standard hardness charts are available where hardness values are provided against load and long diagonal values.

Note that Knoop hardness numbers are not independent of the load used. So, while reporting the hardness numbers, one should also report the load used (especially, when load is less than 300 gf). Knoop hardness technique is much more sensitive to specimen surface preparation than the Vickers hardness in the low load range. Main advantages of Knoop hardness are its ability to measure near-surface hardness, hardness gradients, anisotropy in properties in thin sections, and so forth.

5.1.3
Fracture

This section gives a general discussion of fracture. Fracture is the separation or fragmentation of a solid body into two or more parts under the action of force/stress. The process of fracture consists of two components: crack initiation and crack propagation. Fractures can be classified into two broad categories: ductile and brittle. A ductile fracture is characterized by appreciable plastic deformation with *stable* crack growth. A brittle fracture is characterized by a rapid rate of crack propagation (unstable) with no gross deformation. Stable crack growth implies that once the load is taken off, the crack does not propagate further. Figure 5.13 shows various fracture types observed in metals/alloys subjected to uniaxial tension.

Fractures are classified with respect to different factors, such as strain to fracture, crystallographic mode of fracture, and the appearance of fracture. A shear fracture occurs as the result of extensive slip on the active slip plane. This type of fracture is

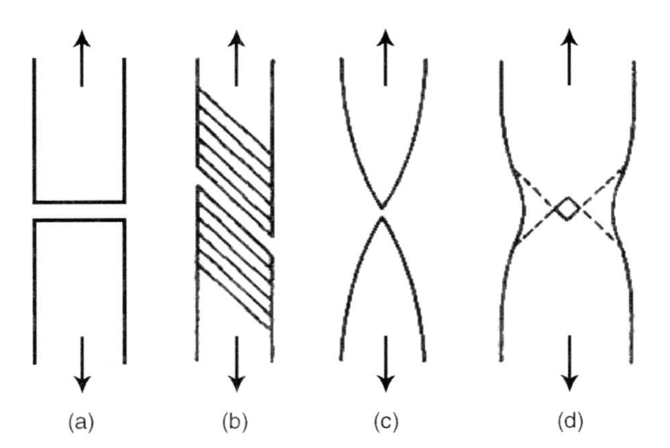

Figure 5.13 Various fracture types observed in metals subjected to uniaxial tension. (a) Brittle fracture of single and polycrystals. (b) Shearing fracture in ductile single crystals. (c) Completely ductile fracture in polycrystals. (d) Ductile fracture in polycrystals. From Ref. [2].

promoted by shear stresses. The cleavage mode of fracture is promoted by tensile stresses acting normal to a crystallographic cleavage plane. In many cases, the fracture surfaces are a mixture of fibrous and granular fracture, and it is customary to report the percentage of the surface area. Fractures in polycrystalline samples are transgranular (the crack propagates through the grains) or intergranular (the crack propagates along the grain boundaries).

5.1.3.1 Theoretical Cohesive Strength

Engineering materials typically exhibit fracture stresses that are 10–100 times lower than the theoretical value. This observation leads to the conclusion that flaws or cracks are responsible for the lower-than-ideal fracture strength. Due to Inglis [4], an approach assumes that the theoretical cohesive stress can be reached locally at the tip of a crack, while the average stress is at much lower value. Then, the nominal fracture stress is given by the expression (for the sharpest possible crack):

$$\sigma_f = \left(\frac{E\gamma_s}{4c}\right)^{1/2}, \tag{5.28}$$

where E is the elastic modulus, γ_s is the surface energy, and c is the crack length.

Microcracks act as precursors for crack propagation in brittle fracture. The process of cleavage fracture involves three steps: (a) plastic deformation to create dislocation pileups, (b) crack initiation, and (c) crack propagation. The initiation of microcracks can be affected by the presence of second-phase particles. Cleavage cracks can also be initiated at mechanical twins.

The ductile fracture starts with the initiation of voids, most commonly at second-phase particles. The particle geometry, size, and bonding play an important role. Dimpled rupture surface (ductile fracture) consists of cup-like depressions that may be equiaxial, parabolic, or elliptical, depending on the precise stress state. Microvoids are generally nucleated at second-phase particles, and the voids grow and eventually the ligaments between the microvoids fracture.

Griffith [5], established the following criterion for crack propagation: "A crack will propagate when the decrease in elastic strain energy is at least equal to the energy required to create new crack surface." The stress required to propagate a crack in a brittle material is a function of crack length (as shown in Figure 5.14) and is given by

$$\sigma_f = \left(\frac{2E\gamma_s}{\pi c}\right)^{1/2}. \tag{5.29}$$

Orowan [6], suggested that Griffith's equation can be applicable to metals (cases where there is some plastic deformation before fracture) with an additional term γ_p (expressing the plastic work needed to extend the crack), and is given by the expression:

$$\sigma_f = \left(\frac{2E(\gamma_s + \gamma_p)}{\pi c}\right)^{1/2} \approx \left(\frac{E\gamma_p}{c}\right)^{1/2}. \tag{5.30}$$

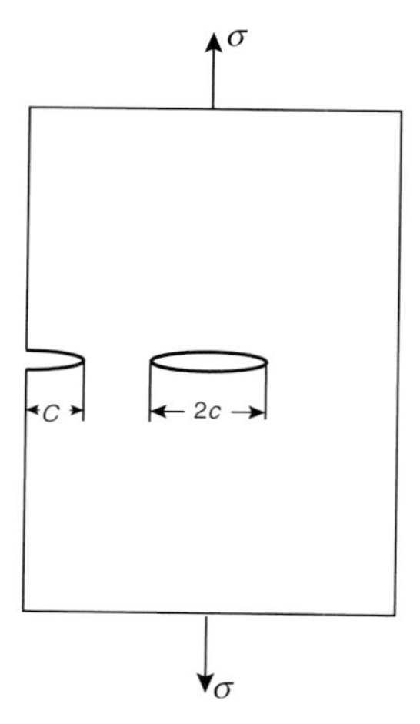

Figure 5.14 The crack configuration used in Griffith's equation.

Sohncke suggested that fracture occurs when the resolved normal stress (σ_c) reaches a critical value (Figure 5.15). The critical normal stress for brittle fracture is

$$\sigma_c = P\cos\phi/(A/\cos\phi) = (P/A)\cos^2\phi. \tag{5.31}$$

5.1.3.2 **Metallographic Aspects of Fracture**

Microcracks act as precursors for crack propagation in brittle fracture. The process of cleavage fracture involves three steps: (a) plastic deformation to create

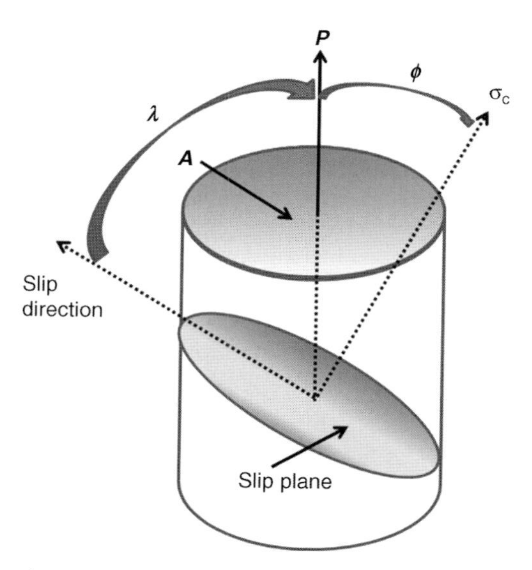

Figure 5.15 The configuration of the sample for determination of the critical normal stress for fracture.

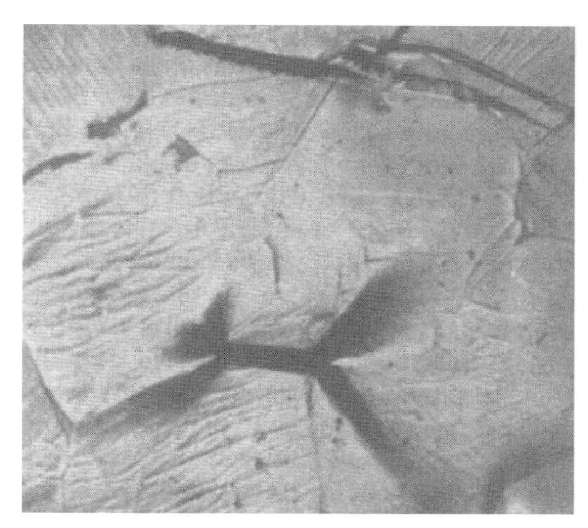

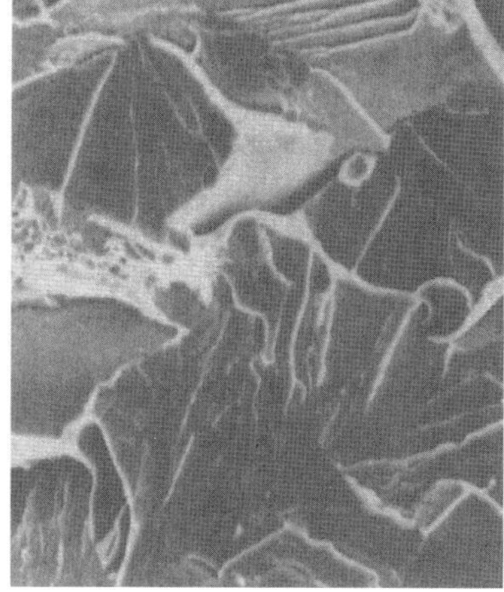

Figure 5.16 (a) Microcracks produced in iron by tensile deformation at −140 °C. (250X-Original *courtesy*: G.T. Hahn; From Ref. [2]. (b) Cleavage in ferrite by overload fracture of hot extruded carbon steel. 2,000X-Original *courtesy*: D.A. Meyn, Naval Research Laboratory; From Ref. [2].

dislocation pileups, (b) crack initiation, and (c) crack propagation. The initiation of microcracks can be affected by the presence of second-phase particles. Figure 5.16a shows one example of microcrack formation in iron after tensile deformation at −140 °C. The appearance of clevage fracture surface in a low carbon steel is shown in Figure 5.16(b). Cleavage cracks can also be initiated at mechanical twins.

The ductile fracture starts with the initiation of voids, most commonly at second-phase particles. The particle geometry, size, and bonding play an important role. Dimpled rupture surface (ductile fracture) contains cup-like depressions (dimples) that may be equiaxial, parabolic, or elliptical, depending on the precise stress state. Microvoids are generally nucleated at second-phase particles, and the voids grow and eventually the ligaments between the microvoids fracture. The different stages of this process are shown in Figure 5.17(a) and an SEM fractograph with dimples shown in Figure 5.17(b).

Ductility (as expressed in true strain to fracture) of a material may depend on the volume fraction of second-phase particles present, as shown in Figure 5.18 with examples in steel.

5.1.4
Impact Properties

In circumstances where safety is extremely critical, the full-scale engineering components may be tested in their worst possible service condition. An example of a full-scale engineering test could be crash of a train carrying the spent fuel casks to see the effect of crash on the integrity of the casks. However, such full-scale tests are extremely expensive and very rarely conducted. Before the advent of fracture mechanics as a discipline, impact-testing techniques were used to determine the

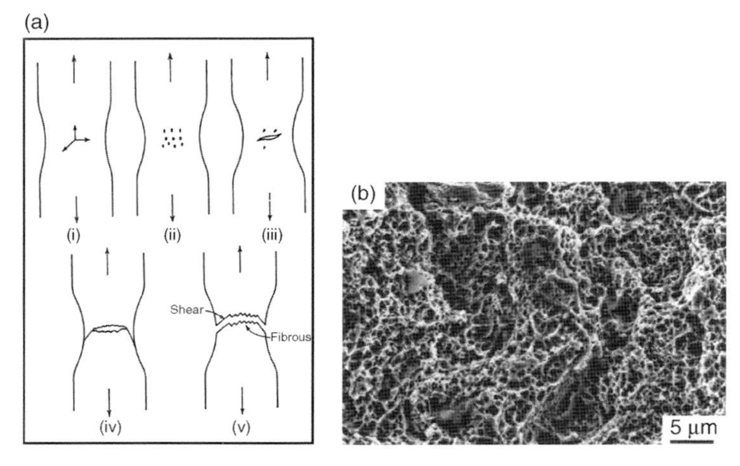

Figure 5.17 (a) Stages in the formation of cup-and-cone fracture. From Ref. [2]. (b) Dimpled rupture surface in a N-80 steel. From Ref. [7].

fracture characteristics of materials. Impact tests are designed to measure the resistance to failure of a material under a sudden applied force in such a way so as to represent most severe conditions relative to the potential of fracture – (i) deformation at low temperatures, (ii) high deformation rate, and (iii) a triaxial state of stress (by introducing a notch in the specimen). The impact tests measure the impact energy or the energy absorbed prior to failure. The most common methods of measuring impact energy are Charpy and Izod (Figure 5.19a). The techniques differ in the manner of specimen support and specimen design. Charpy V-notch test is most commonly used in the United States (Figure 5.19b). In this test, the load is applied as an impact blow from a weighted pendulum hammer that is

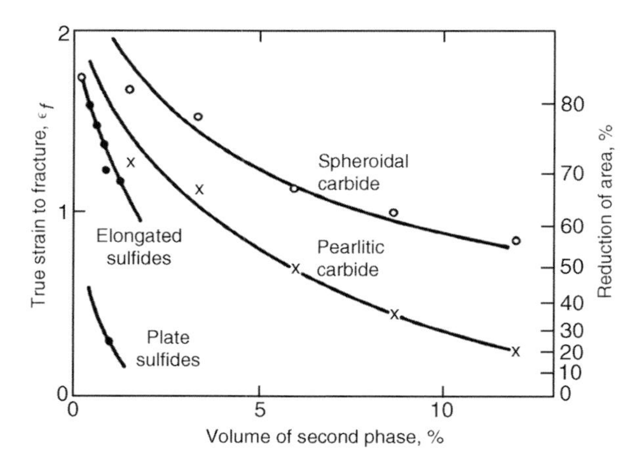

Figure 5.18 Effect of second-phase particles on tensile ductility. From Ref. [2].

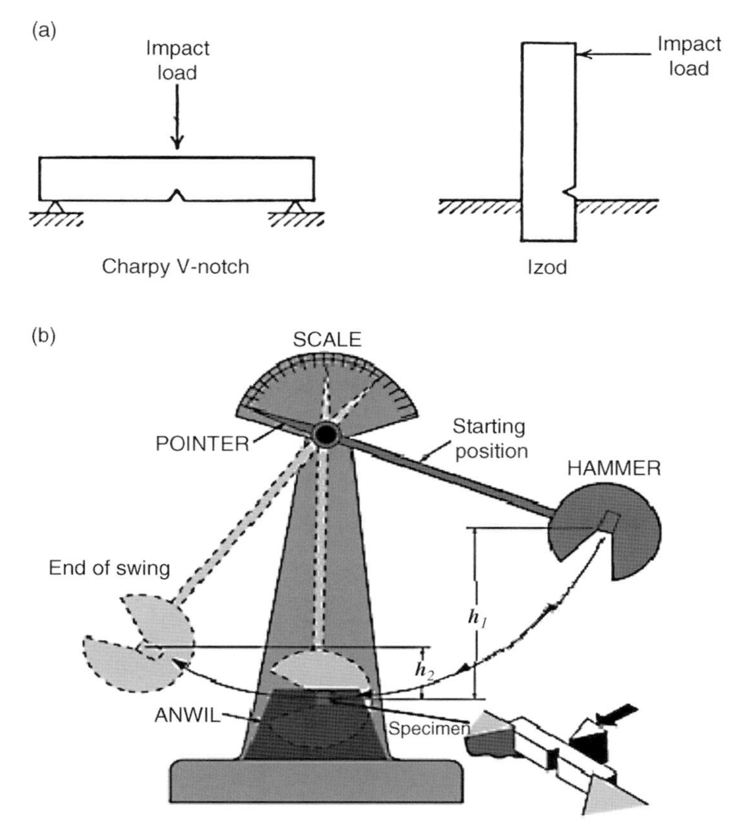

Figure 5.19 (a) The relative specimen configuration of Chapry and Izod tests. (b) A schematic of a Charpy tester.

released from a fixed position at a fixed height (h_1). Upon release, a knife-edge mounted on the pendulum strikes and fractures the specimen at the notch that acts as a stress raiser site for the high-velocity impact blow. After fracturing the specimen, the pendulum continues in its trajectory to reach a height h_2, depending on the absorbed energy during impact. The energy absorption is calculated from the difference between the pendulum's static energies at h_1 and h_2, that is, the impact energy is $Mg(h_1 - h_2)$, where M is the pendulum's mass and g is the acceleration due to gravity. In reality, the machine is equipped with a scale and a pointer that shows the impact energy after the test is done. Nowadays, more sophisticated, instrumented impact testers are used that can follow the load versus time or displacement during the impact event.

Variables including specimen size and shape as well as notch configuration and depth influence the test results. One of the main purposes of the Charpy test is to determine whether or not a material experiences a ductile–brittle transition with decreasing temperature. Figure 5.20 shows a description of ductile–brittle

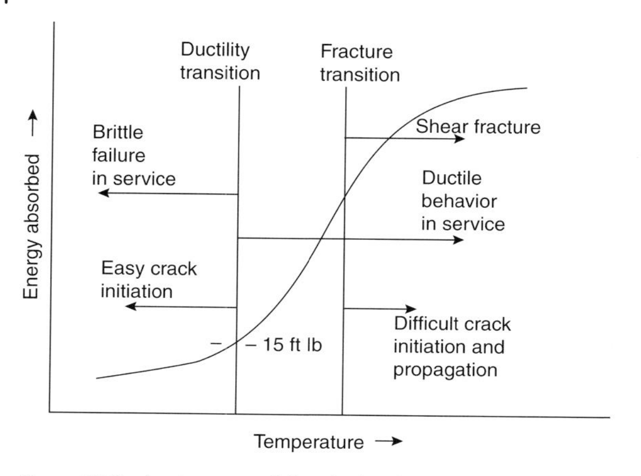

Figure 5.20 A schematic of ductile–brittle transition behavior.

transition behavior. Plane strain fracture toughness is quantitative in nature in that a specific property of the material is determined (i.e., K_{IC}). The results of impact tests are more qualitative in nature and are of little use for design purposes. Impact energies are of interest mainly in a relative sense and for making comparisons – absolute values are of little significance. Attempts have been made to correlate fracture toughness and Charpy impact energy, but with limited success.

5.1.4.1 **Ductile–Brittle Transition Behavior**

For many alloys, there is a range of temperatures over which the ductile–brittle transition occurs. This leads to the difficulty in specifying ductile–brittle transition temperature (DBTT).

a) DBTT is often described as the temperature at which CVN assumes certain value (40 J or 30 ft lb).
b) In some cases, DBTT corresponds to some given fracture appearance (50% fibrous appearance).

However, the most conservative estimate would be at which the fracture surface becomes fully 100% fibrous. Structures constructed from alloys that exhibit the ductile–brittle transition temperature should be used at temperatures above the transition temperature to avoid catastrophic failure. Materials with FCC crystal structures (such as aluminum- and copper-based alloys and austenitic stainless steels) do not exhibit DBTT. BCC and HCP alloys mainly experience this transition, which is mainly due to the highly temperature-sensitive yield stress at lower temperatures. For these materials, the transition temperature is sensitive to both the alloy composition and microstructure. It is to note that decreasing the average grain size of the material decreases DBTT. Figure 5.21 shows the effect of carbon content on the impact energy in plain carbon steels with various carbon concentrations. (0.01–0.67 wt%). Impact energy could also vary depending on the mechanical anisotropy.

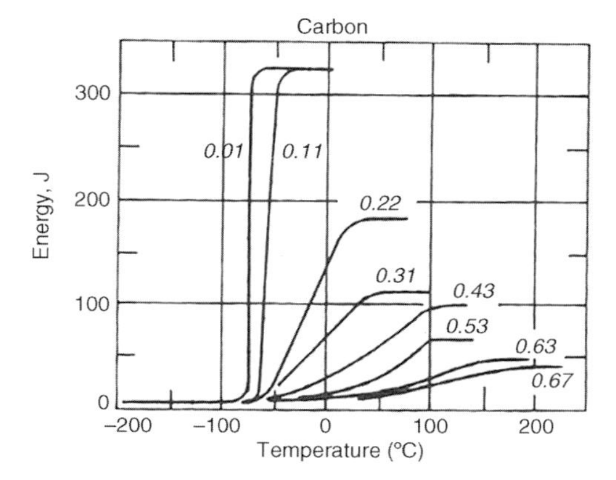

Figure 5.21 Effect of carbon content on the Charpy V-notch energy versus temperature behavior in plain carbon steels with various carbon contents.

5.1.5
Fracture Toughness

It is important to have fracture toughness parameter as an inherent mechanical property of materials just like yield stress. The mode-I fracture toughness (K_{Ic}) meets that definition with certain limitations. As the analysis involved still depends on the linear elastic fracture mechanics theories, the testing procedures followed are applicable to materials with limited ductility, such as high-strength steels, some titanium and aluminum alloys, and, of course, other brittle materials like ceramics.

The elastic stress field around the crack tip can be described by a single parameter known as "stress intensity factor (K)." This factor depends on many factors such as the geometry of the crack-containing solids, the size and location of the crack, and the magnitude and distribution of the loads applied. It can be reasonably assumed that an unstable rapid failure would occur if a critical value of K is reached. There are three modes of testing, as shown in Figure 5.22. In the opening mode (mode-I), the displacement is perpendicular to the crack faces. In mode-II (sliding mode), the displacement is made parallel to the crack faces, but perpendicular to the leading edge. In mode-III (tearing mode), the displacement is parallel to the crack faces and the leading edge.

In reality, opening mode (i.e., mode-I) is most important. That is why the conventional tests for fracture toughness are done under mode-I (i.e., opening mode of loading), and the critical value of K is called K_{Ic}, the plain strain fracture toughness. For a given type of loading and geometry, the relation is

$$K_{\mathrm{Ic}} = Y\sigma\sqrt{\pi a_{\mathrm{c}}}, \tag{5.32}$$

where Y is a parameter that depends on the specimen and crack geometry, a_{c} is the critical crack length, and σ is the applied stress. If K_{Ic} and applied stress are known, one can compute the maximum crack length tolerable. In other words, maximum

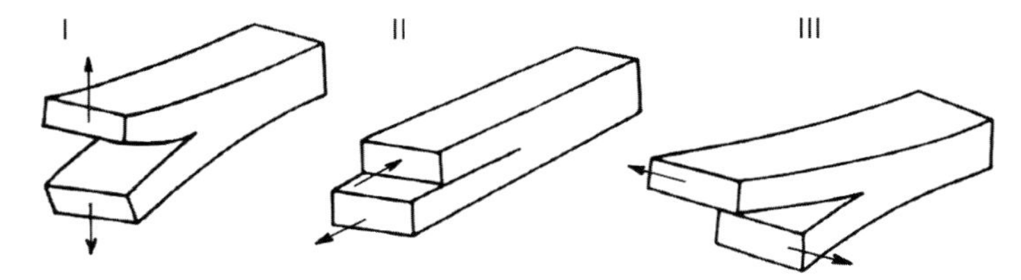

Figure 5.22 Three standard modes of loading: mode-I (opening mode), mode-II (sliding mode), and mode-III (tearing mode). From Ref. [8].

allowable stress can be computed for a given crack size provided one knows the K_{Ic} value of the material. K_{Ic} generally decreases with decreasing temperature and increasing strain rate, and vice versa. It also strongly depends on metallurgical variables such as heat treatment, crystallographic texture, impurities, inclusions, grain size, and so on.

A notch in a thick plate is far more damaging than that in a thin plate because it leads to a plane strain state of stress with a high degree of triaxiality. The fracture toughness measured under plane strain conditions is obtained under maximum constraint or material brittleness. The plane strain fracture toughness is thus designated as K_{Ic} and is a true material property. A mixed mode (ductile–brittle) fracture occurs with thin specimens. Once the specimen has the critical thickness, the fracture surface becomes flat and the fracture toughness reaches a constant minimum value with increasing specimen thickness (Figure 5.23). The minimum thickness (B for breadth) to achieve plane strain condition is given by

$$B \geq 2.5(K_{\mathrm{Ic}}/\sigma_0)^2, \tag{5.33}$$

where σ_0 is 0.2% yield stress.

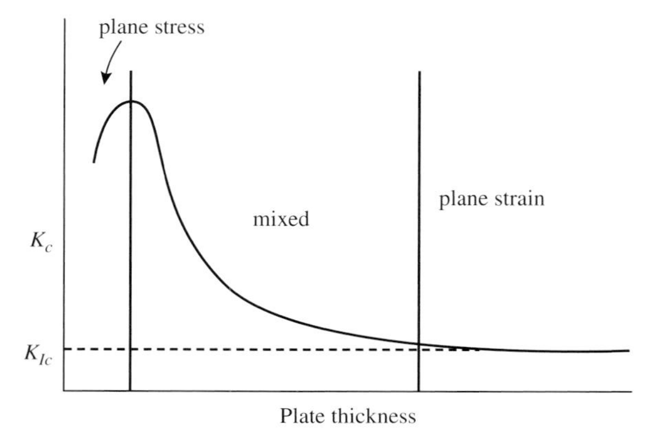

Figure 5.23 The critical stress intensity factor as a function of plate thickness. The thickness must be sufficiently large for achieving plane strain condition.

5.1.5.1 Test Procedure

The following discussion on the fracture test procedure is partly adapted from the book by Dieter [2]. Different types of specimens are used for measuring K_{Ic} (Figure 5.24). The compact test specimen (CT) is quite popular. After the notch is machined into the specimen, the sharpest possible crack is produced at the notch root by fatiguing specimen in low cycle, high-strain mode (typically 1000 cycles). Plane strain fracture toughness is quite unusual in that there is no advance assurance that a valid K_{Ic} can be measured in a particular test. Equation (5.33) is used with an estimate of the expected K_{Ic} to determine the specimen thickness required for plane strain loading condition.

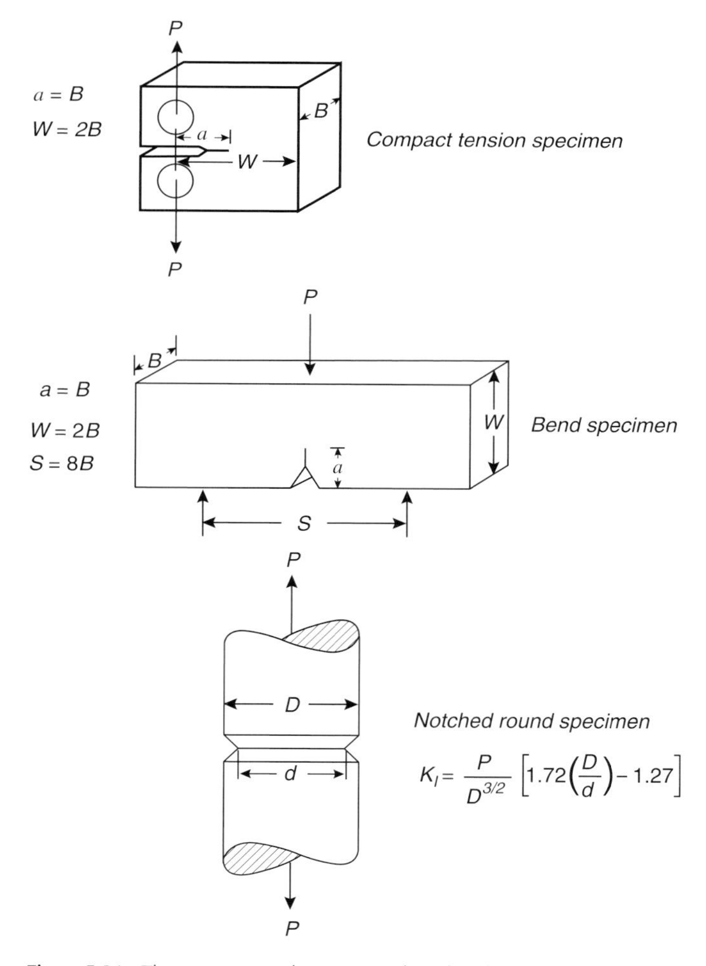

Figure 5.24 Three specimen designs (CT, three-bend, and notched round specimen) for K_{Ic} measurement. From Ref. [2].

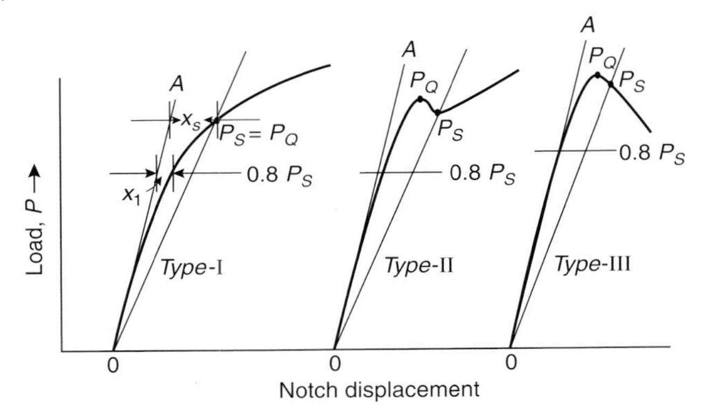

Figure 5.25 Load–displacement curves (type-II and type-III show "pop-in" behavior). From Ref. [2].

The test should be carried out in a testing machine that provides for a continuous autographic record of load (P) and relative displacement across the open end of the notch (proportional to crack displacement). The three types (type-1, type-II, and type-III) of load–crack displacement curves that can be obtained depending on the type of material are shown in Figure 5.25. The ASTM procedure requires to first draw a secant line OP_s from the origin with a slope that is 5% less than tangent OA. This determines the P_s value. Now draw a horizontal line at a load equal to 80% of P_s and measure the distance along this line from the tangent OA to the actual curve. If the value of x_1 exceeds one-fourth of the corresponding distance x_s at P_s, the material is too ductile to obtain a valid K_{Ic}. If the material is not too ductile, the load P_s is then designated as P_Q and used in the calculation.

The value of P_Q determined from the load–displacement curve is used to calculate a conditional value of fracture toughness denoted by K_Q using the following equation (for CT specimen):

$$K_Q = \frac{P_Q}{BW^{1/2}}\left[29.6\left(\frac{a}{W}\right)^{1/2} - 185.5\left(\frac{a}{W}\right)^{3/2} + 655.7\left(\frac{a}{W}\right)^{5/2} - 1017.0\left(\frac{a}{W}\right)^{7/2} + 638.99\left(\frac{a}{W}\right)^{9/2}\right]. \tag{5.34}$$

The crack length (a) used in this equation is measured after fracture. Then, calculate the factor $2.5(K_Q/\sigma_0)^2$. If this quantity is *less* than both the thickness and the crack length of the specimen, then K_Q is actually K_{IC}. Otherwise, it is necessary to use a thicker specimen to determine K_{Ic}. The measured value of K_Q can be used to estimate the new specimen thickness through Eq. (5.33). Table 5.2 shows fracture toughness and tensile strength values of three engineering alloys. It demonstrates that maraging steel that has the highest plain strain fracture toughness would be able to tolerate crack for a given applied stress or with an existing crack of known length and will withstand the highest stress.

Table 5.2 Fracture toughness and tensile strength of two steels and an aluminum alloy.

Property	Ni–Cr–Mo steel	Maraging steel	7075 Al alloy
K_{Ic} (MPa m$^{1/2}$)	46	90	32
Tensile strength (MPa)	1820	1850	560

Example Problem

A tensile stress of 300 MPa acts on a sheet sample of an alloy (FCC: $E = 200$ GPa, $\sigma_y = 350$ MPa, $K_{IC} = 40$ MPa m$^{1/2}$) with a central crack of length 5 mm. Is the sample going to fail in a brittle manner?

Solution

We find K_I for the case and compare it with K_{IC}.

$K_I = \sigma\sqrt{\pi a} = 300\sqrt{\pi \times 2.5 \times 10^{-3}} = 26.59$ MPa m$^{1/2}$ taking $Y = 1$ and this is less than $K_{IC} = 40$ MPa m$^{1/2}$, meaning the sample *does not fail* in a brittle manner.

5.1.6 Creep Properties

At temperatures above 0.4–0.5T_m (T_m is the melting point of the material), plastic deformation occurs as a function of time at constant load or stress. The phenomenon is known as creep "defined as time dependent plastic strain at constant temperature and stress." Note that lead (Pb) may creep at room temperature, but iron (Fe) does not because for lead room temperature represents a higher homologous temperature (i.e., $T/T_m = \sim 0.5$) than that of iron ($\sim$0.16). Because creep occurs as a function of temperature, it is a thermally activated process. Creep must be taken into account when a load-bearing structure is exposed to elevated temperatures for a long duration of time.

A creep test is generally done under uniaxial tensile stress. There are different variants of creep tests, such as compressive creep, double shear creep, impression creep, and indentation creep. However, here we will deal with tensile creep only. A creep curve is basically plotted as a function of creep strain as a function of time at a constant load (or constant stress depending on the availability of such instrument set up to keep stress constant during creep deformation) and temperature. Figure 5.26a shows a creep equipment with a furnace and strain measuring device, a linear variable differential transformer (LVDT). A typical creep curve as shown in Figure 5.26b has three stages: (i) primary stage (transient creep) – work hardening during plastic deformation is more than recovery (softening) exhibiting decreasing strain rate with time; (ii) secondary or steady-state stage (minimum creep rate) – the rate of work hardening and softening balance each other, and (iii) tertiary stage – characterized by an accelerating creep rate where softening mechanisms predominate. The third stage of tertiary creep is often considered as fracture stage rather

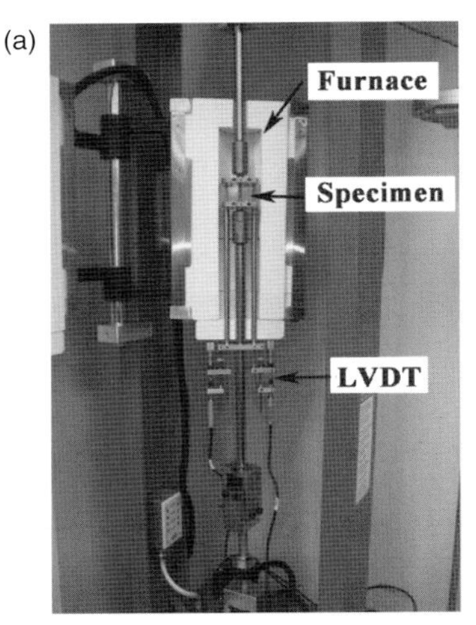

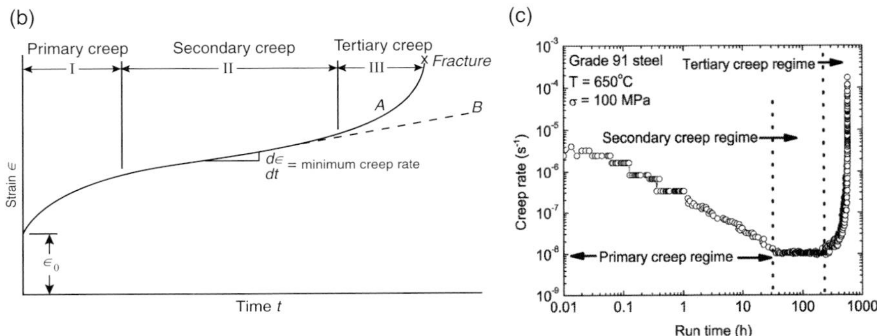

Figure 5.26 (a) A view of a creep tester with specimen loading. (b) A typical creep curve showing different stages of creep. (c) Creep rate versus strain plot for a modified 9Cr–1Mo steel (Grade 91 steel) tested at a temperature of 650 °C and 100 MPa.

than deformation. A real example of creep rate versus time plot for a Grade 91 steel is shown in Figure 5.26c where the steady-state creep rate occurs as a minimum rate that is typical for creep under constant load. Increase in temperature and/or stress tends to enhance creep strains and rates, as illustrated in Figure 5.27 where we note that increased creep rates are accompanied by decreased time to rupture (t_r).

5.1.6.1 Creep Constitutive Equation

Creep constitutive behavior is generally described using the minimum creep rate or in most cases the steady-state creep rate. Norton's law is the equation that is used for describing the dependence of stress on strain rate at a given temperature:

$$\dot{\varepsilon} = A_1 \sigma^n, \tag{5.35}$$

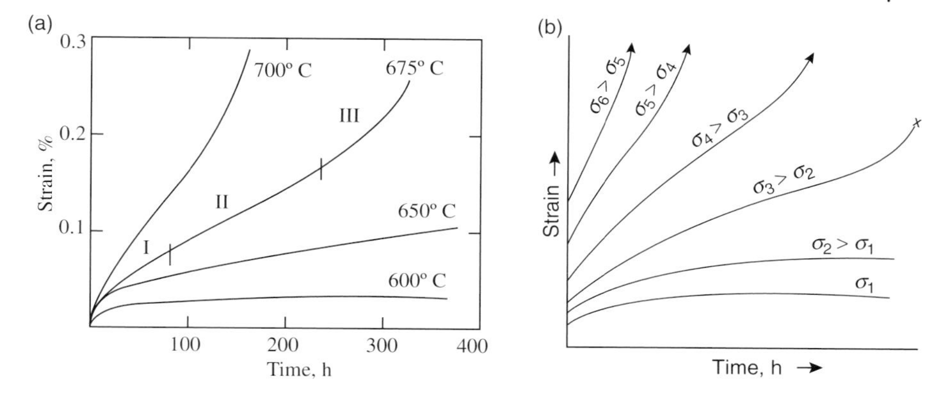

Figure 5.27 The effect of (a) temperature and (b) stress on creep curves.

where A_1 is a constant dependent on the test temperature and n is the stress exponent (basically it is the reciprocal of strain rate sensitivity m, as given in Eq. (5.23)). It is important not to confuse between strain hardening exponent and stress exponent, which have the same symbol, n. The increase of the steady-state creep rate with test temperature (Figure 5.27a) under a given stress follows an Arrhenius equation with a characteristic activation energy for creep (Q_c):

$$\dot{\varepsilon} = A_2\,\mathrm{e}^{-Q_c/RT} \tag{5.36}$$

where A_2 is a constant dependent on the applied stress/load. For creep at high temperatures ($T > 0.4T_m$), the activation energy for creep (Q_c) was shown to be equal to that for self-diffusion (Q_D). Thus, the temperature and stress variations of creep rate can be combined into one equation:

$$\dot{\varepsilon} = A_3\,\mathrm{e}^{-Q_c/RT}\sigma^n, \tag{5.37}$$

where A_3 is a material constant and R is the gas constant ($1.987\,\mathrm{cal\,mol^{-1}\,K^{-1}}$). It has been shown that factors affecting self-diffusivity also influenced the creep rates similarly; for example, pressure dependence, effect of C on self-diffusion and creep of γ-iron, influence of ferromagnetism and crystal structure (diffusion and creep of α-iron versus γ-iron), influence of crystal structure on creep and diffusion in thallium, effect of composition (stoichiometry) on diffusion and creep, and so on. Thus, one can state that the steady-state creep rate is proportional to D_L (lattice or self-diffusion) so that the creep rate equation becomes

$$\dot{\varepsilon} = A_4 D\sigma^n = A_5 D\left(\frac{\sigma}{E}\right)^n. \tag{5.38}$$

Here, A_4 and A_5 are constants dependent on the material and microstructure such as the grain size. Thus, the steady-state creep rate results plotted as $\dot{\varepsilon}/D$ versus σ/E or σ/G in double-logarithmic scale yield a straight line with a slope of n (stress exponent); this is known as Sherby plot and takes care of the temperature variation of the elastic modulus ($E = E_0 - E'(T - T_0)$). At very high stresses ($>10^{-3}E$), however, power-law breakdown is noted with creep rates increasing

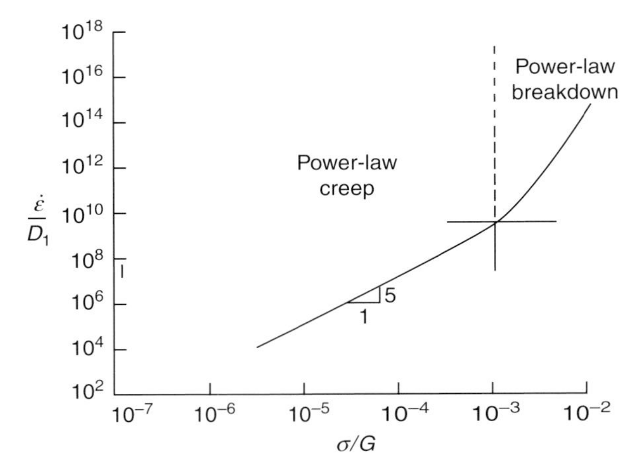

Figure 5.28 Double-logarithmic plot in terms of Sherby parameters exhibiting power-law breakdown at high stresses.

more rapidly (Figure 5.28). This creep regime is due to dislocation glide-climb with climb of edge dislocations being the rate controlling process as exhibited by pure metals (as well as ceramics) and some alloys.

However, these relationships do not show the effect of other parameters in a useful way. In fact, the strain rate is a function of stress, temperature, and microstructural factors. The Bird–Mukherjee–Dorn (BMD) equation is commonly used to express the constitutive behavior during creep deformation:

$$\dot{\varepsilon} = \frac{ADEb}{kT}\left(\frac{\sigma}{E}\right)^n \left(\frac{b}{d}\right)^p, \tag{5.39a}$$

$$\frac{\dot{\varepsilon}kT}{DEb} = A\left(\frac{\sigma}{E}\right)^n \left(\frac{b}{d}\right)^p, \tag{5.39b}$$

where A is a material constant, d is the grain size, and p is known as the grain size exponent. It should be noted that in the above Dorn equation, the strain rate, stress, and grain size are all normalized to dimensionless parameters. The values for A, n, and p along with appropriate diffusivity $\{D = D_0\exp(-Q/RT)\}$ in Eq. (5.39) characterize the underlying creep mechanisms. The underlying creep mechanisms have been dealt with in detail in Section 5.1.6.4.

Example Problem

In a laboratory creep experiment at 1200 °C, a steady-state creep rate of 0.2% h^{-1} is obtained and the specimen failed at 1500 h. The creep mechanism for this alloy is known to be dislocation climb with an activation energy of 25 kcal mol^{-1}. If the service temperature of the alloy is 600 °C, evaluate the creep rate at the service conditions.

Solution

We assume that the stresses are the same in the laboratory tests and in service and use Eq. (5.36).

Given $T_1 = 1473$ K and $\dot{\varepsilon}_1 = 0.2\%\,\text{h}^{-1}$,

$$\dot{\varepsilon}_2 = \dot{\varepsilon}_1 e^{(Q_C/R)((1/T_1)-(1/T_2))} = 0.2 * \exp\left\{\frac{25\,000}{1.987}\left(\frac{1}{1473} - \frac{1}{873}\right)\right\}$$
$$= 0.3 * e^{-5.871} = 5.643 \times 10^{-4}\%\,\text{h}^{-1}.$$

5.1.6.2 **Creep Curve**

The total creep curve from the initial loading time ($t = 0$) to the beginning of the tertiary creep (fracture mode) can be divided into three parts: elastic (ε_0), primary creep strain (ε_p), and steady-state creep strain (ε_s):

$$\varepsilon = \varepsilon_0 + \varepsilon_p + \varepsilon_s. \tag{5.40}$$

In this equation, the elastic strain is given by σ/E, while the steady-state or secondary stage strain is ($\varepsilon_s = \dot{\varepsilon}t$). The transient or primary creep strain follows a trend as shown in the middle part of Figure 5.29. Long time ago, Andrade considered this to follow the following relation known as Andrade β-flow:

$$\varepsilon_p = \beta t^{1/3}. \tag{5.41}$$

While the Andrade $t^{1/3\text{rd}}$ law seems to explain the saturation of the primary creep strain as per the middle part on the right-hand side of Figure 5.29, it predicts an infinite creep rate at loading ($t = 0$). Although the strain rate at loading is relatively high, it cannot be infinite and thus Dorn and Garofalo suggested the following relationship:

$$\varepsilon_p = \varepsilon_t(1 - e^{-rt}). \tag{5.42a}$$

The rate (r) at which the strain reaches the saturation value (ε_t) is related to the ratio of the initial creep rate at loading ($\dot{\varepsilon}_i$) to the steady-state creep rate, which is a function of the applied stress and temperature (Eq. (5.39)). Thus, Eq. (5.42a) may be rewritten in terms of steady-state creep rate:

$$\varepsilon_p = \varepsilon_t(1 - e^{-\beta\dot{\varepsilon}_s t}). \tag{5.42b}$$

If we then plot creep strain as a function of modified time using the steady-state creep rate, we should obtain a single curve for all test temperatures and stresses, as

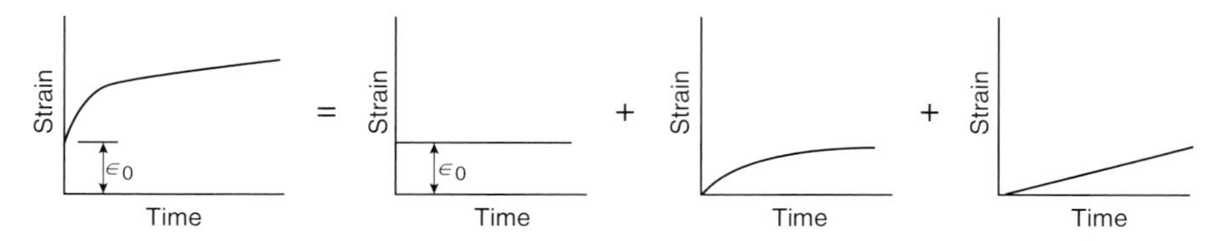

Figure 5.29 Division of the total creep curve into elastic, primary, and secondary creep strains.

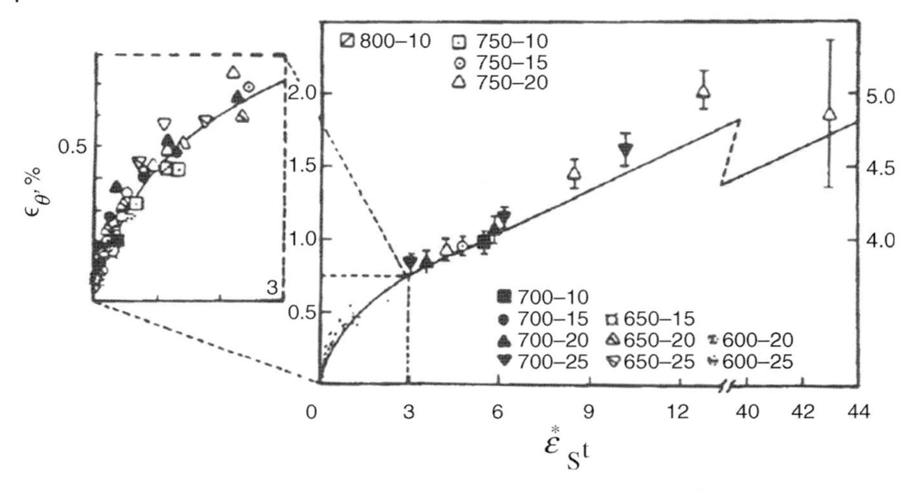

Figure 5.30 Creep strain data for zircaloy at various temperatures (in °F) and stresses (ksi) plotted versus modified time (See the conversion table in the appendix for conversion of temperature and stress units).

shown in Figure 5.30 for Zr-based alloy. Noteworthy feature of this representation lies in the fact that at any given temperature and stress, one can predict the total accumulated creep strain at given time using Eqs (5.39) and (5.42).

5.1.6.3 **Stress and Creep Rupture**

Although it is useful to characterize the creep curve, it is often required to find the rupture time (t_r); the higher the rupture time, the better the life of structures in-service. Stress rupture tests are similar to creep tests; however, the creep strains are not monitored during the test and tests are carried out to fracture and times to fracture are noted at varied temperatures and stresses (loads). The stress–rupture data are plotted as stress against rupture time (Figure 5.31). The basic information obtained in the stress–rupture test can be used for the design of high-temperature components such as jet turbine components and steam turbine blades where the structure cannot undergo more than an allowable amount of creep deformation during the whole service life.

There are empirical relations and functions that can describe stress–rupture behavior of a material and at the same time provide useful design data. Larson–Miller parameter (LMP) is one of them:

$$\mathrm{LMP} = T \log (C + t_r), \tag{5.43}$$

where C is a material constant, which is around 20 when T is in K, and t_r is in hours. Figure 5.32 shows a plot depicting variation of stress against LMP for nuclear grade zirconium alloys.

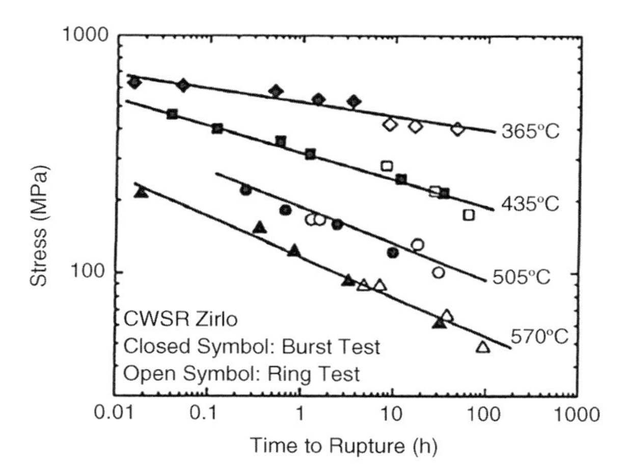

Figure 5.31 Stress versus rupture time for a niobium-bearing zirconium alloy (Zirlo) at different temperatures.

Another important empirical relation is the Monkman–Grant relation that follows from the fact that the time to rupture decreases with stress or minimum creep rate:

$$\dot{\varepsilon}_s \cdot t_r = \text{constant}, \tag{5.44}$$

where $\dot{\varepsilon}_s$ is the steady-state creep rate. Figure 5.33 shows the data for a commercial pure titanium tubing clearly depicting the validity of Monkman–Grant relationship. This has a significant role in predicting the rupture times for service stresses where relatively low stresses and temperatures are appropriate from the short-term

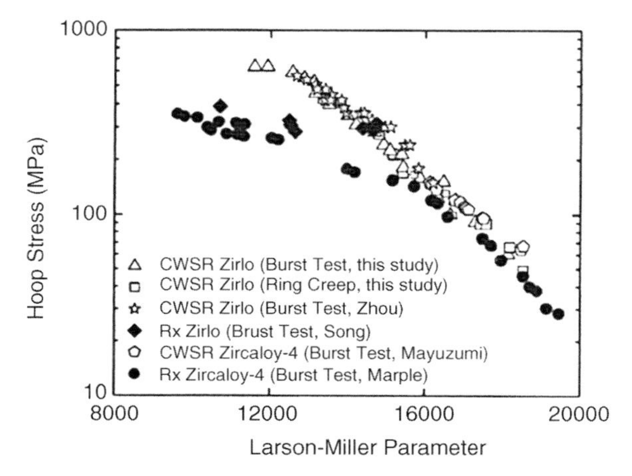

Figure 5.32 Stress versus LMP for two zirconium alloys (Zirlo and Zircaloy-4) under two conditions.

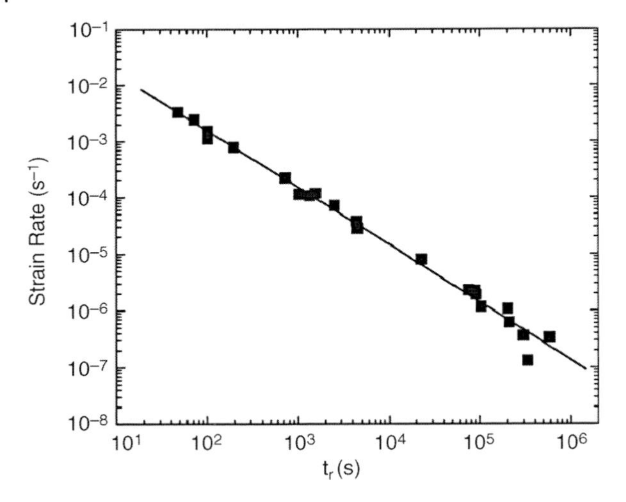

Figure 5.33 Steady-state strain rate versus rupture time for cp-Ti tubing.

laboratory tests performed at high stresses/temperatures. Once the stress and temperature variations of steady-state creep rates are characterized for a given material and the various constants in Eqs (5.36) and (5.37) are determined from laboratory tests, one can first predict the secondary creep rate at the service temperature and stress following which the application of Monkman–Grant relationship can be used to predict the rupture life in-service.

Another useful parameter is known as the Zener–Holloman parameter that is applicable for a given applied stress:

$$Z = \dot{\varepsilon} e^{Q/RT}. \tag{5.45}$$

Figure 5.34 depicts such a correlation for α-iron that shows a single curve describing the primary, secondary (minimum creep rate), and tertiary strains at varied temperatures.

Similarly, Sherby–Dorn parameter involves normalization with temperature-compensated time:

$$P_{SD} = te^{-Q/RT}. \tag{5.46}$$

The creep curves at three different temperatures at a constant stress merge into a single curve when plotted strain versus compensated time, as depicted in Figure 5.35 for aluminum at 3 ksi or ~20.7 MPa.

Example Problem

In a laboratory creep experiment at 1200 °C, a steady-state creep rate of 0.2% h^{-1} is obtained and the specimen failed at 1500 h. The creep mechanism for this alloy is known to be dislocation climb with an activation energy of 25 kcal mol^{-1}. If the

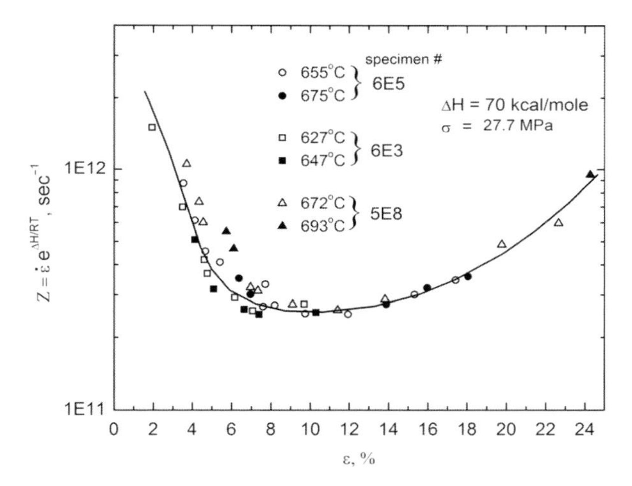

Figure 5.34 Zener–Holloman parameter versus time at varied temperatures at a constant stress in alpha-iron.

service temperature of the alloy is 600 °C, evaluate the service life using Larson–Miller parameter.

Solution

Recalling Eq. (5.43), LMP = 1473 (log 1500 + 20) = 873 (log t_r + 20) so that $t_r = 3.414 \times 10^4$ h.

Note that this can also be found using Sherby–Dorn parameter (Eq. (5.46)) or Monkman–Grant relationship (Eq. (5.44)).

5.1.6.4 **Creep Mechanisms**

As described earlier, majority of pure metals and alloys of class-M (also known as class-II alloys) exhibit power-law creep with stress exponent of around 5 with

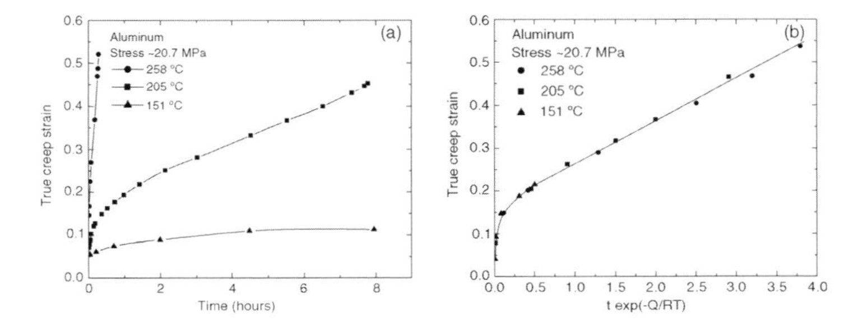

Figure 5.35 Creep data in Al at three different temperatures (a) at ~20.7 MPa stress coalesce into a single curve when plotted versus Sherby–Dorn parameter (b). (t in hours and T in K)

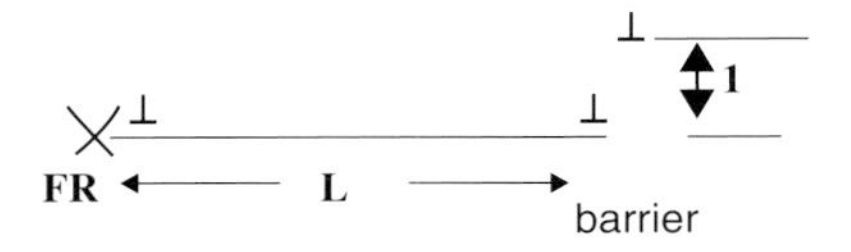

Figure 5.36 Schematic of dislocation glide–climb event.

power-law breakdown at high stresses (Figure 5.28), which is believed to arise from climb of edge dislocations commonly referred to as Weertman dislocation climb model. To describe creep by dislocation climb, Weertman considered in his model the creep processes to be a result of the glide and climb of dislocations; climb being the rate-controlling process. The glide motion of dislocations is impeded by long-range stresses due to dislocation interactions and the stresses are relieved by dislocation climb and subsequent annihilation. The rate of dislocation climb is determined by concentration gradient existing between the equilibrium vacancy concentration and the concentration in the region surrounding the climbing dislocation. Creep strain, however, arises mainly by the glide of dislocations. In the glide–climb model (Figure 5.36), dislocations produced by the Frank–Read source (FR) glide a distance L till it encounters a barrier of height h at which it has to climb so that another dislocation can be generated by the source.

A rather simple way of deriving the relation between the strain rate and the applied stress and temperature is given by the following by referring to Figure 5.36:

$$\Delta\gamma = \text{strain during glide + climb event} = \Delta\gamma_g + \Delta\gamma_c \approx \Delta\gamma_g = \varrho bL, \quad (5.47a)$$

$t =$ time of glide–climb event $= t_g + t_c \approx t_c = h/v_c$, $v_c =$ climb velocity,

$$\dot{\gamma} = \frac{\Delta\gamma}{t} = \frac{\varrho bL}{h/v_c} = \varrho b\frac{L}{h}v_c, \quad (5.47b)$$

where $v_c \propto \Delta C_v e^{-E_m/kT}$, $E_m =$ activation energy for vacancy migration.

$$\dot{\gamma} = \varrho b\frac{L}{h}v_c, \quad \text{where} \quad \Delta C_v = C_v^+ - C_v^- = C_v^0 e^{\sigma V/kT} - C_v^0 e^{-\sigma V/kT}$$
$$= C_v^0 2\,\text{Sinh}\left(\frac{\sigma V}{kT}\right),$$
$$\text{so, } \dot{\varepsilon} = \alpha\varrho b\frac{L}{h}v_c = \alpha\varrho b\frac{L}{h}C_v^0 e^{-Em/kT} 2\,\text{Sinh}\left(\frac{\sigma V}{kT}\right). \quad (5.47c)$$

At low stresses, $\text{Sinh}(\sigma V/kT) \approx \sigma V/kT$, so that $\dot{\varepsilon} = A_1\varrho b(L/h)C_v^0 e^{-Em/kT}(\sigma V/kT)$.

$$\dot{\varepsilon} = A_1\varrho b\frac{L}{h}D_L\frac{\sigma V}{kT} \approx A_2\varrho\sigma\frac{L}{h}D_L. \quad (5.47d)$$

Assuming that the dislocation density (ϱ) varies as the stress raised to power 2 (σ), we find that

$$\dot{\varepsilon} = A\sigma^3 D_L. \quad (5.48)$$

This is known as "natural creep law" and Weertman showed that L/h in Eq. (5.29) varied as $\sigma^{1.5}$, so

$$\dot{\varepsilon} = AD_{\mathrm{L}}\sigma^{4.5}. \tag{5.49}$$

This equation with $n = 4.5$ and $D = D_{\mathrm{L}}$ agreed closely with the experimental results on pure aluminum. Subsequently, this has been generalized with n close to 5 and is referred to as the power-law creep.

At high stresses, Eq. (5.49) predicts an exponential stress dependence,

$$\mathrm{Sinh}\left(\frac{\sigma V}{kT}\right) \approx \exp\left(\frac{\sigma V}{kT}\right), \quad \text{so} \quad \dot{\varepsilon} = A_1 \varrho b \frac{L}{h} D_{\mathrm{L}}\, \mathrm{e}^{\sigma V/kT}.$$

Thus,

$$\dot{\varepsilon} = A\sigma^2 D_{\mathrm{L}}\, \mathrm{e}^{\sigma V/kT} \sim A' D_{\mathrm{L}}\, \mathrm{e}^{\sigma V/kT}, \tag{5.50}$$

as is commonly noted in the power-law breakdown regime. Both the power-law and exponential stress regimes can be combined into a single equation as proposed by Garofalo:

$$\dot{\varepsilon} = AD_{\mathrm{L}}(\mathrm{Sinh}\, B\sigma)^n, \tag{5.51}$$

which describes both the power-law creep regime at low stresses and exponential stress dependence at high stresses ($>10^{-3}\,E$).

In the latter modification, Weertman considered dislocation loops generated from Frank–Read sources on parallel slip planes (Figure 5.37) wherein the leading edge components of the dislocation loops climb to get annihilated. This is commonly referred to as Weertman pill-box model and does not require physical obstacles such as Lomer–Cottrell barriers encountered in FCC metals.

In many alloys, the substitutional impurity (alloying) elements migrate to and lock the dislocations at high temperatures leading to slowing down the glide of dislocations. In these alloys of class-A (also known as class-I), glide and climb of dislocations take place as in pure metals and class-M alloys, but since the glide becomes slower controlled by the diffusivity of solute atoms compared to the climb, creep will be controlled by the dislocation glide. In this case the creep rate becomes

$$\dot{\varepsilon} = A\sigma^3 D_{\mathrm{s}}, \tag{5.52}$$

where D_{s} is the solute atom diffusivity ($= D_{\mathrm{s}}^0\, \mathrm{e}^{-Q_{\mathrm{s}}/RT}$, Q_{s} is the activation energy for solute atom diffusion). This is also known as microcreep and class-I alloys exhibit stress exponents of around 3 versus 5 for pure metals and class-M alloys. Moreover, creep characteristics under these circumstances are different from those observed in pure metals in that relatively small or little primary creep is noted with randomly distributed dislocations with no distinctly formed subgrain boundaries. Solute strengthening in this regime is observed with lower creep rates compared to pure metals; for example, pure Al exhibits climb creep with $n \sim 5$, while addition of Mg results in more creep resistance with $n \sim 3$. It is to be noted that the glide and climb

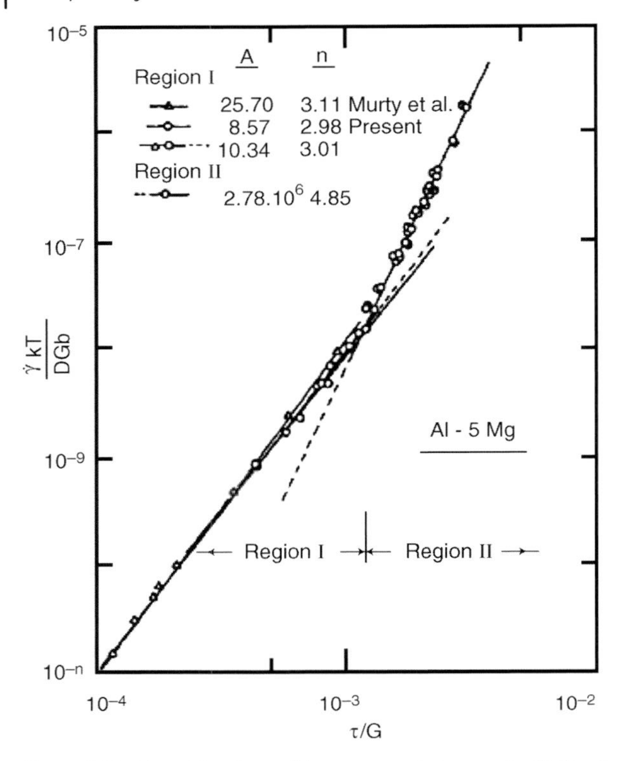

Figure 5.37 Transition from glide creep to climb creep in Al5Mg. From Ref. [9].

processes occur in *series* and in these sequential processes, slower one controls creep. As can be expected, if the applied stress is high enough to make the dislocations free from solute, atom locking the alloy will exhibit climb controlled creep as in pure metals and class-M alloys; Figure 5.37 is an example showing the transition from microcreep to climb creep at high stresses in Al5Mg solid solution.

At relatively high temperatures and/or low stresses, creep can take place by point defects (vacancies) diffusing from tensile boundaries to compressive boundaries for small grain size or thin samples. The possibility of creep occurring by stress-assisted diffusional mass transport through the lattice was first considered by Nabarro in 1948 and Herring in 1950. A few years later, Coble proposed that grain boundaries could also provide an alternative path for stress-directed diffusional mass transport to take place. Figure 5.38 provides schematics of N–H and Coble creep mechanisms. As the figure indicates, under the application of a stress, grain boundaries normal to the applied stress will develop a higher concentration of vacancies. On the other hand, grain boundaries parallel (lateral grain boundaries) to the applied stress will experience compressive stresses and will have reduced vacancy concentration. This causes a concentration gradient between the two boundaries leading to a flux of vacancies diffusing from the normal grain boundaries to the parallel or lateral grain boundaries (atoms diffuse in the opposite

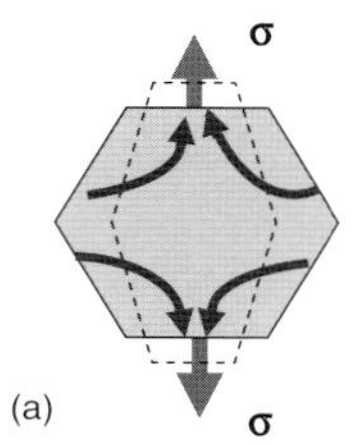

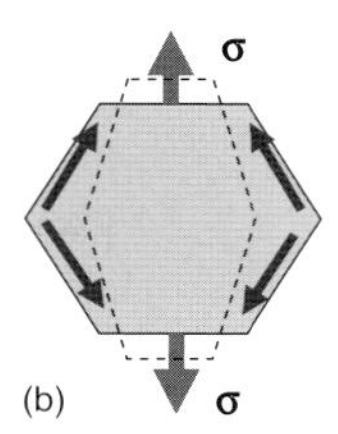

Figure 5.38 Mass transport through the lattice as in N–H model (a) and through grain boundaries (Coble creep).

direction). The diffusion of vacancies can occur through the lattice (N–H) or via grain boundaries (Coble creep) depending on the grain size and temperature. The diffusion of vacancies or the motion of atoms from one grain boundary to another leads to a crystal strain, which in turn contributes to the deformation of the grains and consequently the material.

The calculations of the steady-state flux of vacancies and the corresponding steady-state creep rate lead to the following relations for N–H and Coble creep, respectively:

$$\dot{\varepsilon} = B_{\mathrm{H}} \frac{D_{\mathrm{L}} \sigma \Omega}{d^2 kT}. \tag{5.53}$$

$$\dot{\varepsilon} = \frac{B_{\mathrm{C}}}{\pi} \frac{D_{\mathrm{GB}} \delta_{\mathrm{B}} \sigma \Omega}{d^3 kT}. \tag{5.54}$$

In Eq. (5.53), B_{H} is the N–H constant and has a value of about 12–15, D_{L} is the lattice diffusivity, Ω is the atomic volume, and k is the Boltzmann's constant. In Eq. (5.54), B_{C} is the Coble constant and has a value of 150, D_{GB} is the grain boundary diffusivity, and δ_{B} is the grain boundary thickness. From the above relations, it is understood that the creep strain rate varies linearly with stress and is inversely proportional to the grain size. Usually with decreasing grain size, it is observed that Coble creep dominates N–H creep and vice versa. Since grain boundary diffusion dominates at lower temperatures ($Q_{\mathrm{GB}} \sim 0.6 Q_{\mathrm{L}}$), Coble creep makes major contribution to the total creep strain. However, these two mechanisms operate in parallel so that the strain rate can be expressed by

$$\dot{\varepsilon} = \frac{B \sigma \Omega}{d^2 kT} D_{\mathrm{eff}}, \tag{5.55}$$

where D_{eff} is the effective diffusion coefficient given by

$$D_{\mathrm{eff}} = D_{\mathrm{L}} \left(1 + \frac{\pi}{d} \frac{D_{\mathrm{B}} \delta_{\mathrm{B}}}{D_{\mathrm{L}}} \right). \tag{5.56}$$

In the derivation of the N–H and Coble creep equations, the following assumptions were made:

i) The grain boundaries are perfect sources and sinks of vacancies.
ii) The initial dislocation density of the crystal is low.

This implies that only the sources and sinks for vacancies are the grain boundaries. Since their discovery, both N–H and Coble creep mechanisms have been found to take place in many different materials and experimental results agreed with the model predictions extremely well. These creep mechanisms are known as *viscous* creep mechanisms where strain rate is proportional to the applied stress and attain importance for materials with relatively small grain sizes. Large-grained polycrystalline and bulk single-crystalline materials do reveal viscous creep, but are insensitive to the grain size in contrast to N–H and Coble creep mechanisms and is known as Harper–Dorn (H–D) creep. H–D creep follows the following creep rate equation:

$$\dot{\varepsilon}_{\mathrm{HD}} = B_{\mathrm{HD}} \frac{D_{\mathrm{L}} b \sigma}{kT}. \tag{5.57}$$

H–D creep is shown to take place at stresses below the critical stress needed for dislocation multiplication (i.e., the stress needed for Frank–Read source operation). Thus, in H–D creep regime, the dislocation density is unchanged from the initial dislocation density.

At intermediate temperatures and relatively small and stable (during deformation) grain sizes, grain boundary sliding (GBS) dominates such as commonly observed in superplastic materials. A commonly agreed upon GBS model results in the following expression for the creep rate:

$$\dot{\varepsilon} = B_{\mathrm{GBS}} \frac{D_{\mathrm{GB}} b \sigma^2}{kTEd^2}. \tag{5.58}$$

We note that all these above equations follow the Dorn equation (5.39b) with different Q, n, and p parameters. Table 5.3 compiles these various creep mechanisms with corresponding n, p, and Q parameters.

We included in the table another creep mechanism known as low-temperature climb that becomes important when diffusion through dislocation pipes becomes dominant in which case the climb creep rate is proportional to dislocation pipe diffusivity that increases with the dislocation density. Since dislocation density

Table 5.3 A summary of creep mechanisms with relevant parametric dependencies.

Mechanism	Q[a)]	n	p
Climb of edge dislocations	Q_{L}	5	0
Viscous glide (solute drag creep)	Q_{S}	3	0
Grain boundary sliding	Q_{GB}	2	2
Low-temperature climb	Q_{C}	7	0
Harper–Dorn	Q_{L}	1	0
Nabarro–Herring	Q_{L}	1	2
Coble	Q_{GB}	1	3

a) Q_{L}: activation energy for lattice diffusion, Q_{S}: activation energy for solute diffusion, Q_{GB}: activation energy for grain boundary diffusion, Q_{C}: activation energy for dislocation core diffusion.

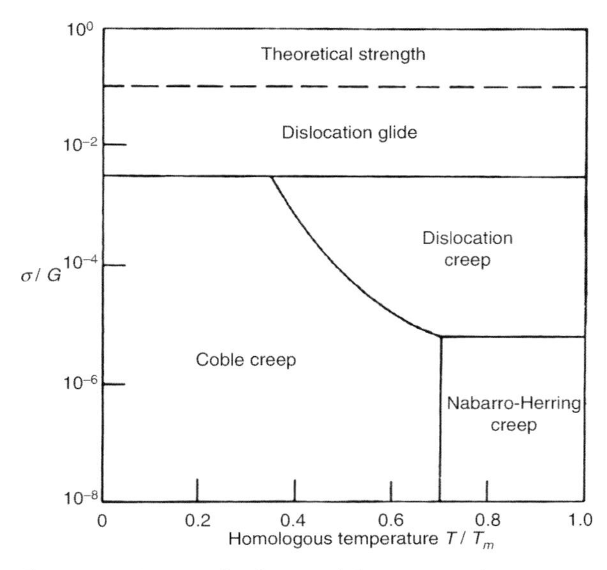

Figure 5.39 An example of a creep deformation mechanism map.

varies as σ^2, the stress exponent becomes $5 + 2$.

$$\dot{\varepsilon}_{\text{LTC}} = B_{\text{LTC}} \frac{D_{\perp}\sigma^5}{kT} = B_{\text{LTC}} \frac{D^0_{\perp}\,\mathrm{e}^{-Q_{\text{C}}/RT}\varrho\sigma^5}{kT} = B_{\text{LTC}} \frac{D^0_{\perp}\,e^{-Q_{\text{C}}/RT}\sigma^7}{kT}. \tag{5.59}$$

A practical way of illustrating and utilizing creep constitutive equations for the various creep mechanisms is with deformation mechanism maps. Ashby and coworkers have developed these maps in stress–temperature space. Figure 5.39 shows such an example and such maps were developed for various metals and alloys as well as ceramics.

Table 5.3 and Figure 5.39 illustrate that various creep mechanisms become dominant at specific loading conditions (stresses and temperatures). As per the examples shown in Figures 5.28 and 5.37, transitions in creep mechanisms occur as stresses are decreased. They can be extended to lower stresses by including GBS and viscous creep mechanisms. Figure 5.40 illustrates such transitions in creep mechanisms in pure metals and class-M alloys where at high stresses climb creep is noted (along with power-law breakdown) followed by viscous creep mechanisms along with GBS at intermediate stresses. This clearly implies that blind extrapolation of high-stress short-term data to lower stresses corresponding to in-service conditions would lead to dangerous nonconservative predictions of creep rates and creep lives.

A creep-resistant material should have the following features:

a) The first requirement for a creep-resistant material is to have a high melting point.
b) Strain hardening cannot be used as a strengthening mechanism at high temperatures as recovery/recrystallization type of softening events may take place. Solid solution strengthening and dispersion strengthening are useful.

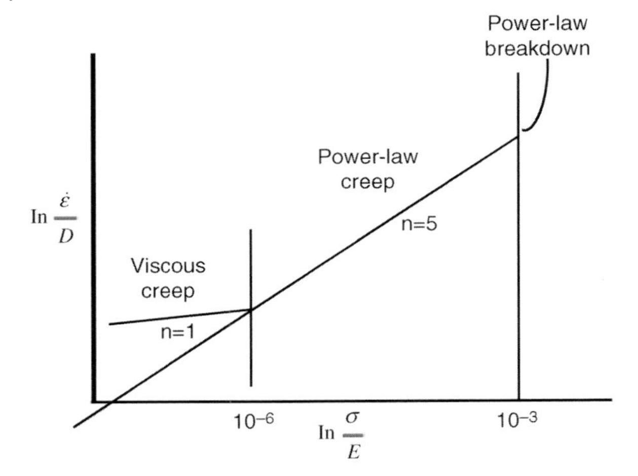

Figure 5.40 Transitions in creep mechanisms in pure metals (and class-M alloys).

c) Precipitation hardening can be applicable when there is no particle coarsening, dissolution, and consequent softening during service.
d) Grain refinement is not applicable under creep conditions. Actually, the opposite is required to minimize the grain boundary sliding effect. A fine grain size means more grain boundary area per unit volume of the material and thus is detrimental to creep resistance. Single-crystalline materials with no grain boundaries are relatively more creep resistant at high temperatures and thus blades for high-temperature gas turbines are made of single crystals.

Example Problem

The following creep data were obtained for a *bulk* aluminum alloy *single crystal* $\{T_M = 660.4\,°C$, Q_L (lattice diffusion) = 34 500 cal mol^{-1}, Q_{GB} (boundary diffusion) = 19 750 cal mol^{-1}, and Q_s (solute diffusion) = 36 000 cal mol$^{-1}\}$:

σ, MPa	T, K	$\dot{\varepsilon}_S$, s^{-1}
150	900	3.32×10^{-7}
150	780	1.70×10^{-8}
300	900	6.64×10^{-7}

i) Using these data, evaluate the parameters n and Q in the power-law creep equation.
ii) Identify the underlying creep mechanism.
iii) If in another example, a *thin polycrystalline* sample exhibited essentially the same n and Q, what would be the controlling creep mechanism?

Solution

i) Find the n value from the data at 900 K as

$$\frac{\log\left((6.64\times10^{-7})/3.32\times10^{-7}\right)}{\log(300/150)}=1$$

and find the activation energy from data at 150 MPa

$$Q=-R\frac{\ln\left((3.32\times10^{-7})/(1.7\times10^{-8})\right)}{(1/900)-(1/780)}=35\ \text{kcal mol}^{-1}.$$

ii) Since $Q_{creep}=Q_L$ and $n=1$, either Nabaroo–Herring or Harper–Dorn creep should be operative. Since the sample is a *bulk single crystal*, it must be Harper–Dorn creep.

iii) If the sample is a polycrystalline material with small grain size (or thin), the underlying mechanism must be Nabarro–Herring due to vacancies migrating from tensile grain boundaries to the compressive ones.

5.1.7
Fatigue Properties

About 90% of engineering failures are attributed to fatigue of materials under cyclic/fluctuating loading. Fatigue failure occurs after a lengthy period of stress or strain reversals and is brittle-like in nature even in normally ductile metals. Generally, the fracture surface turns out to be perpendicular to the applied stress. An example of fatigue failure in a shaft keyway is shown in Figure 5.41.

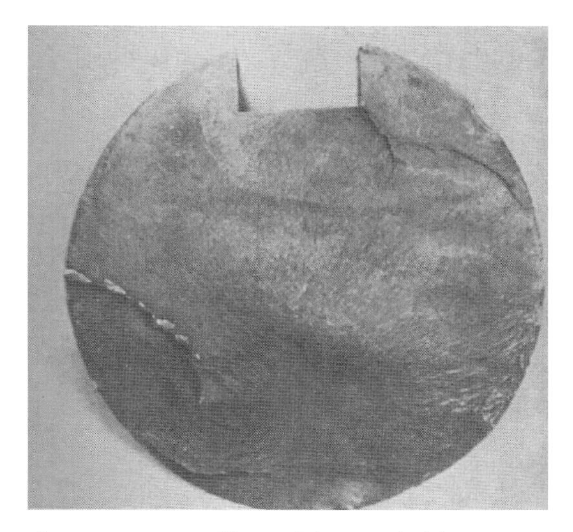

Figure 5.41 A case of fatigue failure in a shaft keyway. From Ref. [2].

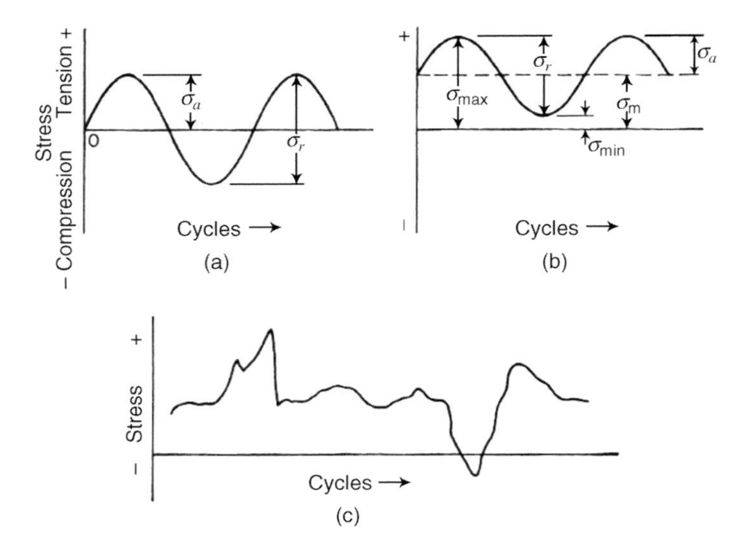

Figure 5.42 Tension–compression (a), tension–tension (b), and irregular stress (c) cycles.

Various types of fatigue loading are shown in Figure 5.42. Figure 5.42a shows a fully reversed tension–compression stress cycle, while Figure 5.42b depicts a tension–tension cycle. No compression–compression cycle is shown since fatigue cracks do not open up under compression; so no fatigue failure is possible under such loading. However, in reality, such a regular loading cycle is often not the norm; the fluctuating stress cycle may look like a random or irregular stress cycle, as shown in Figure 5.42c.

Let us now start defining some common terms used in fatigue. Mean stress (σ_m) is the algebraic mean or average of the maximum and minimum stresses in the cycle:

$$\sigma_m = (\sigma_{max} + \sigma_{min})/2. \tag{5.60}$$

The range of stress (σ_r) is just the difference between σ_{max} and σ_{min}, that is,

$$\sigma_r = \sigma_{max} - \sigma_{min}. \tag{5.61}$$

Stress amplitude or alternating stress (σ_a) is just half the stress range, that is,

$$\sigma_a \quad \text{or} \quad S = \sigma_r/2 = (\sigma_{max} - \sigma_{min})/2. \tag{5.62}$$

The stress ratio (r) is the ratio of minimum and maximum stresses, that is,

$$r = \sigma_{min}/\sigma_{max}. \tag{5.63}$$

$$\text{Amplitude ratio}: A_r = \sigma_a/\sigma_m = (1 - r)/(1 + r). \tag{5.64}$$

We note that Figure 5.42a is a case with $r = -1$.

There are many different ways to carry out fatigue testing. One of them is a cantilever beam fatigue test, as illustrated in Figure 5.43. In this setup, the specimen is

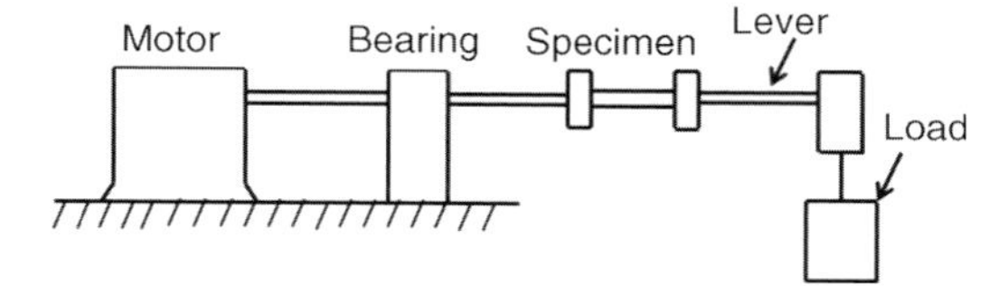

Figure 5.43 A cantilever fatigue beam testing facility.

in the form of a cantilever loaded at one end. It is rotated at the same time by means of a high-speed motor to which it is connected. At any instant, the upper surface of the specimen is in tension and the lower surface is in compression with the neutral axis at the center. During each revolution, the surface layers pass through a full cycle of tension and compression. Other types of fatigue testing include rotating–bending test and uniaxial tension–compression test in universal test machines.

5.1.7.1 **Fatigue Curve**

If the stress amplitude (S) is plotted against the number of cycles to failure (N), S–N curves are created (Figure 5.44). The S–N curve indicates that the higher the magnitude of the stress, the smaller the number of cycles that material is capable of sustaining before failure and vice versa. For some ferrous metals and titanium alloys, the S–N curve becomes horizontal at higher N values or there is a limiting stress level called the "fatigue limit" or "endurance limit," which is the largest value of fluctuating stress that will not cause failure for essentially infinite number of cycles. For many steels, fatigue limits range between 35% and 60% of the tensile strength. However, most other metals/alloys show a gradually sloping S–N curve. For these materials, fatigue limit is not clearly delineated and fatigue strength in these cases is obtained at an arbitrary number of cycles (e.g., 10^8 cycles).

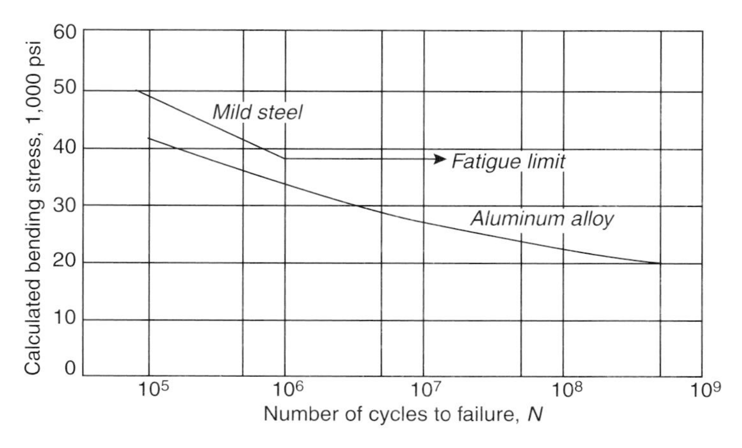

Figure 5.44 S–N curves for mild steel and an aluminum alloy.

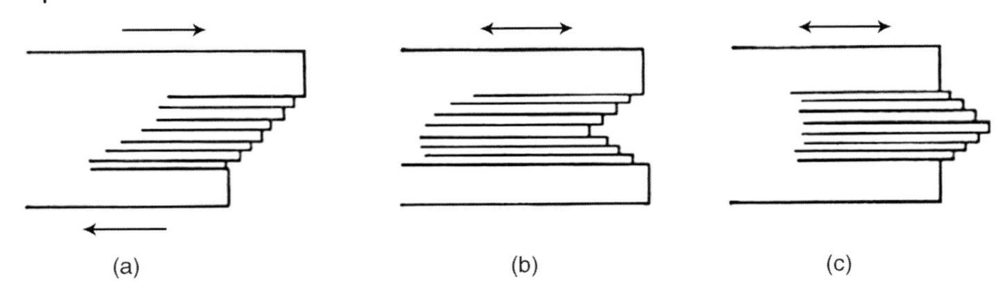

Figure 5.45 During forward loading, slip steps appear (a). But during reverse loading, they tend to take the form of (b) intrusions or (c) extrusions.

The process of fatigue failure consists of three distinct steps: (a) crack initiation, (b) crack propagation, and (c) final failure. The fatigue life (N_f) is thus the sum of the number of cycles for crack initiation (N_i) and the number for the crack propagation (N_p):

$$N_f = N_i + N_p. \tag{5.65}$$

Even when the cyclic stress is less than the yield strength, microscopic plastic deformation on a localized scale can take place. The cyclic nature of the stress causes slip to appear as extrusions and intrusions on the surface, as shown in Figure 5.45. Fatigue cracks can also initiate at other surface discontinuities or stress raisers. During the tensile cycle, slip occurs on a plane with maximum shear stress on it. During the compression part of the cycle, slip may occur on a nearby parallel slip plane with the slip displacement on the opposite direction. These act as the nucleation sites for the fatigue cracks.

The S–N curves as depicted in Figure 5.44 are classified into low and high cycle fatigue with low cycle fatigue (LCF) for $N < 5 \times 10^5$ cycles and high cycle fatigue (HCF) for $N > 5 \times 10^5$ cycles. These tests are usually performed on smooth specimens in strain-controlled mode and the total strain range is divided into elastic and plastic regions:

$$\frac{\Delta\varepsilon}{2} = \frac{\Delta\varepsilon_P}{2} + \frac{\Delta\varepsilon_E}{2}. \tag{5.66}$$

Basquin equation is usually considered to describe the HCF with stress range ($\Delta\sigma = E\Delta\varepsilon$):

$$N\sigma_a^p = C, \tag{5.67}$$

where N is the number of cycles to failure at stress amplitude σ_a ($\Delta\sigma/2$) and p and C are material constants. Similarly, the LCF is characterized by Coffin–Manson equation:

$$\frac{\Delta\varepsilon_P}{2} = A(2N)^c, \tag{5.68}$$

with c ranging from -0.5 to -0.7 and A being a material constant that is proportional to the tensile ductility (total strain to fracture in a tensile test). HCF region

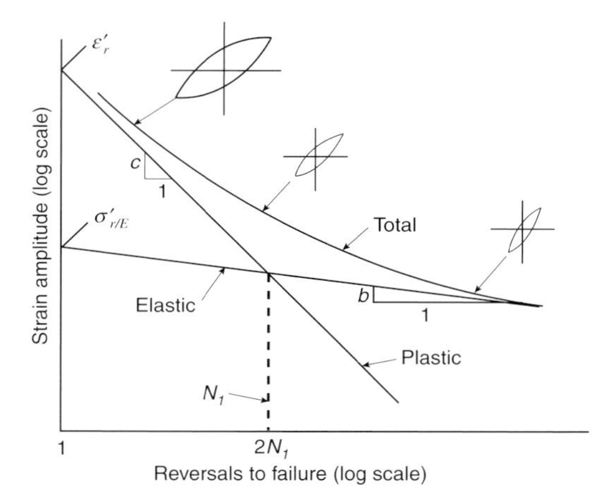

Figure 5.46 Typical fatigue life plot along with corresponding stress–strain loops.

can also be represented in a similar fashion:

$$\frac{\Delta\varepsilon_{\mathrm{E}}}{2} = B(2N)^{b}, \tag{5.69}$$

where b has a value between -0.05 and -0.12 and B is proportional to the tensile or fracture strength of the material. Thus, we note that LCF is controlled by ductility and HCF by strength so that cold working (or exposure to intense neutron irradiation) resulting in higher strength with corresponding decreased ductility will exhibit detrimental effect (i.e., lower N values) during LCF, while being beneficial to HCF.

Figure 5.46 shows typical fatigue life plot as strain range ($\Delta\varepsilon$) against number of failure cycles (N_{f}) along with the corresponding stress–strain loops (broad in LCF and narrow in HCF). The complete fatigue curve can be described by combining the LCF and the HCF formulations by either the universal slopes (Eq. (5.70a)) or the characteristic slopes (Eq. (5.70b)):

$$\frac{\Delta\varepsilon}{2} = 3.5\frac{S_{\mathrm{u}}}{E}N^{-0.12} + \varepsilon_{\mathrm{f}}^{0.6}N^{-0.6}, \tag{5.70a}$$

$$\frac{\Delta\varepsilon}{2} = \frac{\sigma_{\mathrm{f}}}{E}(2N)^{b} + \varepsilon_{\mathrm{f}}(2N)^{c}, \tag{5.70b}$$

where S_{u} is ultimate tensile strength, ε_{f} is true fracture strain, σ_{f} is the true fracture stress, and b and c are material constants. In terms of the characteristic slopes, the value of fatigue life at which the transition from low cycle (plastic) to high cycle (elastic) occurs is given by

$$2N_{\mathrm{tr}} = \left(\frac{\varepsilon_{\mathrm{f}}E}{\sigma_{\mathrm{f}}}\right)^{1/b-c}. \tag{5.71}$$

Example Problem

A steel specimen was subjected to two fatigue tests at two different stress ranges of 400 and 250 MPa and failure occurred after 2×10^4 and 1.2×10^6 cycles, respectively. Determine the fatigue life at 300 MPa stress range.

Solution

First find p and C in Basquin equation (Eq. (5.67)) and use them to find N for 300 MPa range to be 2.54×10^5 cycles.

Example Problem

A steel with an yield strength (S_y) of 30 000 psi, tensile strength (S_u) of 50 000 psi, and true fracture strain of 0.3 is to be used under cyclic loading at a constant strain range. Evaluate the limit on the total cyclic strain range if the steel is to withstand 4.9×10^5 cycles; $E = 30 \times 10^6$ psi.

Solution

Let us use the universal slopes equation: $\Delta\varepsilon = 3.5 \dfrac{S_u}{E} N_f^{-0.12} + \varepsilon_f^{0.6} N_f^{-0.6}$.

$$\Delta\varepsilon = 3.5 \frac{50000}{30 \times 10^6} \left(4.9 \times 10^5\right)^{-0.12} + (0.3)^{0.6} \left(4.9 \times 10^5\right)^{-0.6} = 1.378 \times 10^{-0.3}$$
$$= 0.1378\%.$$

In strain-controlled fatigue tests for life evaluation, it may be noted that the cyclic stress–strain curve leads to a hysterics loop, as depicted in Figure 5.47a where *O–A–B* is the initial loading curve and on unloading the yielding occurs at lower stress (point *C* compared to *A*), which is known as Bauschinger effect. The material may undergo cyclic hardening or softening and in rare cases it remains stable (Figure 5.47b) and this behavior depends on the initial metallurgical condition of the material. According to Figure 5.47b, cyclic hardening leads to decreasing peak strains, while the peak strains increase in the case of cyclic softening with increasing number of cycles. In general, the hysteresis loop stabilizes after about 100 cycles and the stress–strain curve obtained from cyclic loading will be different from that of monotonic loading (Figure 5.47c), but the stress–strain follows a power-law relationship similar to that in monotonic loading,

$$\Delta\sigma = K'(\Delta\varepsilon)^{n'}, \tag{5.72}$$

where the cyclic hardening coefficient n' ranges from 0.1 to 0.2 for many metals and is given by the ratio of the parameters (b/c) (Eq. (5.70b)). In some cases, fatigue ratcheting occurs with a resulting increase in strain as a function of time when tested under a constant strain range that is often referred to as cyclic creep. In a stress-controlled test with nonzero mean stress, the shift in the hysteresis loop along the strain axis, as depicted in Figure 5.47d, is attributed to thermally activated

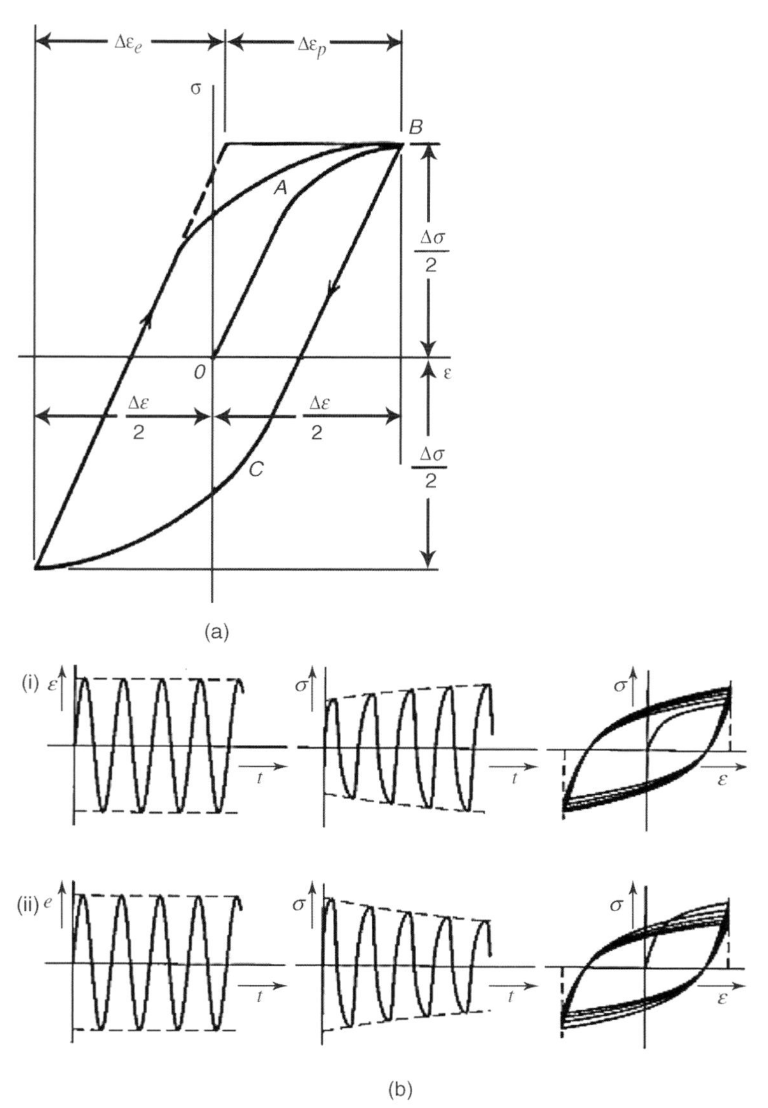

Figure 5.47 (a) Cyclic stress–strain curve illustrating hysteresis loop. (b) Hysteresis loops during cyclic hardening (i) and cyclic softening (ii). (c) Comparison of cyclic stress–strain curve for cyclic hardening and stress–strain curve under monotonic loading. (d) An example of ratcheting fatigue. From Ref. [10].

dislocation movement at stresses well below the yield stress and/or due to dislocation pileup resulting in stress enhancement. Fatigue ratcheting may also occur in the presence of residual stress and in cases where microstructural inhomogeneities exist such as in welded joints.

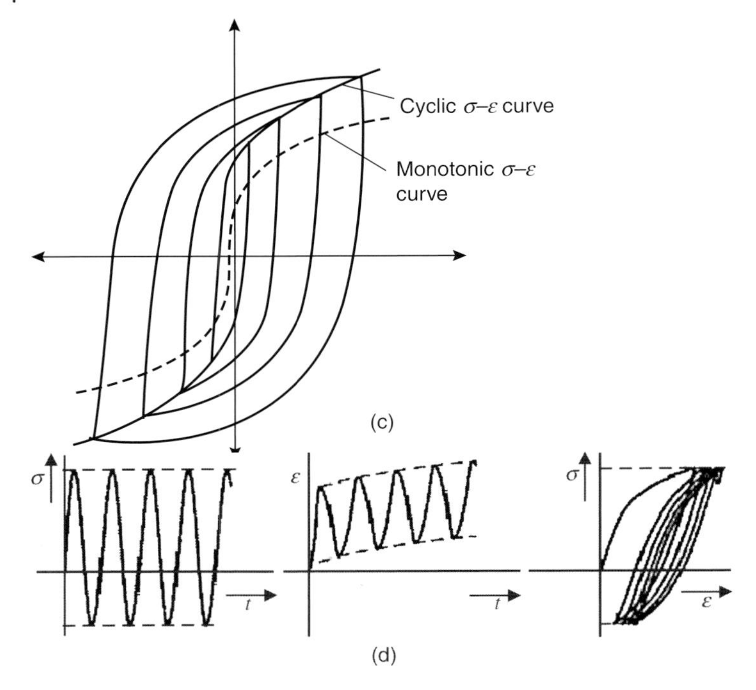

Figure 5.47 (*Continued*)

5.1.7.2 **Miner's Rule**

In real situations, stresses change at random frequencies and, in general, the percentage of life consumed in one cyclic loading depends on the magnitude of stress in subsequent cycles. However, the linear cumulative damage rule, known as Miner's rule, assumes that the total life of a component can be estimated by adding up the life fraction consumed by each of the loading cycles. If N_i is the number of cycles to failure at ith cyclic loading and n_i is the number of cycles experienced by the structure,

$$\sum \frac{n_i}{N_i} = 1. \tag{5.73}$$

Miner's rule is too simplistic and fails to predict the life when notches are present. Furthermore, it fails to predict the life when mean stress and temperature are high enough or cyclic frequency is low where creep deformation dominates over fatigue loading. It turns out that many materials exhibit deviations from this linear addition depending on whether it is cyclically hardening or softening. In particular, the predictions tend to be highly nonconservative for cyclically softening materials. However, it is a very useful rule in fatigue life prediction and has pedagogic advantages.

5.1.7.3 **Crack Growth**

Crack growth tests are generally performed under stress control tests with $r = 0$ using CT specimens with surface notches, while fatigue life tests are performed

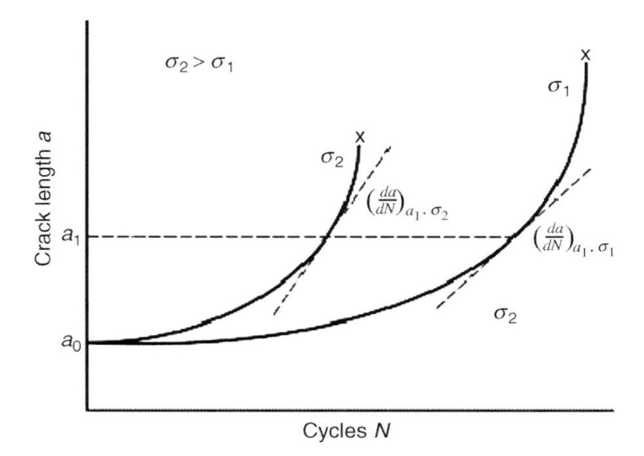

Figure 5.48 Crack length versus number of loading cycles at two stresses ($\sigma_2 > \sigma_1$).

under $r = -1$ condition with smooth specimens. Crack propagation generally occurs in two stages (Figure 5.48):

Stage-1 Propagation: After initiation of a stable crack, the crack then propagates through the material slowly along crystallographic planes of high shear stresses. The fatigue fracture surface has a flat and featureless appearance during this stage.
Stage-2 Propagation: Crack extension rate increases dramatically. At this point, there is also a change in propagation direction to one that is roughly perpendicular to the applied tensile stress. Crack growth occurs through repeated plastic blunting and sharpening. After reaching a critical crack dimension that initiates the final failure step, catastrophic failure starts. The fracture surface may be characterized by two markings: beach marks/clamshell marks (macroscopic), as shown in Figure 5.49a schematically, and striations (microscopic), as illustrated by a SEM image in Figure 5.49b. The presence of beach marks and/or striations on a fracture surface confirms that the cause of failure is fatigue. However, the absence of either or both does not exclude fatigue as a cause of fatigue-type failure.

5.1.7.4 **Paris Law**

The kinetics of crack growth in the subcritical stage can be experimentally determined. One of the commonly used empirical equations correlates the crack growth rate per cycle, $\mathrm{d}a/\mathrm{d}N$, with the range of the stress intensity factor. The intermediate region of crack growth is governed by what is known as Paris law:

$$\frac{\mathrm{d}a}{\mathrm{d}N} = A(\Delta K)^m, \tag{5.74}$$

where A and m (Paris exponent) depend on the particular material, as well as on environment, frequency, and the stress ratio, and ΔK is the range of stress intensity factors $[=\Delta\sigma\,\sqrt{(\pi a)}]$. Figure 5.50 shows a log–log plot of $\mathrm{d}a/\mathrm{d}N$ versus ΔK. For tension–tension loading (with minimum stress being more than zero), the range

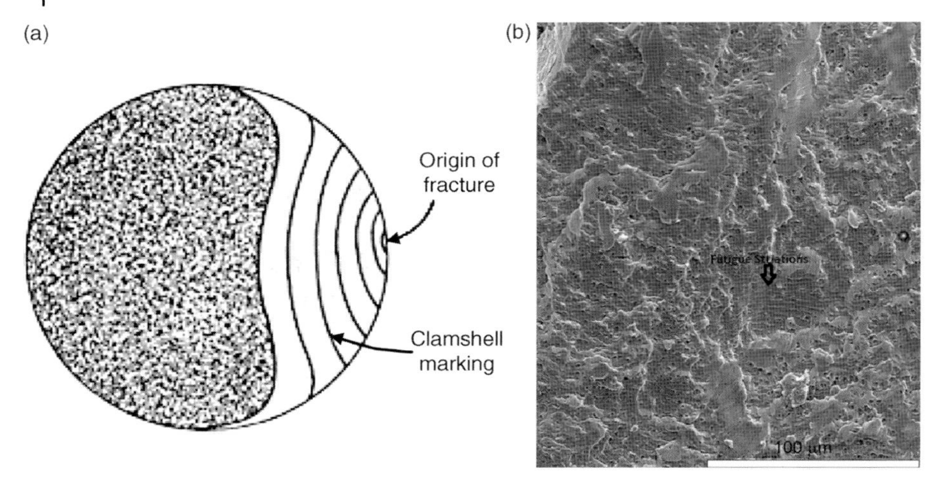

Figure 5.49 (a) A schematic fatigue fracture surface showing clamshell markings. (b) An SEM image showing the striations in a fatigued 304L austenitic stainless steel.

of both K and maximum K are important. However, for tension–compression-type loading, only maximum K is relevant as no crack growth takes place under compressive part of the fatigue cycle. In stage-II known as Paris region, cracks propagate in stable manner exhibiting fatigue striations until a critical crack length (a_c) is reached at the beginning of the stage-III during which unstable crack growth takes place with eventual fracture.

Paris law makes it possible to estimate the number of fatigue cycles (N_f) and fatigue life:

$$N_f = \int_{0}^{N_f} dN = \int_{a_0}^{a_f} \frac{da}{A(\Delta K)^m} = \frac{1}{AY^m(\Delta\sigma)^m \pi^{m/2}} \int_{a_0}^{a_f} \frac{da}{a^{m/2}}, \tag{5.75}$$

where the critical crack length is given in terms of the plane strain critical fracture toughness (K_{IC}):

$$a_f = \frac{K_{IC}^2}{Y^2\pi^2\sigma_{max}}. \tag{5.76}$$

Thus, knowing a_f along with the parameters in Paris law, one can calculate the number of cycles to failure. If the frequency (f) of cycling is known, the time to failure (t_f) can be obtained:

$$t_f = \frac{N_f}{f}. \tag{5.77}$$

The Paris exponent m has values from 2 to 4, and Eq. (5.67) will have a special case for $m = 2$.

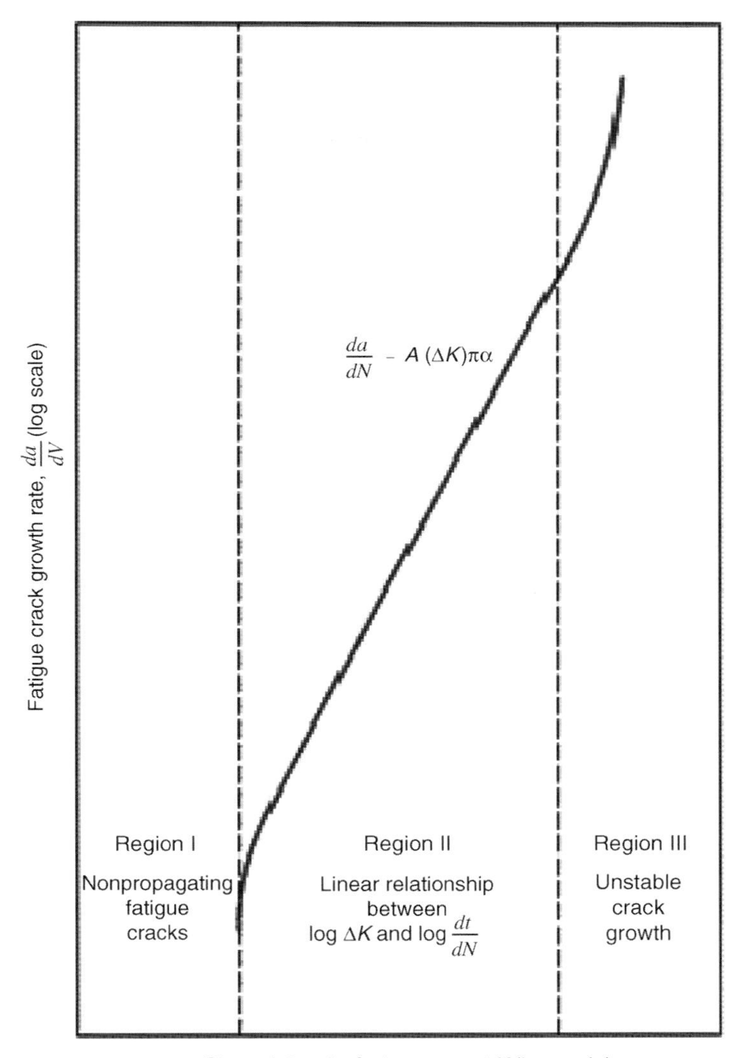

Figure 5.50 The crack growth rate during fatigue as a function of the stress intensity factor range shows a threshold stress intensity factor, a regime of subcritical crack growth, and a fast fracture regime.

Example Problem

A mild steel plate ($K_{IC} = 40\,\text{MPa}\,\text{m}^{1/2}$) contains an initial through thickness *edge* crack of 0.5 mm and was subjected to uniaxial fatigue loading: $\sigma_{max} = 180\,\text{MPa}$ and $R = 0$. If the material follows the Paris crack–growth relation, $(\text{d}a/\text{d}N)$ (m per cycle) $= 5.185 \times 10^{-13}\,(\Delta K)^2$, with K in $\text{MPa}\,\text{m}^{1/2}$. Evaluate the number of cycles required to break the plate (assume $Y = 1$)?

Solution

$da/dN = 5.185 \times 10^{-13}(\Delta K)^2$, where $\Delta K = Y\sigma_{max}\sqrt{\pi a}$, so
$da/dN = 5.185 \times 10^{-13}\sigma_{max}^2 \pi a$.
Thus,

$$\int_0^{N_f} dN = \int_{a_0}^{a_f} \frac{da}{5.185 \times 10^{-13}\sigma_{max}^2 \pi a} = \frac{1}{5.185 \times 10^{-13}\sigma_{max}^2 \pi} \int_{a_0}^{a_f} \frac{da}{a}$$

$$= \frac{1}{5.185 \times 10^{-13}\sigma_{max}^2 \pi} \ln\left(\frac{a_f}{a_0}\right),$$

where $a_0 = 0.5$ mm and $a_f = K_{IC}^2/\pi\sigma_{max}^2 = \frac{40^2}{\pi \times 180^2} = 15.72$ mm, so $N_f = 6.534 \times 10^7$ cycles.

Another important aspect of considering the crack growth versus ΔK is to examine the effects of superimposed environment such as corrosion and radiation. The variation of da/dN with ΔK in these cases would be to shift the threshold stress intensity range to lower values and the critical crack length at fracture is now indicated by K_{ISCC} instead of K_{IC}.

5.1.7.5 Factors Affecting Fatigue Life

1) The frequency of cyclic loading has only a small effect on the fatigue strength.
2) The form of stress cycle such as square, triangular, or sinusoidal wave has no effect on the fatigue life.
3) The environment in which the component undergoes stress reversals has a marked effect on fatigue life. The fatigue life in vacuum could be some 10 times more than that in moist air. Also note the effect of corrosion fatigue.
4) The thickness of the test specimen has an effect on the fatigue properties. Thinner samples show a decrease in the crack growth rate.
5) Stress concentrators such as key ways with sharp corners promote crack initiation, thereby decreasing the number of cycles to failure.
6) Surface smoothness (or roughness) has strong effect on fatigue life and polished surfaces will have improved fatigue life.

5.1.7.6 Protection Methods against Fatigue

1) Shot peening introduces compressive residual stresses near the surface layers. The surface plastically deformed by repeated impingement of hard steel balls can improve fatigue life since the compressive residual stresses first need to be compensated before crack propagation can take place.
2) Carburizing and nitriding introduce compressive residual stresses at the surface while increasing also the strength/hardness. Decarburization lowers the fatigue strength.
3) A fine grain size improves fatigue life.

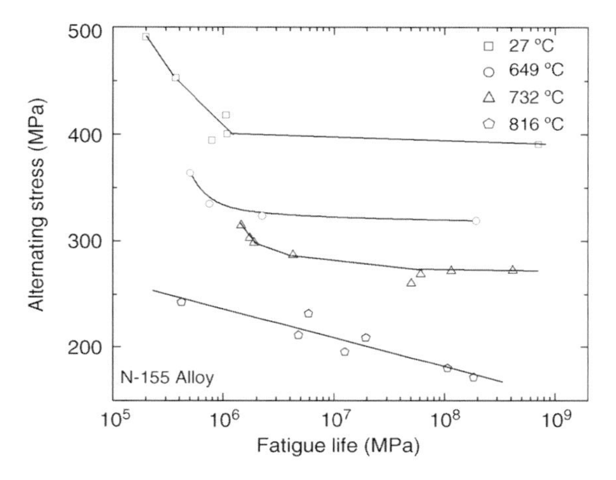

Figure 5.51 Data pertaining to fatigue life of N-155 alloy subjected to various temperatures and reversed bending stresses.

4) The crack initiation almost always happens from a site of stress concentration. In design, fillets of adequate radius of curvature should be provided at places where a sudden change of cross section exists. Shafts are usually highly polished to reduce the chance of surface irregularities.

5.1.8 Creep–Fatigue Interaction

Failure by fatigue is a possibility at elevated temperatures (less than melting point). This type of failure can occur with less plastic deformation and will have the characteristics of a fatigue failure. However, it is generally observed that both fatigue and static strength properties get reduced with increasing temperature. Figure 5.51 illustrates the stress versus fatigue life curve at various temperatures, including room temperature for N-155 alloy (a high temperature alloy of Fe-21Cr-20Ni-20Co-3Mo-2.5W-1.5Mn-1Nb, wt.%). As already discussed, application of a load particularly at higher homologous temperatures would produce time-dependent plastic deformation or creep. The creep–rupture strength decreases with increasing temperature. Usually, the alloys that are creep resistant are also found to be fatigue resistant. However, it does not mean that a material with best creep strength will also provide the best fatigue strength. Therefore, it becomes necessary to design against both creep and fatigue, and testing needs to be carried out in a state where both fatigue and creep loading are applied. The main method of investigating creep–fatigue properties is to conduct strain-controlled fatigue tests with variable frequencies with and without intermittent holding period (hold time) during the test. A lower frequency ($\leq 10^4$ cycles s^{-1}) and hold times allow creep effects to take place. At higher frequencies and short hold times, the fatigue mode predominates and failures just as pure fatigue failure (cracks start at surface and propagates transgranularly inside the bulk material). At longer hold times or decreased frequencies, the creep effect starts playing a growing role and creep–fatigue interaction becomes important. In this regime, the mixed mode fracture is observed, that is, both fatigue cracking and creep cavitation. At another extreme, when the holding time is extended considerably with cyclic loads occurring only sparsely, the situation resembles pure creep deformation. However, where oxidation effects are present, the creep–fatigue interaction becomes much more complex. The ASTM E2714-09 standard (Standard Test Method for Creep–Fatigue Testing) describes the specifics of a creep–fatigue test.

Note that creep–fatigue and thermomechanical fatigue are not the same phenomena. Creep–fatigue is carried out at constant nominal temperatures, whereas thermomechanical fatigue involves thermal cycling (i.e., fluctuating temperatures).

5.2 Thermophysical Properties

The materials characteristics that are not governed directly by mechanical forces and chemical environment are known as physical properties. There are many physical properties, such as electrical conductivity, magnetic susceptibility, specific heat, thermal conductivity, thermal expansion coefficient, and so forth. These properties are usually structure-insensitive and intrinsic properties of materials. There are, however, a number of exceptions. Thermophysical properties (i.e., physical properties affected by heat) are of great interest for nuclear reactors. We focus here exclusively on the thermophysical properties of solid materials, even though these properties pertaining to liquids and gases that may be used as coolants and moderators in nuclear reactors are also important. Here we attempt to understand the basic thermophysical properties along with some specific examples.

5.2.1 Specific Heat

The specific heat (molar heat capacity) is the amount of heat needed to raise the temperature of a g mol of material by 1 degree at constant pressure (C_P) or constant volume (C_V). Thermodynamically, they can be written as

$$C_V = \left(\frac{\partial U}{\partial T}\right)_V, \tag{5.78}$$

$$C_P = \left(\frac{\partial H}{\partial T}\right)_P, \tag{5.79}$$

where U and H are the internal energy and enthalpy of the system, respectively. It is experimentally more suitable to evaluate the specific heat at constant pressure C_P (atmospheric). However, it is theoretically easier to predict C_V. C_P is always greater than C_V; at room temperature or below; the difference is very small for solids

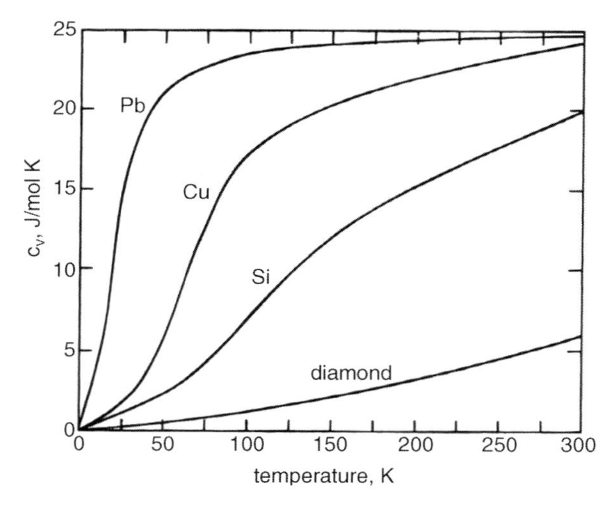

Figure 5.52 The specific heat (at constant volume) for lead, copper, silicon, and diamond as a function of temperature. From Ref. [11].

(whereas for ideal gas, $C_P - C_V = R$). The SI unit of specific heat is generally given in J mol^{-1} K^{-1}.

Petit and Dulong [12] conducted experiments and suggested an empirical rule that states that all solid elements have C_V of $3R$ (where R is the universal gas constant, that is, ~24.9 J mol^{-1} K^{-1}). Kopp [13] introduced an empirical rule that the molar heat capacity of a solid chemical compound is approximately equal to the sum of molar heat capacities of its constituent elements. Even though the molar heat capacity values for elements tend to be close to $3R$ near room temperature, they increase with temperature and become smaller than $3R$ at lower temperatures. Figure 5.52 shows C_V for four pure elements (lead, copper, silicon, and diamond) as a function of temperature. Only metals, lead, and copper show specific heat close to $3R$. However, for other two elements, they are quite different.

The classical physics could not explain why the specific heat changes with increasing temperature. During the early part of the twentieth century, the concept of specific heat was explained with the help of quantum theory. The problem was first treated by Einstein in 1907 by assuming that atoms in a crystal (called Einstein crystal) behave like independent harmonic oscillators. But Einstein's derivation could not explain the experimental data near the absolute zero. Later in 1912, Debye assumed that the range of frequencies available to the oscillators is the same as that available in a homogeneous elastic continuum. Basically, Debye assumed that the vibration occurs in a coordinated way (because of the presence of atomic bonding) as opposed to Einstein's assumption of independent oscillation. They were considered as elastic lattice waves (much like sound waves) that have high frequency with small amplitude. The lattice vibrational energy can be quantized (phonon) in a similar way as done for light waves (photon). An expression for constant volume specific heat developed by Debye provided good agreement with

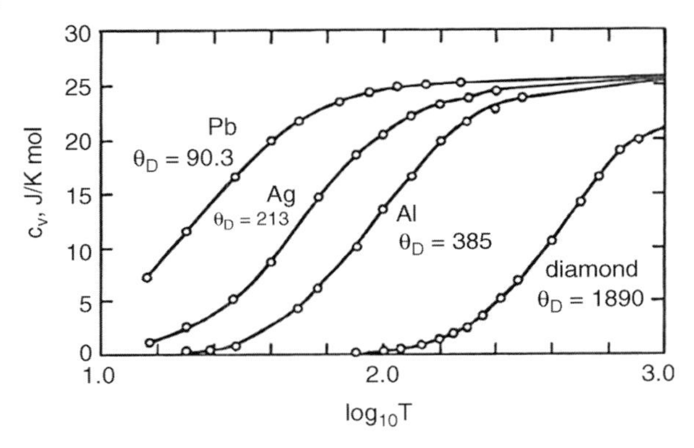

Figure 5.53 The constant volume specific heat (C_V) as a function of temperature for four solid elements. Debye temperatures (θ_D) in K are shown in the figure. From Ref. [11].

experimental data:

$$C_V = 9R\left[4\left(\frac{T}{\theta_D}\right)^3 \int_0^{\theta_D/T} \frac{x^3}{e^x - 1}dx - \frac{\theta_D}{T}\cdot\frac{1}{(e^{\theta_D/T} - 1)}\right], \tag{5.80}$$

where $x = h\nu/kT$ is dimensionless, $\theta_D = h\nu_D/kT$ is Debye temperature, T is the temperature in K, R is the universal gas constant, and ν_D is the maximum frequency of atomic vibration (called Debye frequency). We should be cognizant of two facts regarding the function given in Eq. (5.80): (i) At higher temperatures up to room temperature (i.e., a low value of θ_D/T), C_V tends to reach the value as predicted by Dulong and Petit. (ii) The expression also gives good accord to the experimental data at very low temperatures when Debye's original expression is reduced to

$$C_V = 234R\left(\frac{T}{\theta_D}\right)^3. \tag{5.81}$$

This equation correctly captures the rapid approach toward zero as the temperature reaches the absolute zero. Figure 5.53 shows the C_V values as a function of temperature for four different elements. If the data are normalized with respect to θ_D, all data come together and fall on a single master curve.

It should be noted that C_P rather than C_V is measured more often experimentally. In order to relate the theories to experiments, $(C_P - C_V)$ needs to be known from an expression as shown in Eq. (5.82):

$$C_p - C_V = \frac{\alpha^2 VT}{\beta}, \tag{5.82}$$

where α is the coefficient of thermal expansion, $((\partial \ln V)/\partial T)_P$, and β is the coefficient of compressibility, $-((\partial \ln V)/\partial P)_T$.

However, at a temperature much above the room temperature, C_V gets larger than that predicted by Dulong and Petit. The nonapplicability of Debye's theory in this temperature range stems from the fact that electrons of the atoms not only absorb energy but also increase their energy and this electronic contribution was not considered in Debye's analysis. However, this is only possible for free electrons; that is, the electrons that have been excited from the filled states above the Fermi energy level. This type of situation contribution is certainly valid for metals/alloys that have free electrons. However, they are still a small fraction of total electrons. In insulating and semiconducting materials, the electronic contribution is quite insignificant. The electronic contribution to specific heat bears a linear proportionality with temperature (i.e., $C_V \propto T$). There are also few other contributions to specific heat depending on the system, such as randomization of electron spins in a ferromagnetic material (like alpha-iron) as it is heated through the Curie temperature. A large spike in the specific heat can be observed at that temperature in a differential scanning calorimetry (DSC) curve.

Often the experimentally measured variation in the constant pressure specific heat (C_P) is described by empirical fitting of the form shown below:

$$C_P = a + bT + cT^{-2}, \tag{5.83}$$

where a, b, and c are curve fitting constants. Figure 5.54 shows a number of C_P versus temperature curves for various elements and compounds that exhibit polymorphism. Note the sharp changes at the temperature where the polymorphic transformations take place.

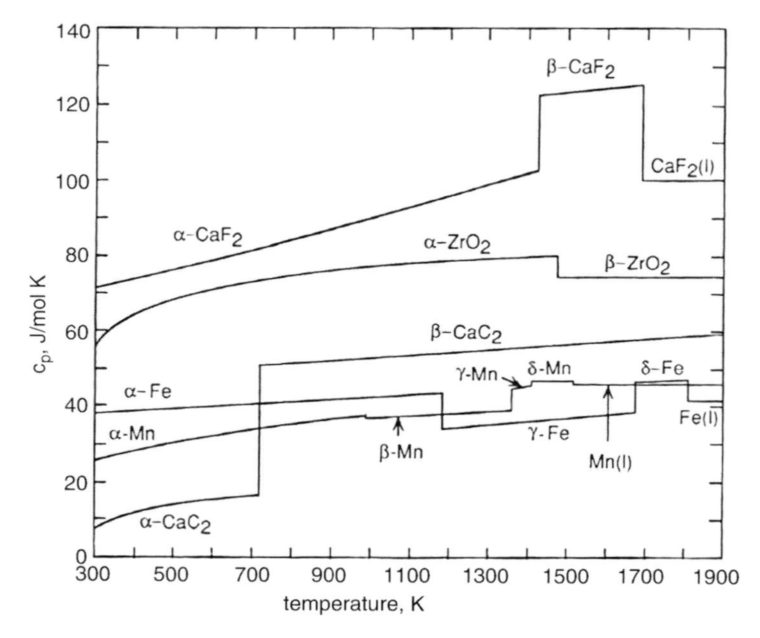

Figure 5.54 The constant pressure specific heat (C_P) versus temperature for various elements and compounds. From Ref. [11].

5.2.2 Thermal Expansion

Almost all solid materials exhibit expansion when heated, and contraction when cooled. Generally, the length change (ΔL) in a material of original length (L_0) is proportional to the temperature change ($\Delta T = T_f - T_i$, where T_f is the final temperature and T_i is the initial temperature). The proportionality constant α_l is the linear coefficient of thermal expansion. Mathematically, it is written as

$$\frac{\Delta L}{L_0} = \alpha_l \Delta T. \tag{5.84}$$

In a similar way, the volumetric coefficient for thermal expansion (α_v) can also be defined by the following relation:

$$\frac{\Delta V}{V_0} = \alpha_v \Delta T, \tag{5.85}$$

where ΔV is the change in volume and V_0 is the original volume. In many materials, α_v is anisotropic and may depend on the crystallographic direction along which the thermal expansion measurement is being done. However, for materials exhibiting isotropic thermal expansion, α_v should be equal to $3\alpha_l$. Dilatometer is generally used to obtain coefficient of thermal expansion.

Thermal expansion occurs due to the change in the balance of attractive and repulsive forces between atoms. Figure 5.55 shows a plot of potential energy of a crystal against the interatomic distance. This plot appears as a potential energy trough. In the figure, the equilibrium interatomic spacing is given by r_0 at 0 K.

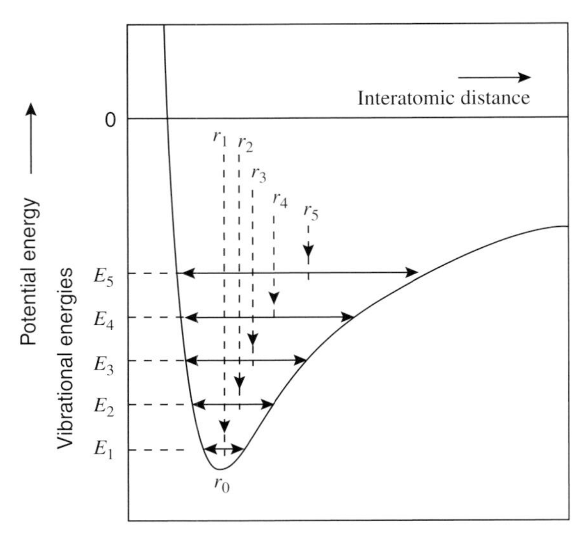

Figure 5.55 A plot of potential energy versus interatomic distance in a crystal showing that the interatomic separation is increasing with temperature.

Now heating the crystal to higher temperatures (T_1, T_2, etc. > 0 K) makes the vibrational energy increase correspondingly from E_1 to E_2, and so on. It is to be noted that the trough width at a particular energy level represents the average vibrational amplitude of an atom, whereas the mean position represents the average interatomic distance. From the viewpoint of thermal expansion, mean interatomic distance (in other words, lattice constant) is important. Basically, the mean interatomic distance increases because of the asymmetry in the potential energy curve, and that is what causes thermal expansion. If the potential energy curve were symmetrical, it would not have led to thermal expansion.

Each class of materials shows different range of values of thermal expansion coefficient. Generally, the stronger the atomic bonding, the deeper and narrower the potential energy trough. Ceramic materials with ionic or covalent bonding show limited thermal expansion as a function of temperature. The values range between 0.5×10^{-6} and $15 \times 10^{-6}\,K^{-1}$. Ceramic materials with low thermal expansion coefficient are susceptible to *thermal shock*. However, some of the common metals show linear coefficients of thermal expansion in the range of 5×10^{-6} to $25 \times 10^{-6}\,K^{-1}$. There are some families of iron–nickel and iron–nickel–cobalt alloys that show quite small linear coefficient of thermal expansion ($\sim 10^{-6}\,s^{-1}$) for applications requiring high degree of dimensional stability. Kovar and Invar are two of the low-expansion alloys.

The magnitude of thermal expansion coefficient usually increases with increasing temperature. The coefficient expansion and the specific heat have interesting similarity in their dependence on temperature including near absolute zero. An example of this resemblance is shown for platinum in Figure 5.56a. There is also an approximate relationship between the melting points and coefficients of thermal expansion for various materials, as shown in Figure 5.56b.

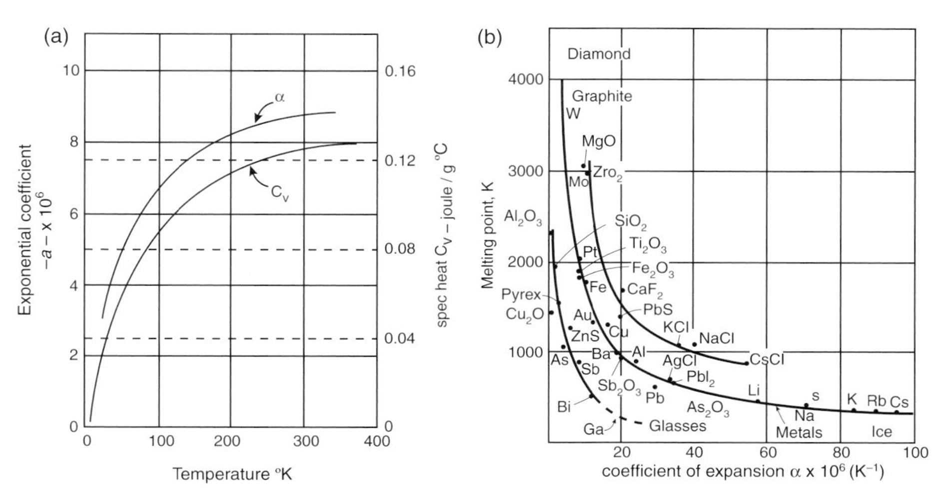

Figure 5.56 (a) Similarity in temperature dependence of thermal expansion coefficient and specific heat (C_V) in platinum. (b) Relation between melting point and thermal expansion coefficient of several elements and compounds.

5.2.3
Thermal Conductivity

Heat is transported through solid materials from the high-temperature region to the low-temperature region. Thermal conduction is a principal mode of heat transfer in solid materials. In nuclear reactors, the heat is conducted away by the cladding materials from the fuel interior. Thermal conductivity (κ) is defined by the following equation known as Fourier's law:

$$q = -\kappa \frac{dT}{dx}, \tag{5.86}$$

where q is the heat flux (the heat flow per unit perpendicular area per unit time) and dT/dx is the thermal gradient. This equation is applicable for the steady-state heat flow. The minus sign comes due to the heat being conducted from the hot region to cold region, that is, down the temperature gradient. The SI unit of thermal conductivity is $W\,m^{-1}\,K^{-1}$. The above equation is much similar to Fick's steady-state flow (Eq. (2.21)). Thermal diffusivity is another term that is often used. It is given by the following expression:

$$D_T = \frac{k}{C_P \varrho}, \tag{5.87}$$

where C_P is the constant pressure specific heat and ϱ is the physical density.

Heat conduction takes place by both phonons and free electrons. Thus, the thermal conductivity (κ) is given by

$$\kappa = \kappa_l + \kappa_e, \tag{5.88}$$

where κ_l is the phonon contribution and κ_e is the electronic contribution to thermal conductivity.

In high-purity metals, thermal energy transport through free electrons is much more effective compared to the phonon contribution as free electrons are readily available as they are not easily scattered by atoms and imperfections in the crystal and they have higher velocities. The thermal conductivities of metals can vary from 20 to 400 $W\,m^{-1}\,K^{-1}$. Silver, copper, gold, aluminum, and tungsten are some of the common metals with high thermal conductivities. Metals are generally much better thermal conductors than nonmetals because the same free electrons that participate in electrical charge transport also take part in the heat conduction. For metals, the thermal conductivity is quite high, and those metals that are the best electrical conductors are also the best thermal conductors. At a given temperature, the thermal and electrical conductivities of metals are proportional, but interestingly the temperature increases the thermal conductivity while reducing the electrical conductivity. This behavior can be explained with the help of Wiedemann–Franz law:

$$L_{WF} = \frac{\kappa}{C_e T}, \tag{5.89}$$

where T is the temperature in K, C_e is the electrical conductivity, L_{WF} is a constant (Lorenz number) that is $\sim 2.44 \times 10^{-8}\,\Omega\,W\,K^{-2}$. The above relation is based on the

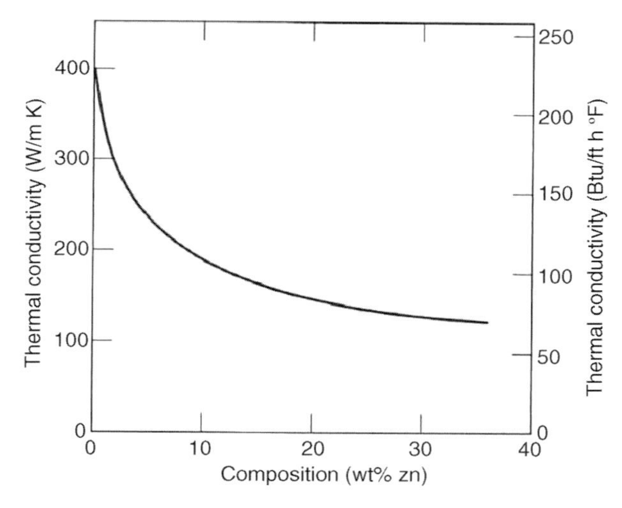

Figure 5.57 The variation of thermal conductivity with respect to zinc content in Cu–Zn alloys.

fact that both heat and electrical (charge) transport are associated with free electrons in metals. The thermal conductivity in metals increases with the average electron velocity as that increases the forward transport of energy. However, the electrical conductivity decreases with increasing electron velocity because the scattering or collisions divert the electrons from forward transport of charge. For more details on thermal properties, readers may consult an excellent text by Ziman [14]. The presence of grain boundaries and other crystal defects reduces the thermal conductivity. Researchers have observed a marked reduction in nanocrystalline materials (i.e., with grain size less than 100 nm).

Alloying generally acts upon the thermal conductivity of metals. The alloying atoms act as scattering centers for free electrons and thus reduce the effective thermal conductivity. Brass has a lower thermal conductivity compared to pure copper for that reason. Specifically, for a 70Cu–30Zn, thermal conductivity at room temperature is $120\,\mathrm{W\,m^{-1}\,K^{-1}}$, whereas the thermal conductivity of pure copper at the same temperature is $398\,\mathrm{W\,m^{-1}\,K^{-1}}$. Figure 5.57 illustrates the point. For the same reason, stainless steels are generally poor conductors of heat compared to pure iron.

The thermal conductivity of glass and ceramics is generally smaller compared to that of metals. They range between 2 and $50\,\mathrm{W\,m^{-1}\,K^{-1}}$. As free electrons are not available in these materials, phonon is the main mode of heat transport (at least at lower temperatures). Glass and other noncrystalline ceramics have much lower thermal conductivity compared to crystalline ceramics as phonons are more susceptible to scattering in the materials lacking definite atomic order. With increasing temperature, thermal conductivity of materials decreases; but at higher temperatures, another mode of heat transport known as infrared radiant heat transport becomes active and its contribution increases as the temperature increases, especially in transparent ceramics. Porosity in ceramics also contributes to the reduction in thermal conductivity. Figure 5.58 shows the variation of thermal

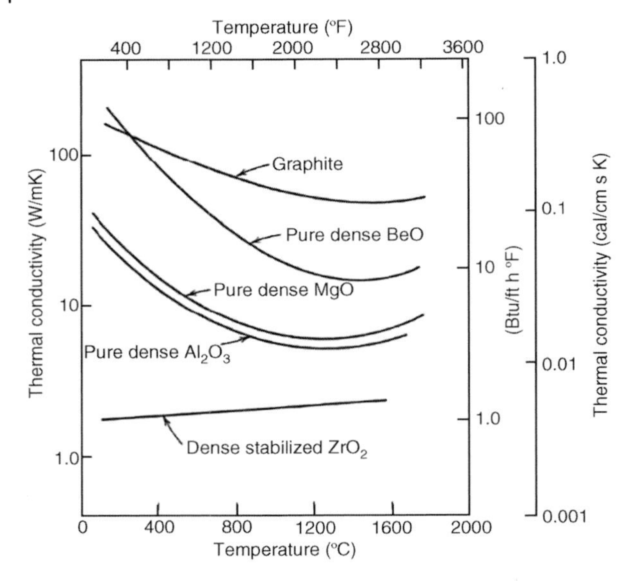

Figure 5.58 Dependence of thermal conductivity of different ceramic materials on temperature. From Ref. [1].

conductivity of certain ceramics as a function of temperature, and compared against that of graphite. The still air generally present in the pores has extremely low thermal conductivity, on the order of 0.02 W m^{-1} K^{-1}, thus giving the porous material low thermal conductivity. That is why thermal insulating materials are made porous. Figure 5.59 shows thermal conductivity of some oxide ceramics (some of nuclear importance) as a function of temperature.

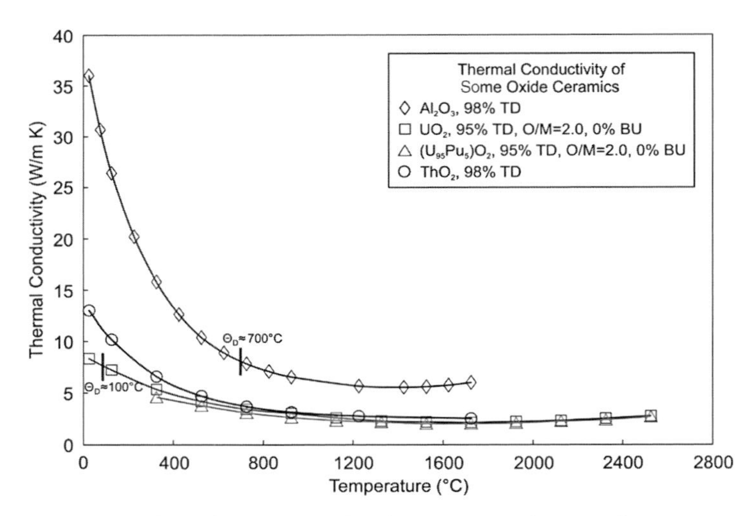

Figure 5.59 Thermal conductivity of oxide ceramics as a function of temperature *Courtesy*: Dr. Jack Henderson, Netzsch.

5.2.4
Summary

Thermophysical properties play an important role in the selection of materials as well as in their service performance in nuclear reactors. Here, three thermophysical properties (specific heat, thermal expansion coefficient, and thermal conductivity) are discussed. The effects of structure and composition on these properties are also highlighted. It also elucidates the effect of temperature on thermal conductivity, specific heat, and coefficient of thermal expansion in metals and nonmetals. However, it should be noted that there are various specific exceptions where the foregoing discussion may not apply.

5.3
Corrosion

Corrosion is a form of surface degradation of metallic materials via electrochemical means. Corrosion properties are not always an intrinsic property of the material since it is influenced by the chemical environment in which the material or the materials system exists. Corrosion is regarded as a life-limiting issue for nuclear reactor components that remain in contact with some kind of fluid for a long duration of time. For example, in LWRs, the core materials remain submerged under the coolant (pressurized water in PWRs or water plus steam environment in BWRs). In modern reactors, coolant chemistry is carefully controlled in order to cause minimum disruption coming from corrosion. However, corrosion is a natural process in a harsh environment such as a nuclear reactor. The cost of corrosion runs into several billions of dollars in the United States in the form of both direct and indirect costs. Note that although until 1960s corrosion was exclusively considered for surface degradation of metallic materials, the corrosion is now generically applied to describe environmental attacks on a wide variety of materials, including ceramics, polymers, composites, and semiconductors.

5.3.1
Corrosion Basics

We are already familiar with the concept of electrolytic cell (recall electrolysis of water into hydrogen and hydroxyl ions). Corrosion occurs in a reverse way. In an electrolytic cell, a voltage is applied between the anode and the cathode to dissociate the electrolyte. In order for corrosion to occur, an electrochemical cell needs to be set up and corrosion current is produced between the electrodes (anode and cathode). A schematic electrochemical cell (also known as galvanic cell) is shown in Figure 5.60, showing various cell components. The components of the cell are anode, cathode, electrolyte, and an external conductive circuit (or path) between the anode and the cathode. At anode, metal M loses electron(s) via the half-cell reaction, $M \rightarrow M^{+n} + ne^{-}$, where n is considered the valence of M. However, for reaction to proceed, both the electrons and the ions need to be removed. The

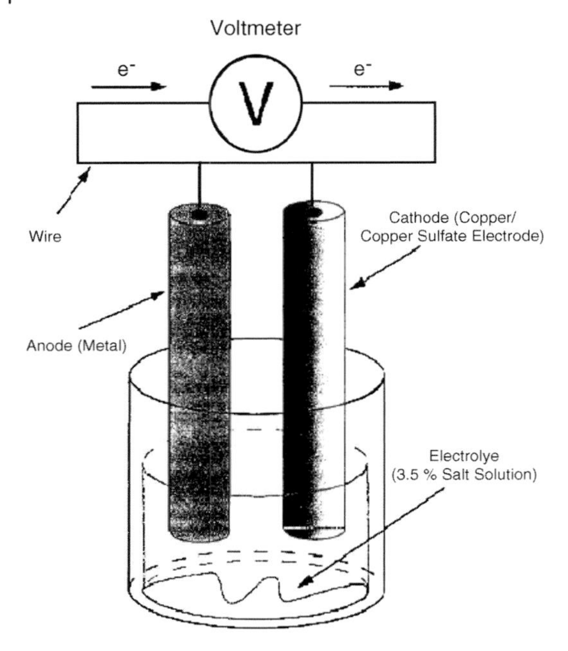

Figure 5.60 A schematic of a basic electrochemical cell.

electrons moving through the conductive path go to the cathode and get accepted by it following the half-cell reaction of $N^{+m} + me^{-} \rightarrow N$ (m is the valence of N) or by other metal ions. The metal ions thus created either remain dissolved in the electrolyte or react together to form insoluble surface deposit.

Basically, an oxidation reaction (i.e., loss of electrons) occurs at the anode and a reduction reaction (gain in electrons) occurs at the cathode with the acceptance of electrons. All metals are from their origin at a higher energy level and they are subject to oxidation reaction. In real world, one does not need to have the exact form of an electrochemical cell, as shown in Figure 5.60, in order for corrosion to occur. Anode and cathode can be set up on the same metal piece depending on different conditions. The metal piece itself can provide the conductive path. The chemical environment in which the metal piece exists will serve as the electrolyte. A schematic example of such a situation is shown in Figure 5.61. Generally,

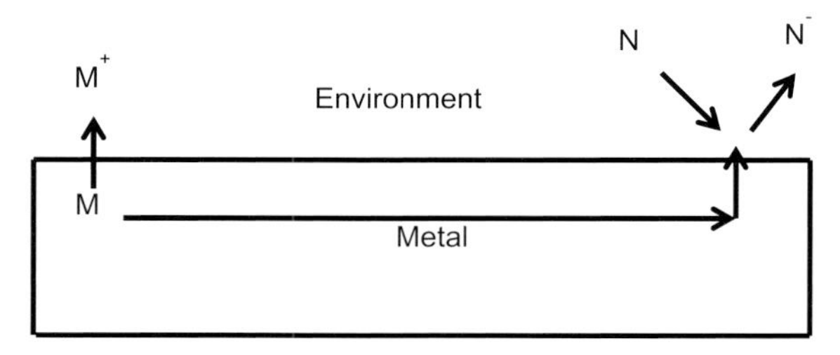

Figure 5.61 A schematic of the operation of a corrosion cell.

Table 5.4 Electrochemical series at 25 °C and 1 M solution.

Anode half-cell reactions	**Electrode potential (V)**
$Au \rightarrow Au^{3+} + 3e^-$	+1.50
$2H_2O \rightarrow O_2 + 4H^+ + 4e^-$	+1.23
$Pt \rightarrow Pt^{4+} + 4e^-$	+1.20
$Ag \rightarrow Ag^+ + e^-$	+0.80
$Fe^{2+} \rightarrow Fe^{3+} + e^-$	+0.77
$4(OH)^- \rightarrow O_2 + 2H_2O + 4e^-$	+0.40
$Cu \rightarrow Cu^{2+} + 2e^-$	+0.34
$H_2 \rightarrow 2H^+ + 2e^-$	0.00
$Pb \rightarrow Pb^{2+} + 2e^-$	−0.13
$Sn \rightarrow Sn^{2+} + 2e^-$	−0.14
$Ni \rightarrow Ni^{2+} + 2e^-$	−0.25
$Fe \rightarrow Fe^{2+} + 2e^-$	−0.44
$Cr \rightarrow Cr^{2+} + 2e^-$	−0.74
$Zn \rightarrow Zn^{2+} + 2e^-$	−0.76
$Al \rightarrow Al^{3+} + 3e$	−1.66
$Mg \rightarrow Mg^{2+} + 2e^-$	−2.36
$Na \rightarrow Na^+ + e^-$	−2.71
$K \rightarrow K^+ + e^-$	−2.92
$Li \rightarrow Li^+ + e^-$	−2.96

corrosion can be stopped if the electrical connection is interrupted or anode reactants are depleted, or cathode products are saturated.

Depending on the relative tendency of metals to lose electrons, one can measure the electrode potential of such a half-cell reaction. The electrode potential of any specific half-cell reaction is generally measured with respect to hydrogen reaction ($H_2 \rightarrow 2H^+ + 2e^-$) that is regarded as a reference and considered as 0 V under standard conditions (1 molar solution and 25 °C temperature). Iron is more active (or anodic) than hydrogen and its electrode potential (for the reaction, $Fe \rightarrow Fe^{2+} + 2e^-$) is −0.44 V under standard conditions. Based on the tendency of these reactions, electrochemical series as shown in Table 5.4 is developed. Note that gold is on the top of the chart, that is, at the noble end, and lithium is at the bottom of the chart, that is, at the active end. From the chart, one can predict which metal will corrode (i.e., act as anode) preferentially and which will be protected (act as cathode) when they are electrically connected.

As already stated, the electrochemical series given in Table 5.4 has been created based on standard conditions, that is, 1 molar solution at 25 °C (298 K). This chart is used by both electrochemists and corrosion engineers alike. There is another variation of this chart where the sign of electrode potential is reversed due to convention. However, the latter form of chart is mostly used by physical chemists and thermodynamists. In this chapter, we base our discussion on the former form of the electrochemical series. To calculate the actual electrode potential E (i.e., under nonstandard conditions), Nernst equation is used. The relevant expression is given as

$$E = E_{298} + \left(\frac{kT}{n'}\right)(\ln X), \tag{5.90}$$

Table 5.5 A condensed version of a galvanic series applicable in seawater.

	Galvanic series in seawater
Noble end	18-8 Stainless steel (type 316) (passive))
	18-8 Stainless steel (type 304) (passive))
	Titanium
	Nickel (passive)
	Copper
	Brass
	Tin
	Lead
	18-8 Stainless steel (type 316) (active))
	18-8 Stainless steel (type 304) (active))
	Pb–Sn solder
	Case iron
	Mild steel
	Alclad
	Aluminum
	Zinc
Active end	Magnesium

where E_{298} is the electrode potential under standard condition, X is the concentration in moles per liter of solution (or molar concentration), k is the Boltzmann's constant ($86.1 \times 10^{-6}\,\text{eV}\,\text{K}^{-1}$), and n' is the valence of the metal or species involved. In practice, the concentration of the electrolyte remains quite dilute ($X \ll 1$). For this reason, the electrode potentials tend to shift more to the anodic end. For example, at room temperature (25 °C), if the concentration of Fe^{2+} in solution is 0.01 M, the electrode potential for the anodic reaction would be −0.47 V compared to −0.44 V under the standard condition (i.e., 1 M).

In practice, we are interested to know the relative corrosion tendency of alloys in different environments. For this, the galvanic series is the more practical chart. One example of a galvanic series in seawater is given in Table 5.5. Magnesium is at the end of the series being the most active and 18/8 stainless steels (in passive condition) is on the top of the chart due to their ability to form chromium oxide-based passive film. Note that titanium and nickel that are more active in the electrochemical series are near the noble end of the galvanic series, by virtue of their ability to form passive films.

As already noted, the anodic reaction is the main reaction for corrosion, which is of the general form $M \rightarrow M^{+n} + ne^{-}$. However, electrons released during anodic reaction need to go to the cathode. So, we need to know some specific cathodic reactions, which are of interest. They are given in Table 5.6. Among them, the reaction for hydroxyl ion formation has some important implications. Hydroxyl ion formed would react with Fe^{3+} ion and form $Fe(OH)_3$ deposit, which is known as the red rust that forms on iron in moist, oxygen-rich environment.

Table 5.6 Some important cathodic reactions.

Type of reaction	Equation of cathodic reaction
Hydrogen evolution	$2H^+ + 2e^- \rightarrow H_2$
Hydroxyl formation	$O_2 + 2H_2O + 4e^- \rightarrow 4(OH)^-$
Water formation	$O_2 + 4H^+ + v4e^- \rightarrow 2H_2O$

5.3.2 Types of Corrosion Couples

Corrosion couples can be categorized into three broad categories: (i) composition cells, (ii) concentration cells, and (iii) stress cells.

5.3.2.1 Composition Cells

A composition cell can be established between two dissimilar metals. One example of it is the galvanized steel in which zinc coating is applied to carbon steel to protect it from rust and other environmental attacks. Here, zinc coating acts as a *sacrificial anode*, which means that the steel underneath is protected while zinc coating gets corroded. Thus, a metal higher on the electrochemical series can be used to reduce corrosion to a less active one. Note that from the relative positions of iron and zinc on the electrochemical series in Table 5.4, zinc is more anodic than iron. The unique aspect of sacrificial anode is that even if there is an accidental scratching on the zinc coating exposing the steel surface, the steel will still be protected until all zinc coating gets corroded. However, tin coating on iron or steel sheets is also used for protection in which case protection remains only up to the point when the tin coating acts as a barrier and is not compromised. However, if tin coating gets scratched or punctured, the corrosion effect will be concentrated in that punctured region (anode) with respect to remaining tin coating (cathode).

In the galvanized steel example, we found out benefits of the composition cell in corrosion protection. However, there are many other engineering examples of composition cells wherever dissimilar metals/alloys are made to work in contact and that situation is not desirable at all from an engineering point of view. A couple of examples include steel bolts in brass naval components and steel piping connected to a copper valve. Whenever a materials selection decision is made to use dissimilar metals/alloys in engineering applications, due care must be exercised. If not, composition cells will be set up and would cause undue galvanic corrosion problems. Galvanic cells can also be created at microlevel. Usually the single-phase metallic alloys are more corrosion resistant. As more second-phase particles are formed through precipitation, galvanic microcorrosion cells are set up between the particle and the matrix metal and lead to a poor corrosion resistance.

5.3.2.2 Concentration Cells

Concentration cells are formed due to the variation in the electrolyte composition. As noted before, according to the Nernst equation (Eq. (5.90)), an electrode in

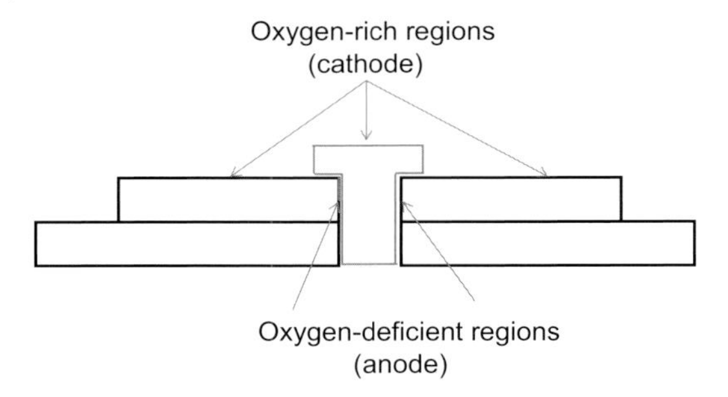

Figure 5.62 A schematic configuration of an oxidation cell involving a bolt-plate assembly: an example of crevice corrosion.

contact with a dilute electrolyte is more anodic compared to that in a concentrated one. Concentration cell aggravates corrosion where the electrolyte concentration is dilute. This concept can be explained by taking a simple example that involves a cell setup where one copper electrode is submerged in a dilute copper sulfate solution while another copper electrode in a concentrated solution. In this case, at the concentrated solution end, a higher concentration of Cu^{2+} will drive the reaction, $Cu^{2+} + 2e \rightarrow Cu^0$, with copper plating the electrode. Thus, that electrode becomes cathode. The electrode in the dilute solution corrodes and acts as anode following the reaction: $Cu^{2+} + 2e^- \rightarrow Cu$. This type of concentration cells is of interest in chemical plants and some flow–corrosion problems.

Oxidation-type concentration cells are of more importance. The oxidation cell aggravates corrosion where the oxygen concentration is low, for example, *crevice corrosion*. An example would be as simple as leaving a stone on an iron plate under a moist environment. After a few days, if you come back and check, you will find a rust formation in a much more aggressive manner under the region that the stone occupied. This happens because the cathodic reaction $O_2 + 2H_2O + 4e^- \rightarrow 4(OH)^-$ occurs due to the preponderance of oxygen and moisture outside the region blocked by the stone. But such cathodic reactions would not happen until electrons are supplied from the anode. The region underneath the stone does not have enough oxygen for the above reaction to happen and thus acts as anode and supplies electrons to the cathode regions. Thus, the area below the stone corrodes more. Similar oxidation cells can be formed in situations with steel bolts holding steel plates. The regions inside bolt head and shank do not get exposed to oxygen as much as the plate surface as shown in Figure 5.62. Thus, the region deficient in oxygen acts as anode and gets corroded.

5.3.2.3 **Stress Cells**

Stress cells do not involve compositional differences, but involve regions with stress gradients such as dislocations, grain boundaries, and highly stressed areas that become prone to corrosion. One example is the chemical etching of as-polished surface of a metallographic sample. The grain boundaries being the stressed regions act as anodes and get attacked by the etchant preferentially.

Another example would be a cold bent rod when submerged in water gets attacked more at the bent region, which acts an anode because of the high residual stress present in it. There are corrosion mechanisms (stress corrosion cracking (SCC) and corrosion fatigue) that are aided by the presence of stress.

Specific Examples of Corrosion Mechanisms

Intergranular Corrosion Intergranular corrosion occurs when corrosion attack develops along the grain boundaries. One important example is the sensitization of 18-8 austenitic stainless steel (SS 304). The steel contains 18 wt% Cr and 8 wt% Ni with a low carbon content. But the weldability characteristics of these stainless steels are poor because the carbon content is enough to form deleterious carbides at the grain boundary areas of the heat-affected zone upon welding as observed in the quasi-binary phase diagram (Figure 5.63). When the material is fusion welded, the area near the weld zone (i.e., the *heat affected zone* or HAZ) experiences a high temperature, but does not melt the metal. As the welded steel is cooled down, chromium carbide ($Cr_{23}C_6$) formation takes place at the grain boundaries, as illustrated in Figure 5.64a. Because these particles contain a high amount of chromium, they make the regions adjacent to the grain boundary lean in chromium (<12 wt%), as illustrated in Figure 5.64b. In order for the stainless steel to retain its corrosion-resistant (stainless) property, it must have at least 12 wt% Cr. Even though the grain interiors contain higher Cr contents, the areas surrounding grain boundaries in the HAZ become lean in Cr. This condition is known as *sensitization*. This type of sensitized steel becomes susceptible to intergranular corrosion. This is a serious problem for this type of stainless steels.

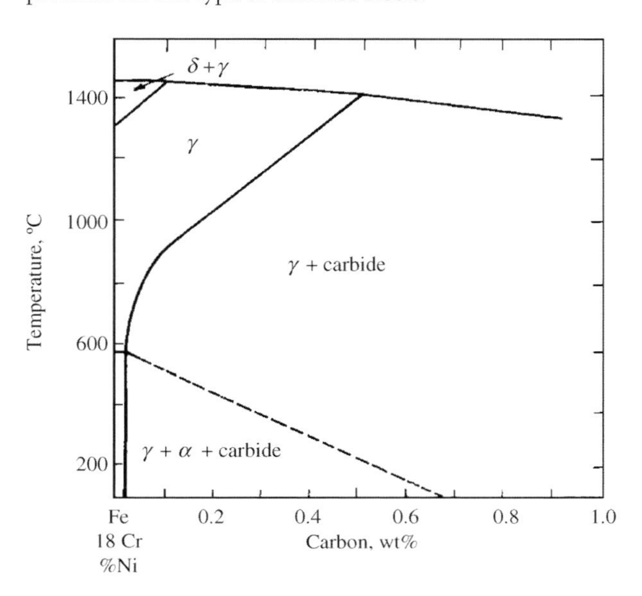

Figure 5.63 A quasi-binary phase diagram for a 18-8 stainless steel. Adapted from Ref. [8].

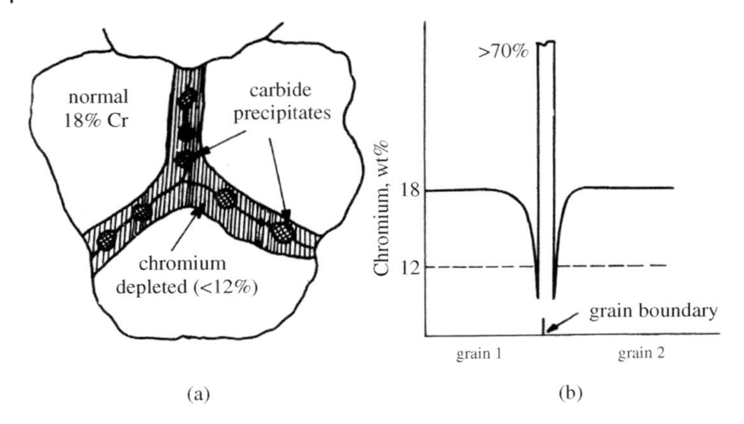

Figure 5.64 (a) A schematic of sensitized microstructure of a 304-grade stainless steel. (b) The chromium content at the grain boundary becomes high compared to the adjacent grain boundary areas, thus making it prone to the intergranular corrosion attack. From Ref. [8].

Sensitization problem can be solved in three different ways: (i) After welding, the plate is cooled rapidly across the temperature range where chromium carbide forms. If the chromium carbide formation can be avoided, there will be no problem of sensitization. However, quenching the plate may have other undue consequences of residual stresses or distortion. (ii) Special stainless steel grades with very low carbon content (~0.01 wt% C) have been developed such as in 304L stainless steel grades. If there is not much carbon present, there will be less amount of chromium carbide formation. (iii) Stabilized grades of austenitic stainless steels (like 316, 324, etc.) have been developed. Strong carbide-forming elements like molybdenum, titanium, niobium, and so on are added to the steel composition and during welding, corresponding carbides (TiC etc.) are preferentially formed throughout the bulk of the material (not just the grain boundaries) with not much carbon left for chromium carbide formation.

Stress Corrosion Cracking Stress corrosion cracking involves cracking of a material under static load by the combined action of stress and a chemical environment. SCC is generally found in alloys, not in pure metals. SCC occurs only in a specific environment for a given alloy. The presence of a tensile component of stress is necessary for SCC to take place. In majority of cases, the crack path is intergranular; transgranular cracking is rare. Cracking occurs in two stages: crack initiation and crack propagation. It has been noted that titanium alloys are immune to crack initiation in a chloride environment, but a precracked material shows susceptibility to crack propagation. Season cracking and caustic embrittlement are two examples of SCC. We will discuss this topic in detail in Section 6.4.4.

The crack growth mechanism that is involved is thought to be by anodic dissolution at the crack tip. However, in materials that form passive films, the crack tip does get exposed to the corrosion medium as the plastic deformation at the crack tip exposes fresh metal surface. Thus, an active–passive cell gets created between the

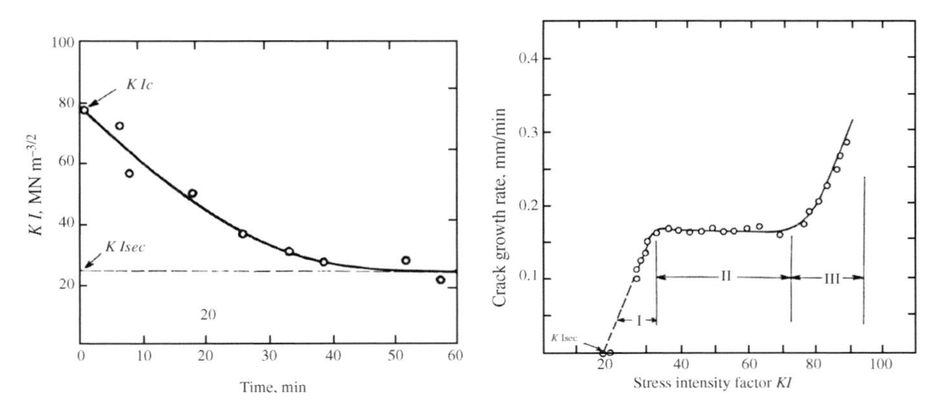

Figure 5.65 (a) The stress intensity factor versus time. (b) Crack growth rate in a Ni–Cr–Mo steel in a sodium chloride solution as a function of stress intensity factor, depicting three distinct stages of SCC. From Ref. [8].

crack tip and the crack faces. Since only a small area at the crack tip gets exposed, a very high current density is generated and corrosion occurs. In a laboratory environment, slow strain rate tensile testing in the chosen chemical environment is used to determine the SCC properties of a material. Another way to study the time-dependent fracture in a corrosive environment is to precrack a specimen and surround it in the corrosion environment and keep the specimen under constant load. The measured time to fracture is plotted as a function of mode-I fracture toughness (K_I) in Figure 5.65a. If the applied stress corresponds to the critical fracture toughness (K_Ic), fracture occurs without any delay time. When $K_\mathrm{I} < K_\mathrm{Ic}$, fracture occurs after a delay period implying that crack grows slowly. However, when the stress intensity factor is below a critical level (K_Iscc), no fracture occurs even after very long periods. Figure 5.65b shows the measured crack growth rates as a function of K_I. Below K_Iscc, the crack growth rate is essentially zero. Above K_Iscc in region-I, the crack propagates at an accelerating rate till it reaches region-II in which crack moves at a constant speed independent of the stress intensity factor. In region-II, the crack propagation is plausibly controlled by electrochemical factors. Finally, crack moves again at an accelerated rate resulting into fracture. The energy balance equation used to develop Griffith's equation needs to be modified as electrochemical energy is released due to anodic dissolution at the crack tip. The final equation for K_Iscc is given by

$$K_\mathrm{Iscc} = \left[2\gamma_\mathrm{p}E - \left(\frac{n'F\varrho\delta}{M}\right)V\right]^{1/2}, \tag{5.91}$$

where γ_p is the plastic work term at the crack tip, E is the Young's modulus, n' is the number of electrons taking part in the anodic dissolution process, F is Faraday's constant, ϱ is the density of the alloy, δ is the opening at the crack tip, M is the atomic weight, and V is the electrode potential. Following Eq. (5.91), K_Iscc decreases with an increase in yield strength (through decreasing γ_p) and with an increase in propensity toward corrosion (via electrode potential V).

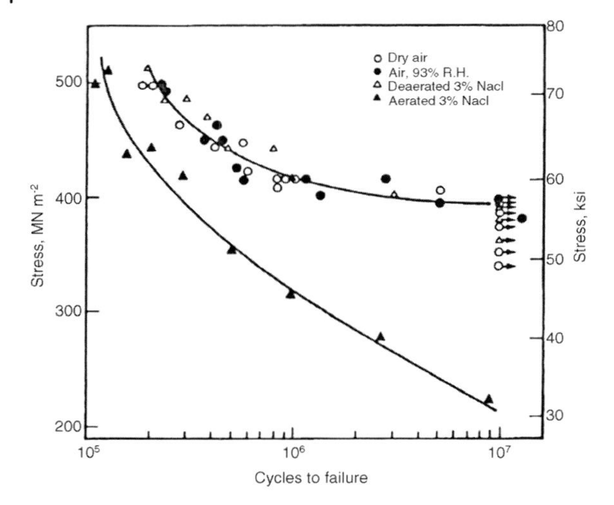

Figure 5.66 The deleterious effect of aerated aqueous chloride solution on the high cycle fatigue life of smooth AISI 4140 steel. From Ref. [15].

Corrosion Fatigue

Corrosion fatigue occurs when the simultaneous action of cyclic stresses and chemical environment is in play. Corrosive environment generally shorten the fatigue life and reduce the endurance limit, as shown in Figure 5.66. There are various ways in which corrosive environment can affect the fatigue behavior. One could be generating pits (pitting corrosion) that can lead to crack nucleation sites. Also, the crack growth rate is increased as a result of the chemical environment. The mode of load application also affects the corrosion fatigue behavior, such as reduction in loading frequency during fatigue testing may result in longer period of time through which the crack stays in opening mode in contact with the corrosive medium, leading to a reduction in fatigue life.

Corrosion Prevention Methods There are a host of corrosion prevention methods that can be introduced. The first choice of prevention would be to develop an intrinsically corrosion-resistant alloy if it is economical. One way is to create passive films naturally on the material. There are yet several other ways that can prevent or minimize corrosion. For example, even though aluminum appears as very active on the electrochemical series, it forms a thin, impervious layer of alumina, thus protecting it from a variety of corrosive environments. Anodizing is a commercial process by which a much thicker alumina film is developed on aluminum to have usefulness in marine environment. Generally, the thin passive film of aluminum oxide gets destroyed under the exposure of chloride ions in seawater. As already noted, stainless steel is mainly stainless because of the formation of a thin, impervious film of chromic oxide that occurs due to higher chromium content.

Inhibitors are sometimes used to provide passive surface film over surfaces. Inhibitors are highly oxidized chromate and the like solutions that are adsorbed on the metal surface to be protected. The passive layer formed by inhibitors works like

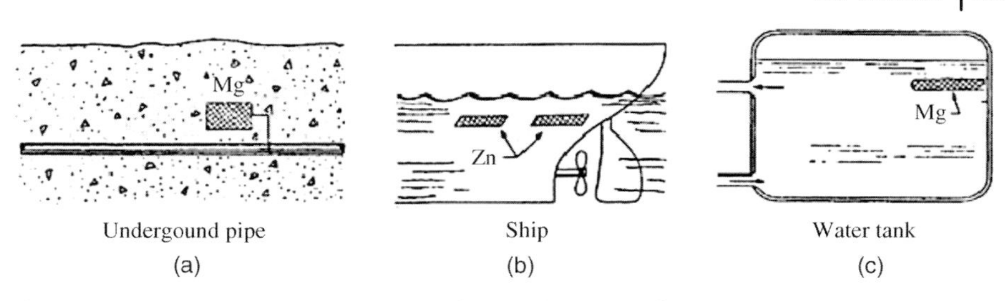

Figure 5.67 Corrosion prevention using sacrificial anode. From Ref. [16].

passive film, but cannot be regenerated. Hence, new inhibitors need to be applied if the layer gets washed out.

Galvanic protection is another way of preventing corrosion. As already mentioned, one of the ways of achieving galvanic protection is to use a more active metal as anode (sacrificial anode). The sacrificial anode needs to be replaced after it gets exhausted (Figure 5.67). Another way of achieving galvanic protection is to apply a direct current (DC) that feeds electrons into the metal part that needs to be protected. An example would include the use of impressed current minimizing or even stopping corrosion occurring at the underground pipeline (Figure 5.68).

Various types of paints, enamel coating, and so on are also used in various applications for prevention of corrosion. However, they need to be reapplied in regular intervals as they may exfoliate under service condition and restart corrosion.

5.3.3 Summary

Here, we have provided a succinct review of corrosion properties without going into details about the aspect. Corrosion is described as a surface deterioration phenomenon via electrochemical means. The importance of electrochemical and galvanic series is covered. Then we discussed about three types of corrosion cells: composition cells, concentration cells, and stress cells. A special case of intergranular corrosion as well as its prevention in austenitic stainless steels has also been discussed. Stress corrosion cracking is introduced. Some general corrosion prevention methods are summarized.

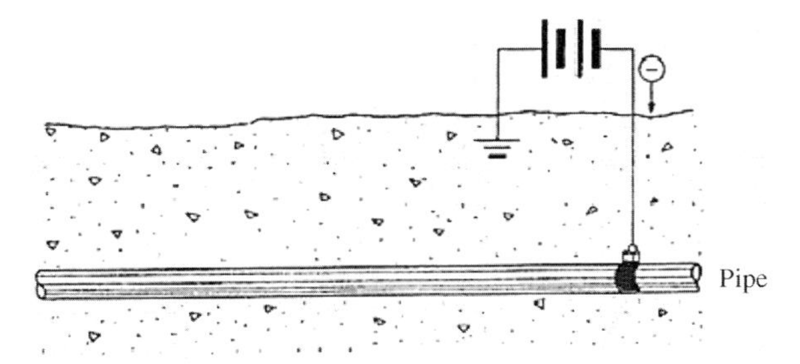

Figure 5.68 Corrosion prevention using impressed current. From Ref. [16].

Appendix 5.A
Typical Room Temperature Mechanical Properties of Some Material in the Annealed Condition (from Ref. [17])

Material	Yield strength (MPa)	Tensile strength (MPa)	Ductility, % elongation (in 50 mm (2 in.) gauge length)	Modulus of elasticity (GPa)	Shear modulus (GPa)	Poisson's ratio
Aluminum	35	90	40	69	25	0.33
Copper	69	200	45	110	46	0.34
Brass (Cu–30 wt % Zn)	75	300	68	97	37	0.34
Iron	130	262	45	207		0.27
Nickel	138	480	40	207	76	0.31
1020 Steel (Fe–0.2 wt% C)	180	380	25	207	83	0.30
Titanium	450	520	25	107	45	0.34
Molybdenum	565	655	35			
Tungsten				407	160	0.28
Magnesium				45	17	0.29
Lead				14		0.45
Al_2O_3		207		380		0.26
Si_3N_4				303		0.24
SiC (dense)		170		470		
PVC		41	2–30	2.8		

Appendix 5.B
K and *n* Values in Eq. (5.17) for Some Metals and Alloys (from Ref. [17])

Material	n	K (MPa)
Low carbon steel (annealed)	0.21 (0.26)	600 (530)
4340 Low alloy steel (annealed)	0.15	640
4340 Low alloy steel (tempered at 315 °C)	0.12	2650
304 Stainless steel (annealed)	0.44 (0.45)	1400 (1275)
Copper (annealed)	0.44 (0.54)	530 (315)
Aluminum (annealed)	0.20	180
Naval brass (annealed)	0.21	585
2024 Aluminum alloy (heat treated – T3)	0.17	780
AZ-31B magnesium alloy (annealed)	0.16	450

Problems

5.1 A tensile test conducted on a 2 in. long metal specimen revealed 0.1in. increase in length. (i) Calculate the strain. (ii) What will be the strain if the original length of the specimen is 3 in.

5.2 If a material exhibited a strain of 0.02% strain under an applied load of 4 kN on a specimen 20 cm long × 2 cm wide × 1 cm thick, calculate the modulus of elasticity of the material. If the Poisson's ratio of the material is 0.3, find the shear and bulk moduli.

5.3 A copper alloy specimen having a diameter of 5 mm and length of 25 mm has been tensile tested that revealed the final length to be 27.2 mm with the final diameter of 4.1 mm. Calculate the ductility.

5.4 (from Ref. [1].) A cylindrical specimen of stainless steel having a diameter of 0.505 in. (12.8 mm) and a gauge length of 2 in. (50.8 mm) is pulled *m* tension. Using the load–elongation data tabulated below, complete the problems (a)–(g).

Load		Length		Load		Length	
lb_f	N	in.	mm	lb_f	N	in.	mm
2850	12 700	2.001	50.825	26 800	119 400	2.030	51.562
5710	25 400	2.002	50.851	28 800	128 300	2.040	51.816
8560	38 100	2.003	50.876	33 650	149 700	2.080	52.832
11 400	50 800	2.004	50.902	35 750	159 000	2.120	53.848
17 100	76 200	2.006	50.952	36 000	160 400	2.140	54.356
20 000	89 100	2.008	51.003	35 850	159 500	2.160	54.864
20 800	92 700	2.010	51.054	34 050	151 500	2.200	55.880
23 000	102 500	2.015	51.181	28 000	124 700	2.230	56.642
24 200	107 800	2.020	51.308	Fracture			

a) Plot the data as engineering stress versus engineering strain.
b) Determine the modulus of elasticity.
c) Determine the yield strength at a strain offset of 0.002.
d) Determine the tensile strength of this alloy.
e) What is the approximate ductility, in percent elongation?
f) Compute the modulus of resilience.
g) Determine an appropriate working stress for this material.

5.5 Evaluate K and n from the data in problem 5.4 by making a double-log plot of true stress versus true strain.

5.6 For a tensile test, it can be shown that $d\sigma/d\varepsilon = \sigma$ at UTS (at P_{max} or at necking where $dP = \sigma dA + Ad\sigma = 0$; also known as tensile instability). Assuming that the standard work hardening law ($\sigma = K\varepsilon^n$) is obeyed, show that the value of the true strain [ε_u] at the onset of necking is given by the work hardening parameter [n].

5.7 The graph shows results of a tensile test of 1040 steel. Specific values at several points are given below (from Ref. [18]).

	Load (kN)	Length (cm)	Diameter (mm)
Initial	0.0	5.08	12.8
a	23.6	5.0848	12.796
b	27.1	5.19	—
c	46.4	5.56	12.4
d	52.5	6.25	10.9

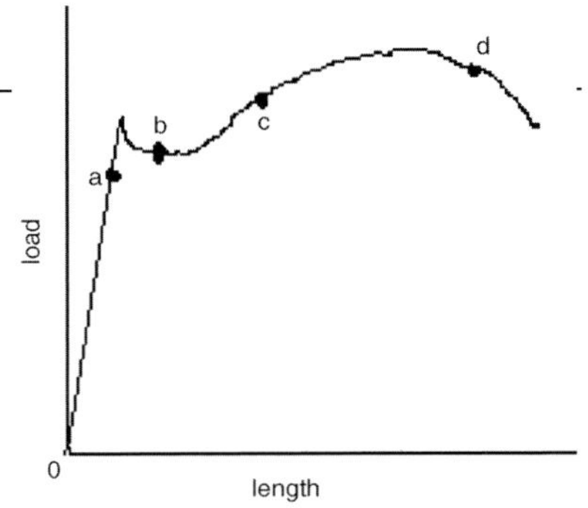

i) Calculate the constants K and n for the plastic strain hardening region.
ii) Calculate the modulus of elasticity.
iii) Estimate the yield strength.
iv) Calculate the maximum stress during the test.
v) Estimate the true uniform and necking strains (if the fracture dia is 10 mm).
vi) Evaluate the toughness of the material.

5.8 A tensile test performed on a copper alloy at 900 K to a strain of 0.15 at a strain rate of $3.32 \times 10^{-7}\,\text{s}^{-1}$ gave a flow stress value of 150 MPa and a sudden change of the strain rate at that strain from (0.15) to $1.98 \times 10^{-6}\,\text{s}^{-1}$ exhibited an increase in the flow stress to a value of 250 MPa. Calculate the strain rate sensitivity (m) of the material.

5.9 Show that the magnitude of the maximum stress that exists at the tip of an internal crack length of 0.001in. with crack tip radius of 10^{-5} in. under a tensile stress of 25 ksi is 354 ksi?

5.10 Show that the critical crack size for an edge crack in a plate is 4.66 in. if the stress is 14 ksi, the width of the plate is 70 in., the critical fracture toughness is 60 ksi in.$^{1/2}$, and the geometry factor Y is 1.12.

5.11 A sample of an Al alloy with an edge crack of length $a = 1.5$ mm fractures at a tensile stress of 364 MPa. If $E = 200$ GPa and $\gamma_s = 4\,\text{J m}^{-2}$ for the alloy, show that the critical fracture toughness (K_{Ic}) is given by 25 MPa m$^{1/2}$.

5.12 An aluminum alloy that has a plane strain fracture toughness of 25 ksi in.$^{1/2}$ fails when a stress of 42 ksi is applied. Observation of the fracture surface indicates that fracture began at the surface of the part. Estimate the size of the flaw that initiated the fracture to be 0.093 in. given $Y = 1.1$.

5.13 Fatigue life is usually estimated from Manson's method of universal slopes relating the fatigue life to the tensile parameters [1]: $\Delta\varepsilon = (S_u/E)N_f^{-0.12} + \varepsilon_f^{0.6}N_f^{-0.6}$. For a steel, $S_y = 30$ ksi, $S_u = 50$ ksi, $\varepsilon_f = 0.30$, and

$E = 30$ Mpsi. Estimate the limit on the total cyclic strain range if the steel is to withstand 4.9×10^5 cycles.

5.14 A514 steel (with the critical fracture toughness K_{IC} of 165 MPa m$^{1/2}$) containing an edge crack was tested in fatigue under conditions $a_0 = 2$ mm, $\sigma_{max} = 280$ MPa, and $\sigma_{min} = 140$ MPa. It was observed experimentally that fatigue crack growth in this steel can be described by a Paris law, $da/dN = 0.66 \times 10^{-8} (\Delta K)^2$ where da/dN is in m per cycle and ΔK is in MPa m$^{1/2}$. (a) Find the critical crack size at failure a_f to be 0.11 m. (b) Compute the fatigue life N_f. (c) Compute t_f at a cyclic frequency $\nu = 15$ Hz.

5.15 Suppose that a nickel-based alloy can safely undergo a creep rate of 0.01% h^{-1} at an applied stress of 10 000 psi and at a temperature of 600 °C. If the stress is raised to 15 000 psi, what is the temperature at which the creep rate reaches the same value? (For this alloy, $n = 5$ and $Q_c = 70$ kcal mol^{-1}.).

5.16 The following rupture times were obtained for stress rupture tests on an alloy steel:

Stress (psi)	Temperature (°F)	Rupture time (h)
80 000	1080	0.43
80 000	1030	6.1
80 000	1000	22.4
80 000	975	90.8
10 000	1400	1.95
10 000	1350	6.9
10 000	1300	26.3
10 000	1250	84.7

a) Estimate the value of the activation energy for creep (Q_c).
b) Establish the validity of the Larson–Miller parameter and plot the data appropriately (t_r versus LMP).

5.17 The following data were obtained for a copper alloy:

σ (MPa)	T (°K)	$\dot{\varepsilon}_{ss}$ (s^{-1})
150	900	3.32×10^{-7}
250	900	1.98×10^{-6}
150	780	1.70×10^{-8}

Evaluate the parameters A, n, and Q in the power-law creep equation: $\dot{\varepsilon} = A\sigma^n e^{-Q/RT}$.

5.18 a) If a copper ($A = 64$) specimen is exposed to monoenergetic (2 MeV) neutrons of flux ϕ ($= 2 \times 10^{15}$ n cm^{-2} s^{-1}) for 6 days continuously, show that the number of atomic displacements per atom (dpa) is 2.856 ($E_d = 22$ eV, $\sigma_{el} = 2$ b).

Assume that only one vacancy survived per billion atoms displaced (due to recombination).

b) What will be the (equivalent) temperature at which the same number of vacancies would exist in unirradiated specimen? ($Q_v = 28\,\text{kcal mol}^{-1}$, $Q_D = 52\,\text{kcal mol}^{-1}$)

c) Determine the enhancement in self-diffusivity at 400 °C due to radiation-produced vacancies?

d) If the steady-state creep rate ($\dot{\varepsilon}$) follows a power law, $\dot{\varepsilon} = AD\sigma^{4.5}$, calculate the %change in the creep rate due to radiation (assume that the same equation is valid before and after irradiation).

e) If the creep rupture life of the unirradiated material is 45 days at a stress (σ_0), estimate the rupture life of the irradiated specimen.

5.19 Describe briefly: galvanic corrosion versus crevice corrosion, stress corrosion cracking versus corrosion fatigue, and galvanized steel versus tin-coated steel.

5.20 Explain clearly why aluminum used as a sacrificial anode needs to be kept in a solution of chloride ions.

5.21 Explain the differences in the electrochemical mechanism of protection offered by (i) zinc (standard potential −0.76 V) and (ii) tin (−0.14 V) applied as coating (assume the substrate material is steel).

5.22 Calculate the electrode potential of a Zn plate immersed in a solution containing 0.01 molar Zn ion solution at a temperature of 80 °C.

5.23 Define sensitization? Explain the available methods for avoiding sensitization in welded austenitic stainless steels?

5.24 Identify what type/types of corrosion cell may be present in the following cases and identify the anode, cathode, and electrolyte in each case with rough sketches:

a) A copper valve in a steel tube exposed to a marine atmosphere.

b) A steel bolt joining two steel plates placed in open air.

c) Revealing the microstructure of a single-phase alloy through chemical etching.

d) A galvanized steel pipe underground.

5.25 A tensile test of a metal was run in the laboratory, but only the maximum load was measured during the test (59 kN). All other data were taken before or after the test. The initial diameter of the gauge length was 13 mm. The minimum diameter after test was 10 mm. Maximum diameter after test away from the fracture point was 11.6 mm.

a) What is the strain hardening exponent for this material?

b) Calculate the load on the specimen just prior to fracture.

5.26 A test bar having a diameter of 12.83 mm and a gauge length of 50 mm is loaded elastically with 156 kN and is stretched 0.356 mm. Its diameter is 12.08 mm under load. What is its shear modulus?

5.27 A sample of an alloy with an edge crack of length $a = 1.5$ mm fractures at a tensile stress of 364 MPa. If $E = 200$ GPa and $\gamma_s = 4\,\text{J m}^{-2}$ for the alloy, show that the critical fracture toughness is given by $25\,\text{MPa m}^{1/2}$.

5.28 A steel sheet is exposed to a cyclic tension–compression loading of magnitude 100 and 50 MPa, respectively. Prior to testing, the length of the largest surface crack is 2.0 mm. Estimate the fatigue life of the sheet if it has K_{Ic} of 25 MPa $m^{1/2}$ and the constant values of Paris law are 3.0 and 10^{-12}, respectively, for $\Delta\sigma$ in MPa and c in m.

5.29 Explain why specific heat does not follow values suggested by Dulong and Petit at elevated temperatures.

5.30 Thermal conductivity of ceramics is lower than most common metals. Why?

5.31 Explain why zinc coating can protect steel sheet.

5.32 Describe briefly: galvanic corrosion versus crevice corrosion and SCC versus corrosion fatigue.

Bibliography and Suggestions for Further Reading

1 Callister, W. (1994) *Materials Science and Engineering*, John Wiley & Sons, Inc., New York.

2 Dieter, G.E. (1988) *Mechanical Metallurgy*, McGraw-Hill.

3 Kehl, G.L. (1949) *The Principles of Metallographic Laboratory Practice*, McGraw-Hill, New York.

4 Inglis, C.E. (1913) Stresses in a plate due to the presence of cracks and sharp corners.

5 Griffith, A.A. (1920) The phenomenon of rupture and flow in solids, *Philosophical Transactions of Royal Society of London*, **221A**, 163–198.

6 Orowan, E. (1952) *Fatigue and Fracture of Metals*, John Wiley & Sons, New York.

7 Xu, T., Jin, Z., Feng, Y., Song, S., and Wang, D. (2012) Study on the Static and Dynamic Fracture Mechanism of Different Casing-Drilling Steel Grades, *Materials Characterization*, **67**, 1–9.

8 Raghavan, V. (1995) *Physical Metallurgy: Principles and Practice*, Prentice Hall, New Delhi.

9 Murty K.L., F.A. Mohamed and J.E. Dorn (1972) Viscous glide, dislocation climb and Newtonian viscous deformation mechanisms of high temperature creep in Al-3Mg, **20**, 1009.

10 Roesler, J., Harders H., and Baeker, M. (2006) *Mechanical Behavior of Engineering Materials*, Springer, Berlin.

11 Gaskell, D.R. (1995) *Introduction to Thermodynamics of Materials (3rd ed.)*, Taylor and Francis, Philadelphia.

12 Petit, A.-T.; Dulong, P.-L. (1819). "Recherches sur quelques points importants de la Théorie de la Chaleur" (in French). *Annales de Chimie et de Physique* **10**: 395–413.

13 Kopp, H. (1865) "Investigations of the Specific Heat of Solid Bodies", *Journal of the Chemical Society*, **4** (19): 154–234.

14 Ziman, J. (1967) *The Thermal Properties of Materials*, Scientific American.

15 Lee, H.H. and Uhlig, H.H. (1972) Corrosion Fatigue of Type 4140 High Strength Steel, *Metallurgical Transactions*, **3** (1972) 2949–2957.

16 Van Vlack, L.H. (1989) *Elements of Materials Science and Engineering*, 6th edn, Prentice Hall.

17 Suryanarayana, C. (2006) *Experimental Techniques in Mechanics and Materials*, John Wiley & Sons, Inc., New York.

18 Felbeck, D.K. and Atkins, A.G. (1984) *Strength and Fracture of Engineering Solids*, Prentice Hall.

19 Darken, L.S. and Gurry, R.W. (1953) *Physical Chemistry of Metals*, McGraw Hill.

20 Rose, R.M., Shepard, L.A., and Wulff, J. (1966) *Structure and Properties of*

Materials, vol. **4**, John Wiley & Sons, Inc., New York.

21 Jones, D.A. (1995) *Principles and Prevention of Corrosion*, 2nd edn, Prentice Hall.

22 Fontana, M.G. (1986) *Corrosion Engineering*, McGraw Hill.

23 Ahmad, Z. (2006) *Principles of Corrosion Engineering and Corrosion Control*, Butterworth-Heinemann.

24 Scully, J.C. (1975) *The Fundamentals of Corrosion*, Pergamon Press, Oxford, UK.

25 Murty, K.L., High Temperature Steady-State Creep ofIron, MS thesis, Cornell University, Ithaca, NY, 1967

26 O.D. Sherby and P.M. Burke, 'Mechanical Behavior of CrystallineSolids at Elevated Temperature,' *Prog. Mat Sci.*, vol. **13**, 1968.

Additional Reading

ASM Handbook, Mechanical Testing and Evaluation, Vol. 8, ASM International, USA.

6
Radiation Effects on Materials

> "Imagination is more important than knowledge. For knowledge is limited to all we now know and understand, while imagination embraces the entire world, and all there ever will be to know and understand."
>
> —*Albert Einstein*

We have learned about some aspects of the primary radiation damage in Chapter 3 and covered on the evaluation of damage in terms of displacements per atom (dpa). Radiation effects generally refer to the behavior (or properties) of materials in the aftermath of the radiation damage. As this book is on nuclear reactor materials, we will focus on the radiation effects caused by neutron irradiation. However, most of the radiation effects are universal in nature regardless of the type of radiation (i.e., proton, heavy ions, electrons) with specific different features. Let us make some general observations on the radiation effects: (i) defect concentration increases with fluence, (ii) nuclear transmutation occurs, (iii) chemical reactivity changes, (iv) diffusion increases, (v) new phases, both equilibrium and nonequilibrium, can take place, and (vi) impurities are produced. The changes of properties are, in general, proportional to neutron flux, irradiation time, and temperature. From the previous chapters, it has been very clear that the essence of materials science as applied to nuclear field is in structure–property performance. Figure 6.1 illustrates this very aspect. Defects formed during primary radiation damage evolve to form clusters and subsequently extended defects and change the microstructural features in various ways. These microstructural changes, in turn, alter the materials properties. That is why materials undergo extensive irradiation testing before they could qualify to be fit for use in nuclear reactors. In this chapter, while we will discuss general aspects of radiation effects, we will give examples from nonfuel (i.e., structural/fuel cladding, etc.) reactor materials. Some specific examples of radiation effects in fuels will be discussed in Chapter 7.

6.1
Microstructural Changes

Following the primary damage induced during a few picoseconds, the irradiated material goes through several stages of evolution, as described in Chapter 3, over a

An Introduction to Nuclear Materials: Fundamentals and Applicatioins, First Edition.
K. Linga Murty and Indrajit Charit.
© 2013 Wiley-VCH Verlag GmbH & Co. KGaA. Published 2013 by Wiley-VCH Verlag GmbH & Co. KGaA.

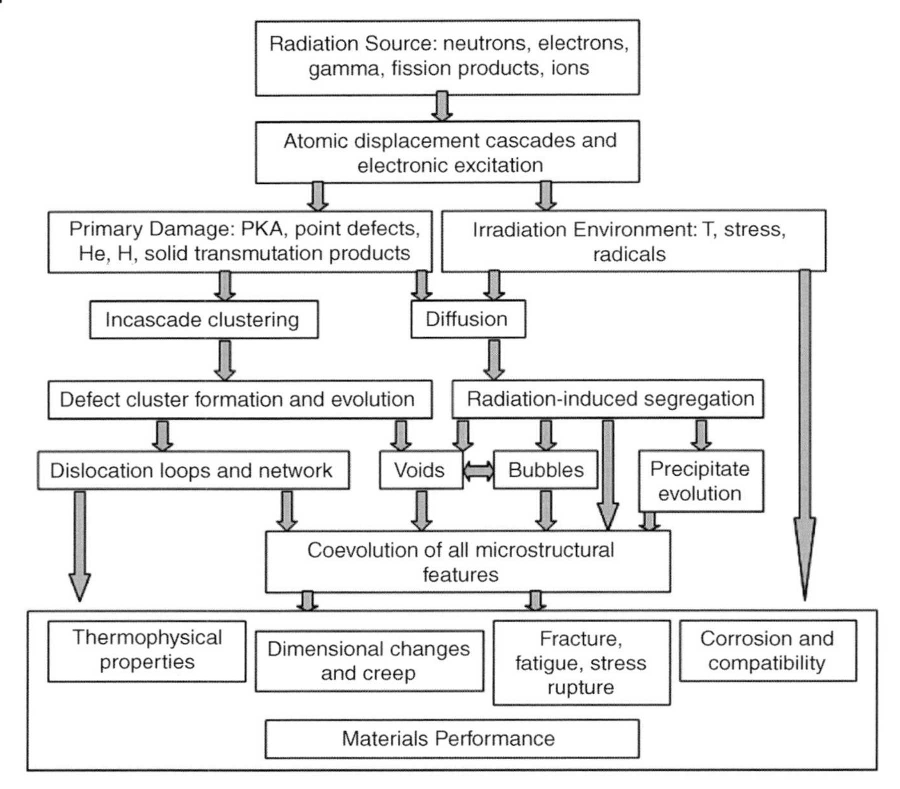

Figure 6.1 A flowchart showing various causes of radiation damage and their consequent effects. Courtesy: US Department of Energy.

long period of time. We need to now discuss how the higher order defects are formed from the primary damage defects in the irradiated materials. For this, let us revisit the specific characteristics of two primary damage defects: vacancies and self-interstitial atoms summarized in Table 6.1.

Table 6.1 Comparison between SIAs and vacancies.

Metric	SIA	Vacancy
Formation energy	Higher (>2 eV)	Lower (<2 eV)
Relaxation volume[a)]	$\sim +1\,\Omega$ to $+2\,\Omega$	$-0.1\,\Omega$ to $-0.5\,\Omega$
Migration energy	Lower (<0.2 eV)	Higher (>0.5 eV)

a) *Relaxation* (or excess) *volume* is generally associated with a defect leading to some form of internal stress and is expressed in terms of atomic volume (Ω). *Relaxation volume* basically measures the distortion volume induced in the lattice due to the insertion of a vacancy or interstitial. Most calculations of relaxation volume assume lattice as an elastic continuum. The relaxation volume for SIA is considered positive (because of the volume increase), while the relaxation volume for vacancies is considered negative for monovacancies (because of the volume shrinkage).

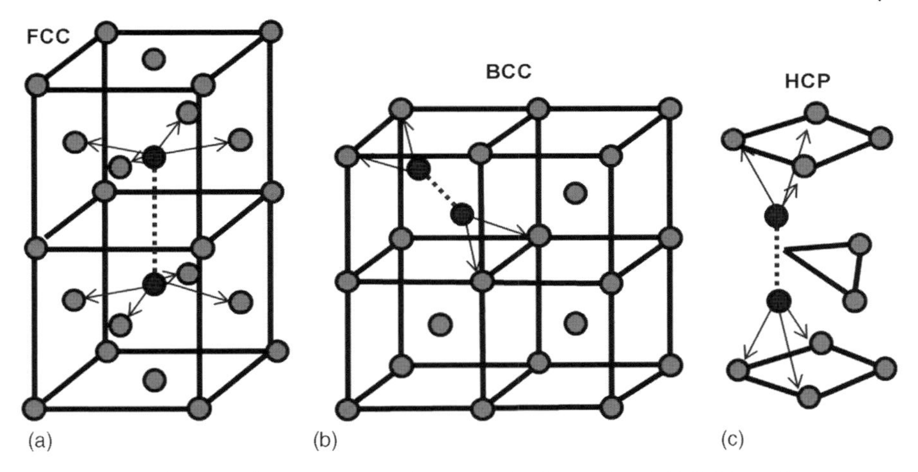

Figure 6.2 Single SIA configuration in (a) FCC metal, (b) BCC metal, and (c) HCP metal.

The high relaxation volume due to SIAs causes large lattice distortions, which create a strong interaction with other SIAs and other lattice defects (dislocations, impurity atoms, etc.). As we noted, the classical picture of self-interstitial like the interstitial impurity atoms is untenable energetically. Thus, the single SIAs become stable only in a dumbbell or split interstitial configuration around a single lattice site, as shown in Figure 6.2. The dumbbell axis is generally found to be along $\langle 100 \rangle$ in FCC metals, $\langle 110 \rangle$ direction in BCC metals, and $\langle 0001 \rangle$ in HCP metals. Multiple interstitials (interstitial clusters) are created by the aggregation of mobile SIAs at higher temperatures. Multiple interstitials have a high binding energy ($\sim$1 eV). Figure 6.3(a) and (b) show two di-interstitial configurations in FCC and BCC lattices, respectively.

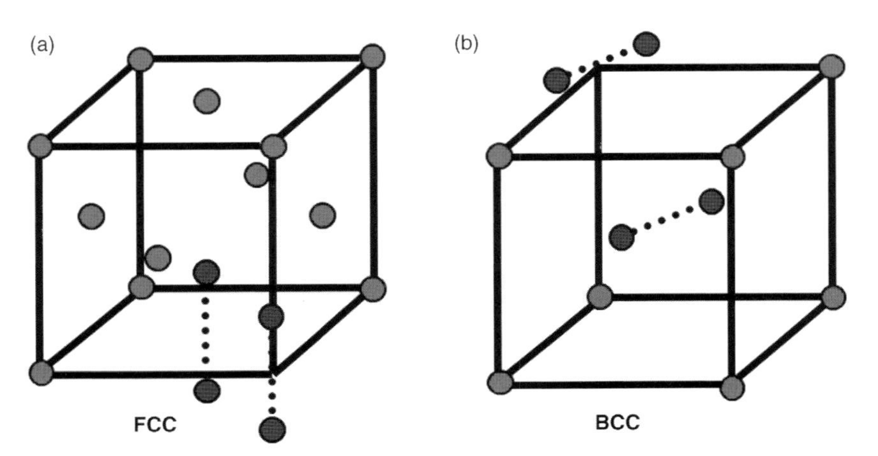

Figure 6.3 Di-interstitial configurations in (a) FCC and (b) BCC metals.

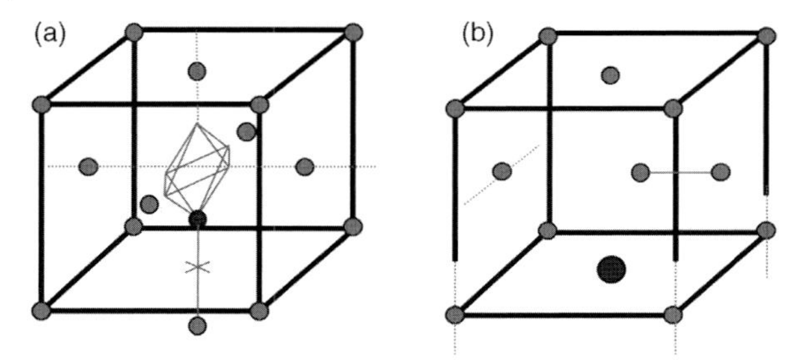

Figure 6.4 (a) Mixed dumbbell configuration of interstitial–impurity (undersized) complex formation. (b) The configuration of interstitial–oversized impurity complex.

Impurity atoms can act as effective traps for SIAs. Stable interstitial–impurity complexes having undersized impurity atoms (with respect to the host lattice atom) do not dissociate thermally under a certain temperature range where vacancies become mobile. Binding energy of the interstitial–impurity complexes vary typically between 0.5 and 1.0 eV. However, weaker trapping is generally observed with oversized impurity atoms. Figure 6.4 shows an interstitial–undersized atom complex. It has got a mixed dumbbell configuration that is stable. This configuration however can reorient itself through jumping of the undersize impurity atom across the vertices of the octahedron (forming a cage-like structure) shown in Figure 6.4. The activation energy associated with this type of motion, known as cage motion, is quite small, on the order of 0.01 eV.

Smaller binding energies (~0.1 eV) are associated with multiple vacancies compared to multiple interstitials, often observed in irradiated metals. Various configurations of vacancies are shown in Figure 6.5. The migration energy of divacancies is

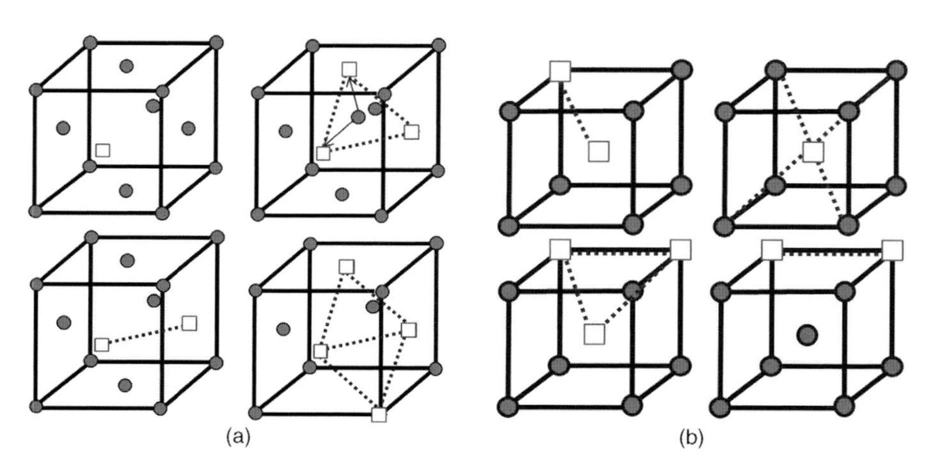

Figure 6.5 Schematics of some vacancy configurations shown in (a) FCC unit cells, and (b) BCC unit cells.

less than that for two monovacancies (0.9 versus 1.32 eV for Ni). We have commented on the energetics of divacancy formation in Section 2.2. Tetravacancies can only migrate by dissociation. Nonetheless, it can act as the first stable nucleus for further clustering.

Vacancies can bind with oversize solute/impurity atoms in order to lower the overall free energy of the solid. Estimates of the binding energy of a vacancy to an oversize solute in an FCC lattice range from ~0.2–1.0 eV. These solutes can act as efficient traps for vacancies in the lattice.

6.1.1 Cluster Formation

The fraction of defects produced in a cascade is between 20–40% of that that is predicted by the NRT model because of intracascade recombination. If the clusters are stable, they may migrate away from the cascade region and can be absorbed at various sinks such as dislocations and grain boundaries. In general, vacancy clusters and interstitial clusters should be treated separately. Interstitial clusters are stable, whereas vacancy clusters are not. Interstitial clusters possess higher mobility than their vacancy counterparts.

Incascade clustering is important as it helps promote nucleation of extended defects. Interstitial cluster occurs either in the transition phase between the collisional and thermal spike stages or during the thermal spike stage. The probability of clustering is enhanced with increase in the PKA energy with interstitial clustering being predominant, as shown in Figure 6.6.

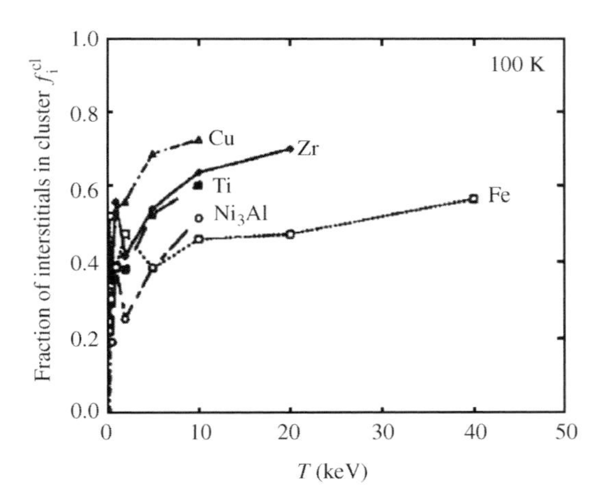

Figure 6.6 The fraction of SIAs that survive as clusters containing at least two interstitials in several metals and Ni_3Al – MD simulation results Ref. [1].

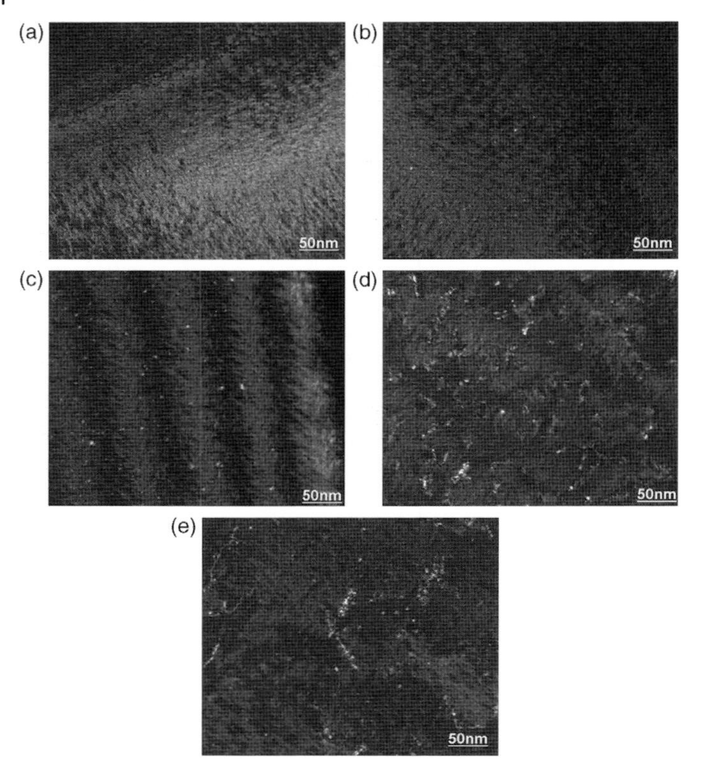

Figure 6.7 Weak beam dark field TEM images of defect clusters in neutron-irradiated molybdenum at different dose levels: (a) $7.2 \cdot 10^{-5}$ dpa, (b) $7.2 \cdot 10^{-4}$ dpa, (c) $7.2 \cdot 10^{-3}$ dpa, (d) 0.072 dpa, and (e) 0.28 dpa Ref. [2, 3].

The structure of clusters is generally a strong function of the crystal structure. In α-Fe (BCC crystal structure), the most stable configuration is small clusters (<10 SIAs), a set of ⟨111⟩ crowdions. The next in stability is the ⟨110⟩ crowdions. As the cluster size grows, only two configurations become stable: ⟨111⟩ and ⟨110⟩. These crowdions can act as the precursor for the formation of perfect interstitial loops. Figure 6.7 shows a few TEM weak beam dark field images of small interstitial type loops in fast neutron irradiated molybdenum (BCC).

In copper (FCC crystal structure), the ⟨100⟩ dumbbell configuration is the stable configuration of the SIA; the smallest cluster may contain only two such dumbbells (di-interstitials). Larger clusters could be a set of ⟨100⟩ dumbbells or a set of ⟨110⟩ crowdions each with {111} habit plane. During growth, the clusters change to faulted Frank loops with Burgers vector (1/3)⟨111⟩ and to perfect loops with (1/2)⟨110⟩. Figure 6.7 shows a few TEM weak beam dark field images of small interstitial type loops in fast neutron irradiated molybdenum.

Note

Crowdion

Crowdion takes place when an atom is added to a lattice plane, yet it does not stay in an interstitial position. To accommodate the atom, lattice atoms numbering over 10 or more in a particular direction are all shifted with respect to their lattice sites. A crowdion configuration is shown in Figure 6.8. Also, see crowdion configuration in Seeger's model as illustrated in Figure 3.1(b). The configuration can resemble a dumbbell spread over 10 atoms along a row. This phenomenon is a regular feature in focusons (i.e., focusing collisions). However, these configurations are not stable and attempt to go back to the original configuration as the knock-on atom energy is dissipated.

Clustering of the vacancies occurs within the core of the cascade and the extent of clustering varies with the host lattice. Based on size and density measurements of vacancy clusters, the fraction of vacancies in clusters is estimated to be less than 15%. The stability of vacancy clusters is low relative to the interstitial clusters.

Alpha-Fe (BCC): A set of divacancies on two adjacent {100} planes (that can transform to a dislocation loop of Burgers vector ⟨100⟩) or a set of first nearest neighbor vacancies on a {110} plane (that can change to a perfect dislocation loop with (1/2)⟨111⟩ Burgers vector).

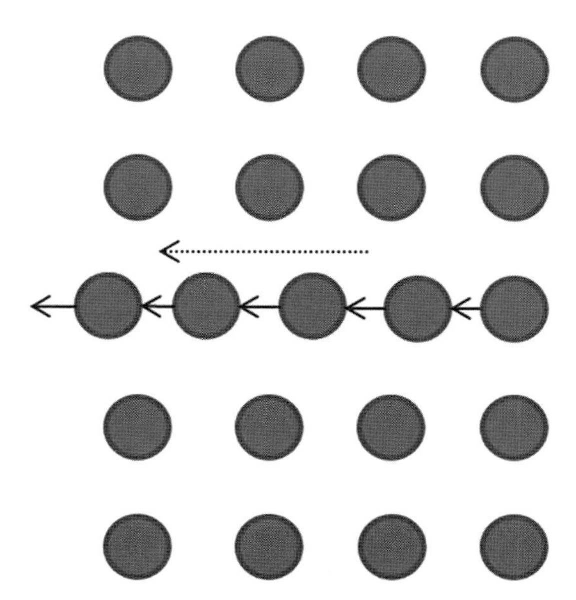

Figure 6.8 A schematic configuration of a crowdion.

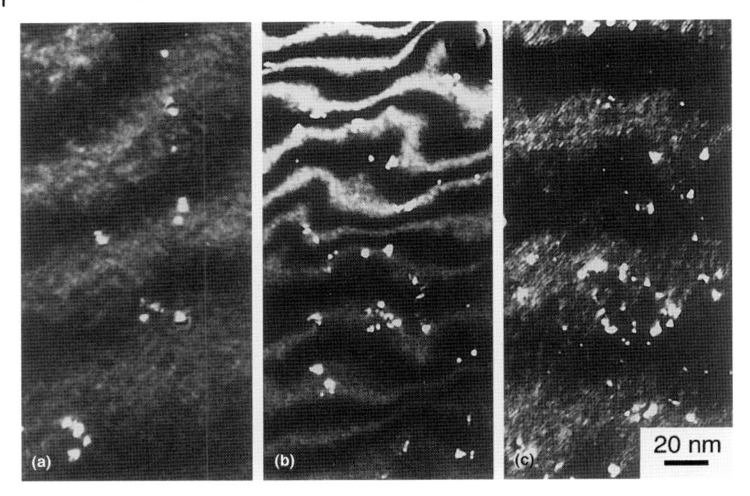

Figure 6.9 Dark field TEM images showing the defect structures of (a) gold at 3.5×10^{20} n/m^2, (b) silver at 2.1×10^{20} n/m^2 and (c) copper at 1.5×10^{21} n/m^2, irradiated as thin foils with 14 MeV fusion neutrons at room temperature Ref. [4].

Copper (FCC): The most stable configurations are the stacking fault tetrahedron (SFT) and faulted clusters on {111} planes that form Frank loops with Burgers vector $(1/3)\langle111\rangle$. The binding energy per defect in vacancy cluster is much less than that for interstitial clusters.

Figure 6.9(a), (b) and (c) show the defect structures (consisting of primarily SFTs) in fusion neutron-irradiated gold, silver and copper (all FCC), respectively. The majority of vacancy clusters in FCC metals and alloys have the shape of stacking fault tetrahedra and they appear as white triangles when viewed along the [110] direction with a proper weak beam dark field imaging condition. SFTs are created from Frank dislocation loops and subsequent dislocation reactions. An SFT consists of four triangular {111} planes as faces and 1/6<110> type stair-rod dislocations as six sides.

6.1.2
Extended Defects

Figure 6.10 illustrates schematically all the point defect and other higher order defects that may occur. Besides stable faulted Frank loops, SIAs may form metastable arrangement of SIAs that do not reorganize into a stable, glissile form by the end of thermal spike. They are important because they do not migrate away from the cascade and act as precursors to extended defects. Mobile clusters can interact with other clusters or with impurity atoms. Vacancy clusters that form perfect dislocation loops are also intrinsically glissile. Formation of dislocation loops and voids occurs from defect clusters. Interstitial-type dislocation loops in deuterium-irradiated molybdenum and tungsten are shown in Figure 6.11.

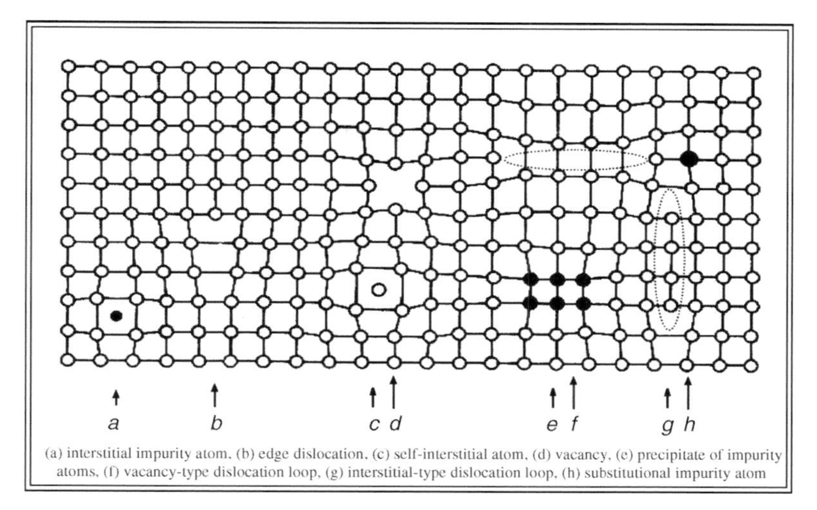

(a) interstitial impurity atom, (b) edge dislocation, (c) self-interstitial atom, (d) vacancy, (e) precipitate of impurity atoms, (f) vacancy-type dislocation loop, (g) interstitial-type dislocation loop, (h) substitutional impurity atom

Figure 6.10 Schematics of various radiation-induced defects in crystals. Courtesy: Professor Helmut Föll, University of Kiel, Germany.

6.1.2.1 Nucleation and Growth of Dislocation Loops

Dislocation loops resulting from vacancy and interstitial condensations are created from clusters of the respective defects, and either shrink or grow depending on the flux of defects reaching the embryo. Once they have reached a critical size, the loops become stable and grow until they unfault by interaction with other loops or with the network dislocations. Russell and Powell (1973) [5] have determined that interstitial loops nucleate much easier than vacancy loops because interstitial loop nucleation is less sensitive to vacancy involvement than is vacancy loop nucleation to interstitial involvement. The nucleation of loops is essentially a clustering process in which enough of one type of defect needs to cluster, in the presence of other types of defects, to result in a critical size embryo that will survive and grow.

The effects of irradiation temperature on the faulted Frank loop size and density in irradiated cubic silicon carbide (SiC) are shown in Figure 6.12(a) and (b), respectively. As the loops grew in size, their number density decreased. Figure 6.13 shows the effect of radiation dose (in dpa) on the faulted Frank loops in ion-irradiated cubic SiC at an irradiation temperature of 1400 °C.

6.1.2.2 Void/Bubble Formation and Consequent Effects

We have, so far, discussed various effects that energetic radiations can produce in reactor materials. Void swelling is one of them. Increase in strength and decrease in ductility are some of the commonly observed effects of void formation, which negatively affect effective life of the reactor components, mostly in-core components (such as fuel cladding). In 1967, using transmission electron microscopy, Cawthorne and Fulton gave the first experimental evidence of radiation-produced voids in LMFBR stainless steel fuel cladding tubes causing void swelling due to the fast neutron irradiation at reactor ambient temperatures (400–600 °C). Figure 6.14a

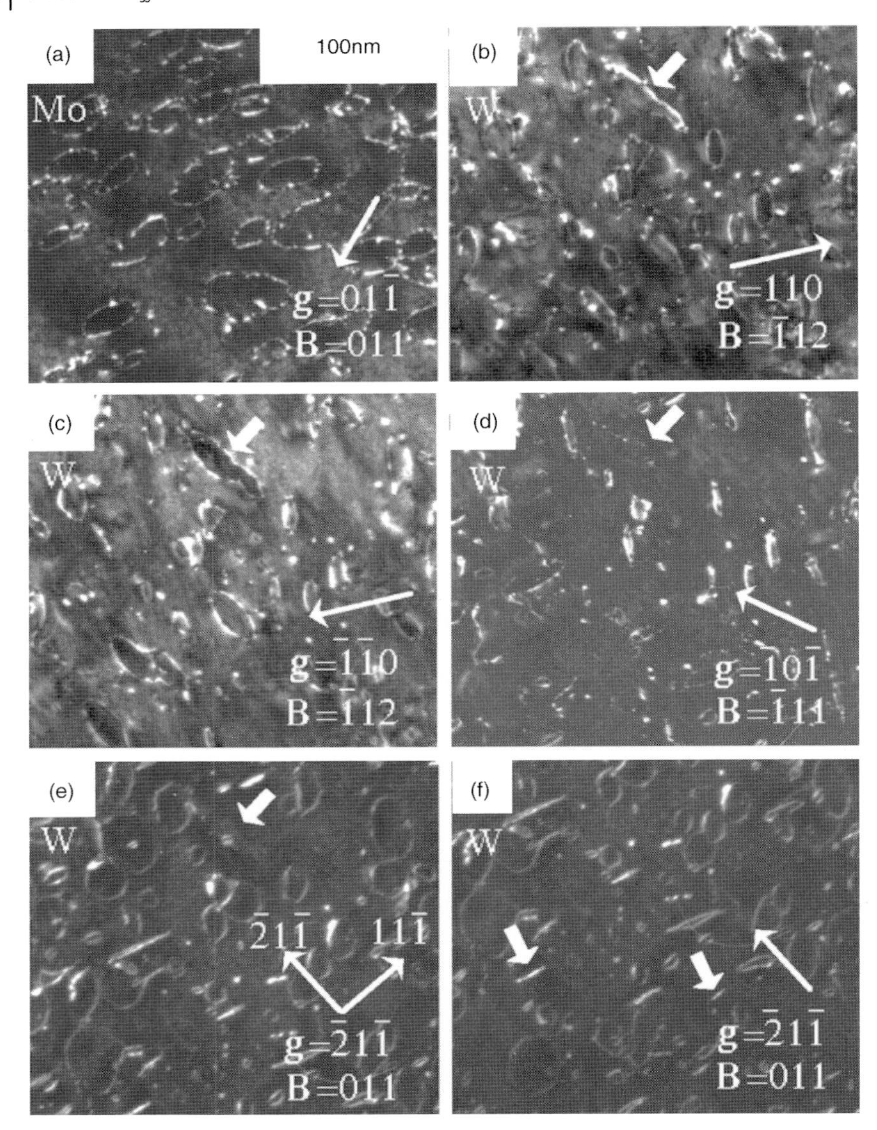

Figure 6.11 Dark field images of dislocation loops produced by 10 keV deuterium ions at room temperature: (a) in molybdenum (at 2×10^{22} ions/m^2) and (b)–(f) in tungsten (at 5.0×10^{21} ions/m^2) Ref. [6].

depicts the microstructure of voids in stainless steel and Figure 6.14b the consequent macroscopic swelling of fuel rods. Voids contain some helium gas (generated due to transmutation reactions), but generally do not contain sufficient amounts to be called "bubbles." But generally their interrelation is very strong, and will be described in the following sections.

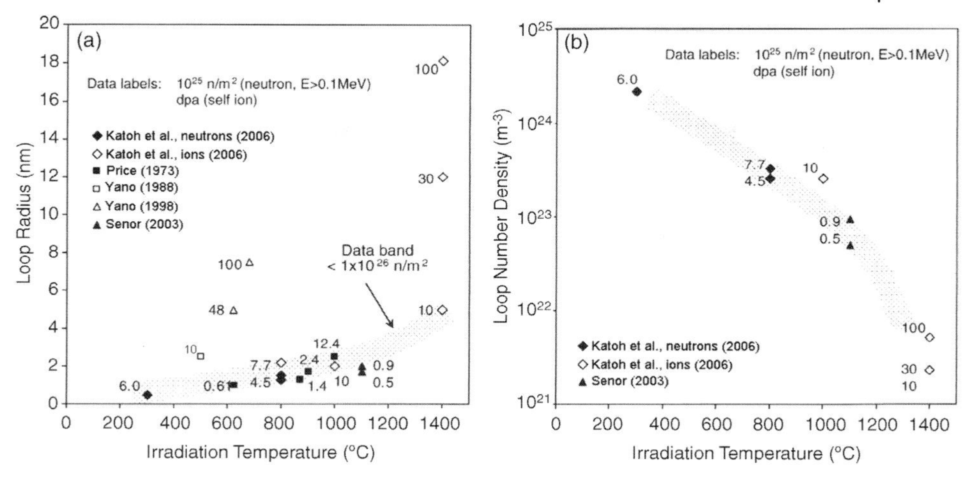

Figure 6.12 The variation of (a) Frank loop size and (b) number density as a function of irradiation temperature in irradiated cubic silicon carbide Ref. [7].

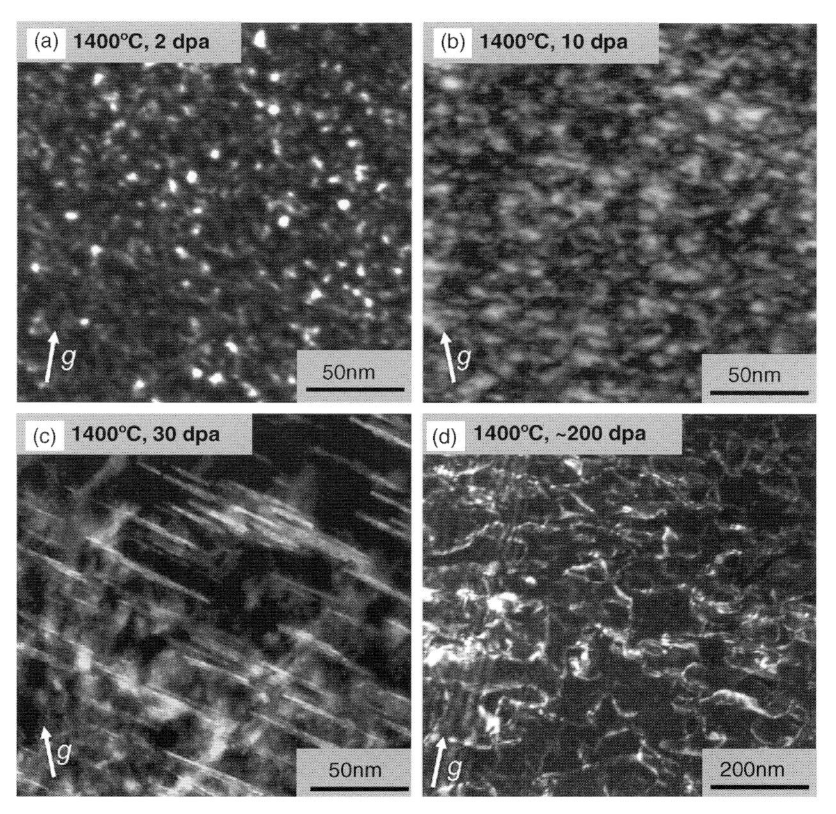

Figure 6.13 Dose dependent evolution of dislocation microstructure (mainly faulted Frank loops) in cubic silicon carbide irradiated by Si^{+2} ion at 1400 °C Ref. [7, 8].

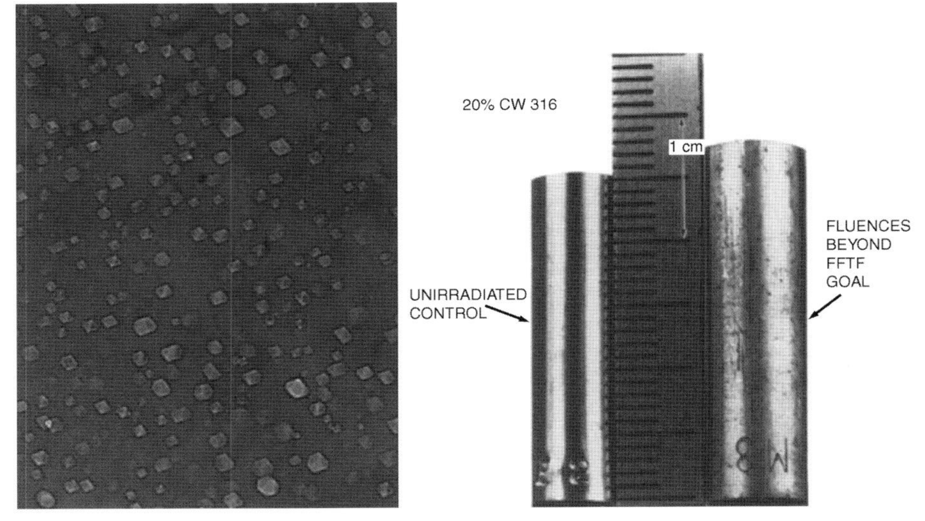

Figure 6.14 (a) Void microstructure in a proton-irradiated Fe–15Cr–20Ni alloy to 10 dpa at 500 °C. (marker scale is not provided: the average diameter is measured to be about 25 nm) Murase et al. [9]. (b) Macroscopic swelling (~10% linear as measured by length change, ~33% volumetric, as measured by density change) observed in unfueled 20% cold worked AISI 316 open cladding tube at $1.5 \times 10^{23}\,\mathrm{n\,cm^{-2}}$ ($E > 0.1$ MeV) or ~75 dpa at 510 °C in EBR-II. Note that in the absence of physical restraints, all relative proportions were preserved [10].

Is Void Swelling Observed Only in Stainless Steels?

Void formation is not unique to stainless steels only; as a matter of fact, all metals and alloys swell through void formation in the homologous temperature range of 0.3–$0.55T_m$ (where T_m is the melting point of the material in Kelvin). There may well be a variation in void swelling behavior depending on the material composition, He content, grain size, temperature, and cold working. (Void formation is also a problem in the fusion technology. Fast neutrons produced from the plasma can cause void formation in the first wall blanket material with (n,α) reactions for neutron energy increasing above 5 MeV. Increase in helium content generally increases the propensity for void swelling. The problem of void swelling may be expected to be more severe in fusion reactors than in LMFBRs due to the relatively higher neutron energies.)

Origin of Void Swelling

Let us be very clear that the origin of void swelling has been a topic of study for many years and still remains an active area of research interest. Fast neutrons tend to readily produce defects like vacancy–interstitial (Frenkel) pairs as a result of their interaction with the lattice atoms. Most of these defects recombine with each other or migrate to the sinks following their formation. The defects remain in a dynamic balance and maintain steady-state concentrations that are more than those of thermal equilibrium. Both vacancies and interstitials cluster together if the temperature is high enough for diffusion, yet not so high that the defect supersaturation is maintained by not allowing point defects to be annihilated by recombination or migration to sinks. A cluster of interstitials forms a dislocation loop, while vacancies can cluster in two different ways:

a) Vacancies can agglomerate into platelets that collapse into dislocation loops
b) Vacancies can form three-dimensional clusters known as voids.

The driving force for void nucleation is given by supersaturation of vacancies due to irradiation defined by

$$S_v = C_v/C_{vo}, \tag{6.1}$$

where C_v and C_{vo} are the total vacancy concentration and thermal equilibrium concentration of vacancies, respectively. If we work on the energetics of void formation, one could see that void is the stable form for small clusters of vacancies. However, as the number of vacancies in the cluster grows, the loops become energetically favorable configurations. But the collapse of void embryo into a vacancy loop is not favored by the presence of inert gas (such as He) in the void and that is why voids can survive and grow. Now the question remains why radiation-produced point defects form separate interstitial loops and voids. Since the vacancies and interstitials formed in equal numbers by fast neutron irradiation, it is expected that point defects of both types would diffuse to voids at equal rates and hence produce no net growth of the voids. Since voids represent accumulated excess vacancies, the interstitials must be preferentially absorbed elsewhere in the solid. The preferential interstitial sink is dislocation because dislocations interact more strongly with interstitials than vacancies because they distort the surrounding lattice more than vacancies. The preferred migration of interstitials to dislocations leaves the matrix metal depleted in interstitials relative to vacancies at somewhat greater rate than interstitials, and growth results.

The following are the conditions necessary for void swelling [11]:

a) Both vacancies and interstitials must be mobile in the solid.
b) At least one type of sink must differentiate between interstitials and vacancies, and should have more interactions with other types of defects.
c) The supersaturation of vacancies must be large enough to permit voids and dislocation loops to be nucleated and to grow.
d) Trace quantities of insoluble gases like He must be present to stabilize the embryo voids and prevent their collapse to vacancy loops.

Void Distribution Function

The presence of voids in irradiated materials can be primarily measured from transmission electron microcopy studies. Only this kind of examination can provide information on the void size and density. The extent of swelling is generally measured by Archimedes principle by immersing the irradiated material in a fluid of known physical density and compared with the original volume (V), that is, it is measured in terms of the ratio of swelling-induced volume change ($\Delta V/V$) to the original volume. For narrow void distribution, the term is given by the following relation:

$$\frac{\Delta V}{V} = \frac{4}{3}\pi \bar{R}^3 N_v, \tag{6.2}$$

where $\bar{R}$ is the average void radius and N_v is the number of voids per cm^3.

Temperature Dependence of Void Swelling

A typical plot of void swelling as a function of temperature in an irradiated 304 austenitic stainless steel is shown in Figure 6.15. It is clear from this figure that the peak in swelling happens at the intermediate temperature. Lower defect mobility limits void growth at lower temperatures, whereas at higher temperatures the irradiation-produced defect concentration reaches the value close to thermal equilibrium value and loses the level of vacancy supersaturation (i.e., a lack of excess vacancy concentration) that is required to sustain void growth. These two opposing effects give the shape of the swelling–temperature curve as shown in Figure 6.15.

With increasing temperature, the void number density falls and the average size increases, which is a typical behavior for a process that is dominated by nucleation at lower temperatures (i.e., the void growth is slow) and by growth at higher

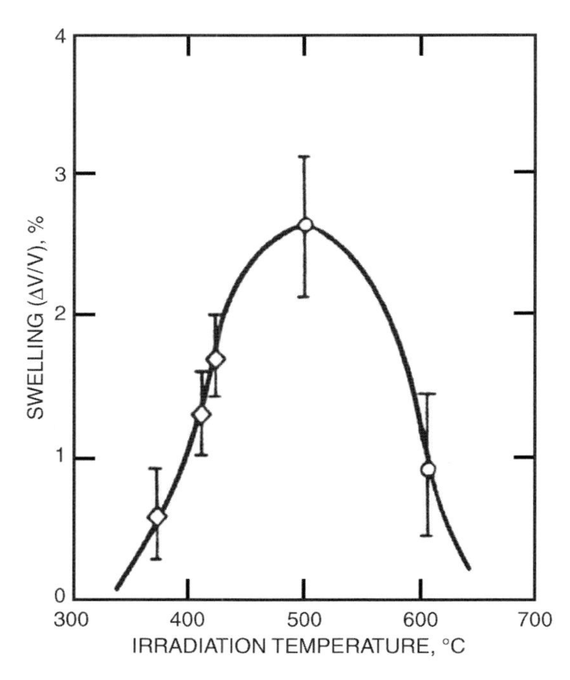

Figure 6.15 Effect of irradiation temperature on the swelling behavior of 304-type stainless steel at a fluence of 5×10^{22} n cm^{-2} ("o" – measured by TEM; "◇" – immersion principle) [12].

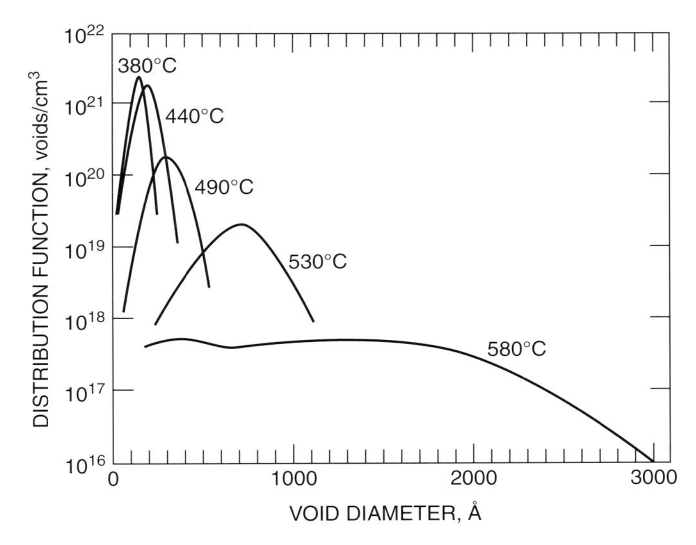

Figure 6.16 Void size distribution in a 316-type stainless steel irradiated to a fluence of 6×10^{22} n/cm^2 at various temperatures ref. [13].

temperatures where the driving force for void growth is small. Figure 6.16 shows the void distribution function as a function of void diameter at different irradiation temperatures in a 316-type stainless steel irradiated to a fluence of $6 \times 10^{22}\,\text{n cm}^{-2}$. The void size distribution function is narrow and average void size small at low temperatures. However, with increasing temperature, the distribution becomes wider and the average void size also increases. This occurs due to the enhanced mobility of point defects at elevated temperatures.

Effect of Cold Work

Cold work changes dislocation network density and has been found to have a significant effect on the swelling characteristics of metallic materials. The higher dislocation density restricts the nucleation and growth of voids. The effects of cold work and fluence reveal that the same amount of cold work provides less effect at higher fluences since cold work enhances the incubation dose for swelling but does not diminish the swelling rate in the linear swelling region at higher doses. This is thought to be a result of the fact that the dislocation density decreases in cold worked material during irradiation, while that in an annealed material increases. This accounts for the decreasing effect at higher doses.

Effect of Grain Size

Guthrie *et al.* [14] have demonstrated the effectiveness of fine grain size in inhibiting void formation by studying a film of sputtered nickel with ~0.5 μm thickness.

Under neutron exposure, the film with the ultrafine grain size did not show evidence of voids, while a nickel foil with a grain size of 30 μm produced voids. Also, similar experiments carried out in finer grain size material have shown similar results. The reason for this behavior is that the grain boundaries present in greater numbers in a fine grain size material act as efficient sinks for point defects. Recently, people have tried to investigate various forms of nanocrystalline material to see whether they are more resistant to radiation effects. However, the issue has not been resolved decisively to date.

Effect of Composition and Precipitates

The swelling behavior in a material depends on composition. An example is shown in Figure 6.17 for nickel-based materials at 425 °C. Nickel with just 0.4% impurities shows greater resistance to swelling than the high-purity nickel, and Inconel-600 (73Ni–17Cr–8Fe, in wt%) does not show any swelling (actually shows densification!) during irradiation. The swelling resistance observed in Inconel is due to the fine coherent precipitates present in this type of material. In coherent precipitates, the precipitate–matrix interface is continuously bonded. Coherent precipitates (variable bias sink) act as recombination sites for vacancies and interstitials, thus reducing swelling. However, precipitates need to be used as direct control of swelling due to instability of the structure in radiation environment.

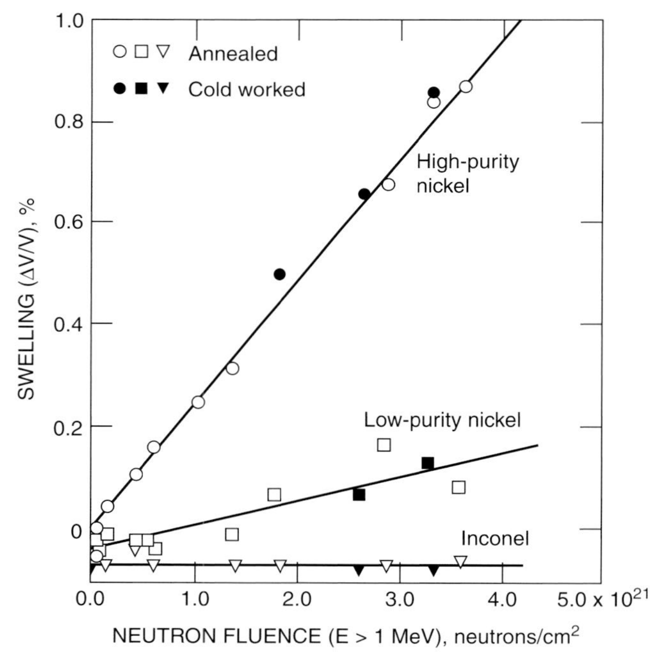

Figure 6.17 Swelling in high-purity nickel, 99.6% purity nickel, and Inconel (73% Ni–17% Cr–8% Fe–rest minor elements) at 425 °C Ref. [15].

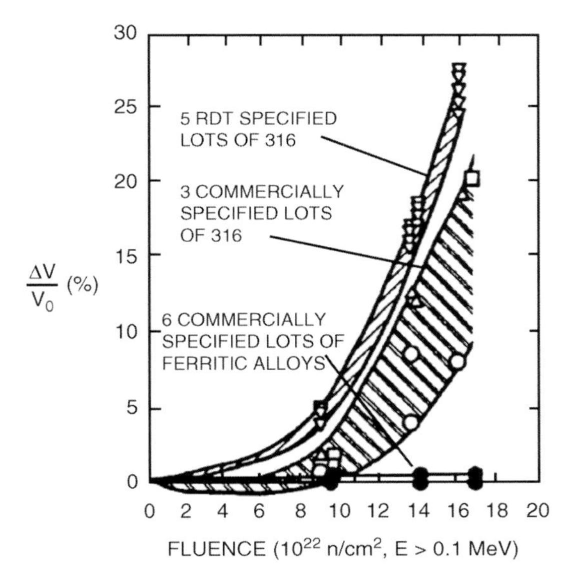

Figure 6.18 Comparison between the swelling behavior of a ferritic/martensitic steel with austenitic stainless steels (316 type) as a function of fast neutron fluence at 420 °C Ref. [16].

It has generally been a common knowledge found out from the EBR-II irradiation testing that ferritic/martensitic (F–M) alloys show greater resistance to void swelling compared to austenitic steels. Swelling behavior of six commercial ferritic/martensitic steels is compared with a 316-type austenitic stainless steel as a function of neutron fluence at ~420 °C in Figure 6.18. At the peak swelling temperature (400–420 °C), only <2% swelling was noted in F–M alloys like HT-9 (12Cr–1MoWV) and T-91 (9Cr–1Mo–V–Nb) steels even after irradiation dose of 200 dpa. The complex microstructure of the ferritic/martensitic steels involving lath boundaries, various types of precipitates, high dislocation density, and so on provides numerous defect sink sites that help in limiting void swelling. On the other hand, in austenitic stainless steels, presence of Cr, Ni, and other elements in higher amounts leads to the formation of helium through transmutation reactions and promote void swelling. However, recent studies have shown that F–M steels also swell at a greater rate at very high radiation doses; the phenomenon is generally not observed in conventional irradiation experiments as the incubation dose required is relatively large compared to that of the austenitic stainless steels.

Void swelling characteristics in a 16Cr–4Al–Y_2O_3 oxide dispersion- strengthened (ODS) steel were studied by Kimura *et al.* [17]. They also compared the results with those of a non-ODS, reduced activation martensitic steel (JLF-1, 9Cr–WVTa alloy). The ODS alloys contain a high number density of nanometric (<5 nm) Y–Ti based oxide precipitates. They used dual ion irradiation at 500 °C (773 K) at a higher dose rate to accumulate extensive displacement damage. The

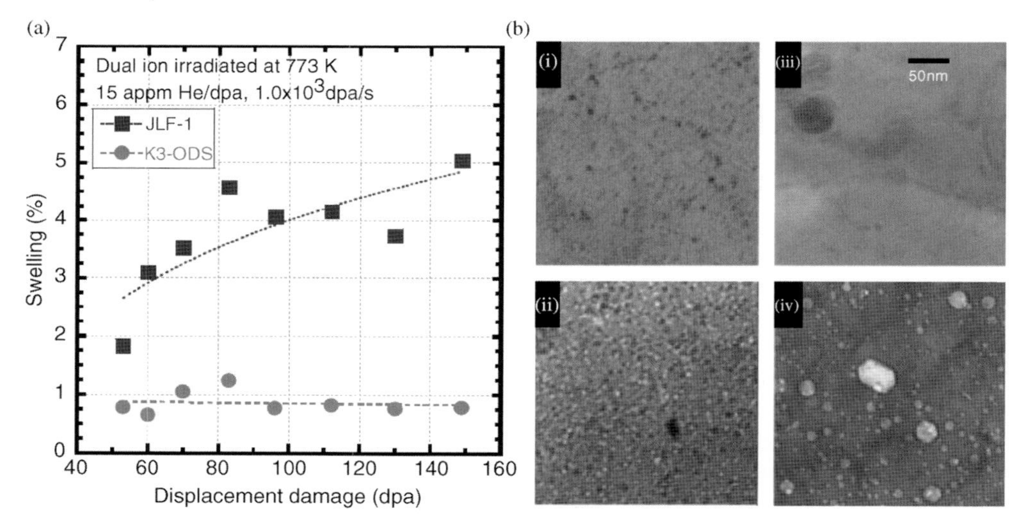

Figure 6.19 (a) Volumetric swelling versus displacement damage level in a dual ion-irradiated ODS steel and a 9Cr F–M steel. (b) TEM pictures of the ODS steel (a) before and (b) after the dual ion irradiation up to 60 dpa, respectively. (c and d) JLF-1 steel before and after the same irradiation condition, respectively [17].

volumetric swelling has been plotted against the displacement damage levels in Figure 6.19a. Interestingly, the ODS steel showed not much void swelling and was almost independent of the dose level. On the other hand, the JLF-1 steel exhibited comparatively higher volumetric swelling levels. They also presented evidence from TEM studies of both alloys irradiated up to 60 dpa, as shown in Figure 6.19b. The helium bubbles formed were much finer and numerous in the ODS alloy compared to the JLF-1 that showed larger bubbles. This microstructural evidence supports the hypothesis that helium bubbles are created at the Y–Ti–O nanoprecipitates, and thus help keep a reduced level of volumetric swelling in the ODS alloys.

Effect of Helium

In certain materials under neutron irradiation, there is no way of turning off the helium production as a means of controlling void formation. So, in materials where helium is produced, helium bubbles are formed and contribute to the swelling behavior. Some details of helium production have been discussed in Section 6.2.3. In reality, void formation takes place under conditions of damage rate, temperature, and sink density at which the vacancy supersaturation is not enough to nucleate voids. This observation led Cawthorne and Fulton [18] to suggest that helium atoms in the metal stabilize the small void nuclei and prevent the collapse of embryo voids into vacancy loops. Helium remains intimately involved in the formation of voids as nucleation process turns heterogeneous. So, voids are essentially partially or fully vacuum. As the radiation dose increases, more helium is produced to fill up the voids fully and then the features are rather called "bubbles." However, one important distinction should be remembered – voids do not need helium gas to grow, but bubbles need gas to grow. The morphology of the bubbles tends to be more spherical compared to voids that assume faceted shape.

Fluence Dependence of Void Swelling

We have already observed the effect of fluence in Figure 6.18 on the void swelling behavior of austenitic stainless steels like 316 type. It has been observed that the swelling extent increases with increasing neutron fluence (or dpa) and empirical relations have also been developed. It has been noted that there exists an incubation period that represents the neutron dose needed to produce enough helium and point defect concentration to make void nucleation possible. The incubation period may also be needed to develop enough number of interstitial dislocation loops to allow the preferred absorption of interstitials by dislocations to sufficiently bias the point defect population in the material in favor of vacancies as to permit vacancy agglomeration into voids. However, incubation period does depend on temperature and material/microstructure. Figure 6.20 shows a generalized form of void swelling behavior as a function of dose or fluence. The initial transient period is followed by a steady-state swelling period. This steady-state period for FCC-based materials continues to proceed with no sign of saturation. However, void swelling in BCC-based metals/alloys (like F–M steels) shows saturation effect. In these systems, at higher radiation doses, voids/bubbles self-organize following the crystallography of the metal leading to void/bubble lattices.

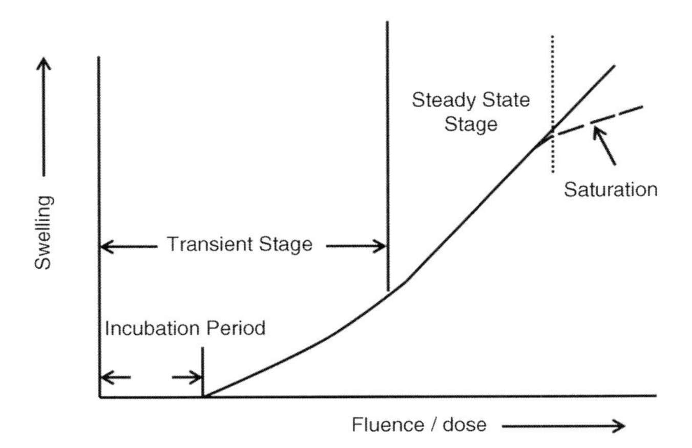

Figure 6.20 Generalized void swelling behavior as a function of radiation dose, showing different stages.

Note

Radiation growth is another radiation effect that occurs in materials with anisotropic structure or strong texture. Radiation growth does not lead to change in volume as the void swelling, but rather conserves the volume. It has been observed in zircaloys (HCP crystal structure) and alpha-uranium (orthorhombic). It is easier to think of a single crystal of alpha-Zr to understand the radiation growth effect. The c/a ratio of the crystal decreases or the single crystal becomes short and fat, thus conserving the volume. The length increase of fuel rod under irradiation can occur if the crystallographic texture of the zircaloy is such that the a-axis of the crystals is oriented near to the length axis. This effect is further discussed in Chapter 5 with respect to radiation growth effect seen in uranium.

6.1.3 Radiation-Induced Segregation

Radiation-induced segregation (RIS) involves segregation of alloying elements under fast particle irradiation to certain microstructural locations leading to a situation where otherwise homogeneous alloys become heterogeneous. During fast particle irradiation, significant diffusion fluxes of point defects (vacancies and interstitials) can be set up in the vicinity of the defect sinks like surfaces or internal grain boundaries. Generally, we have come to know from Section 2.3 that the different atomic species in an alloy migrate at different rates in response to the already set up point defect fluxes so that some species travel toward the said defect sinks while others move away. Thus, RIS can lead to significant alterations in the local composition near the sinks like grain boundaries, and this can have substantial bearing on the macroscopic properties of the materials. This phenomenon has been studied in some detail, particularly in structural materials used in nuclear reactor components. Understanding RIS is of particular importance in chromium containing austenitic stainless steels or nickel base superalloys because these alloys are used in commercial power reactors and potential candidate materials for advanced reactors. RIS can potentially lead to irradiation-induced stress corrosion cracking as chromium segregates away from the grain boundaries where it is most needed. At low temperatures, defect concentration builds up and rather than going to the sinks, point defects tend to recombine. At higher temperatures, thermal diffusion dominates and composition becomes equilibrated or homogeneous. At intermediate temperatures, RIS becomes acute due to the operation of a process known as "inverse Kirkendall" effect. Figure 6.21 shows the composition profile in an irradiated 300 series stainless steel, analyzed by energy dispersive spectroscopic measurement conducted with a JEOL 2010F high-resolution transmission electron microscope, showing depletion of Cr and enrichment of Ni, Si, P at the grain boundary due to RIS.

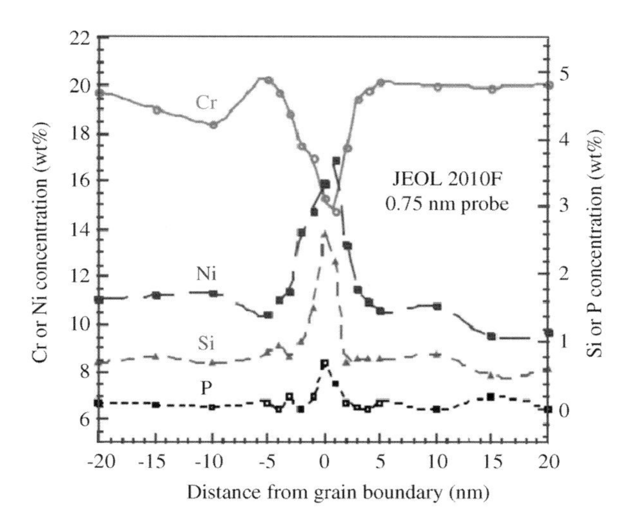

Figure 6.21 Radiation-induced segregation of Cr, Ni, Si, and P at the grain boundary of a 300 series stainless steel irradiated in a LWR core to several dpas at 300 °C [19].

6.1.4 Radiation-Induced Precipitation or Dissolution

At intermediate temperatures and radiation doses of >10 dpa, RIS effect may, in fact, lead to phase instability, involving precipitation of new phases or dissolution of existing phases. In Ni–Si system, Si as an undersize solute (atomic radius: 110 pm) compared to Ni (atomic radius: 135 pm) enriches the sinks, and when conditions are ripe, the solubility of Si in Ni exceeds the level forming new phase Ni_3Si (γ'-type structure). A more illustrative example is shown in Figure 6.22. Nemoto *et al.* [20] showed evidence of radiation-induced precipitation in various irradiated Mo–Re alloys. Precipitates formed from Mo–4Re alloys after fast neutron irradiation consist of chi and sigma phases.

6.2 Mechanical Properties

Mechanical properties are important to varying degrees in a majority of nuclear reactor components that experience a variety of loading conditions under various temperature and irradiation regimes.

6.2.1 Radiation Hardening

The strength (both yield and ultimate tensile stress) of metals/alloys increases along with appreciable reduction in ductility coupled with reduction in strain

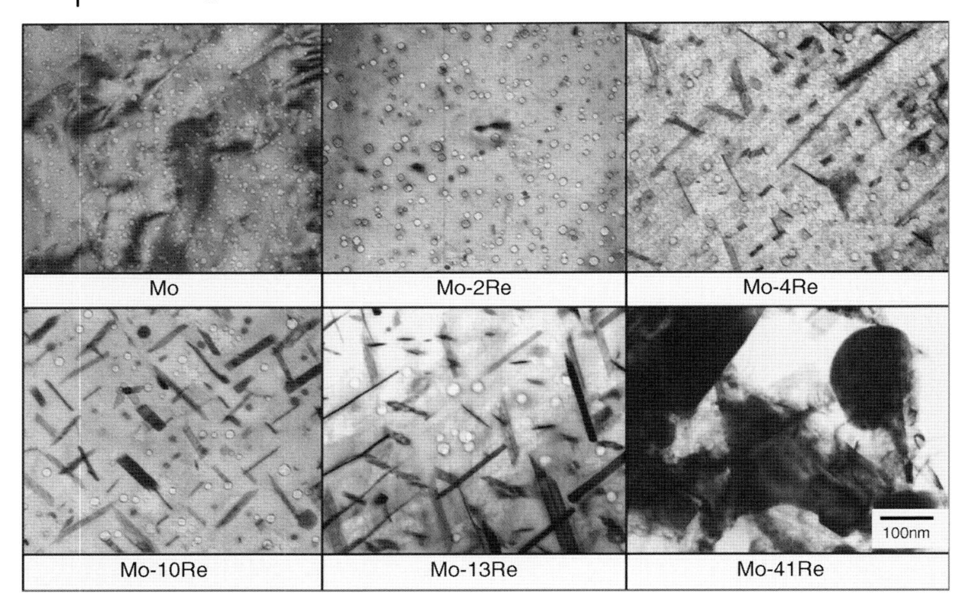

Figure 6.22 TEM Micrographs of recrystallized Mo–Re alloys irradiated at ~800 °C up to 18 dpa [20].

hardening exponent. For example, strain hardening exponent of Zircaloy-2 has been found to vary between 0.1 to 0.15 in annealed condition. However, after irradiation, the exponent may decrease to 0.02 to 0.01 depending on the extent of radiation damage, thus affecting the uniform elongation of the alloy (as described in Chapter 5.1). Radiation hardening can occur due to the multitude of defect creation in irradiated materials: (i) point defects (vacancies and self-interstitials), (ii) impurity atoms, (iii) small defect clusters, (iv) dislocation loops, (v) dislocation lines, (vi) cavities (voids/bubbles), and (vii) precipitates. Generally, radiation hardening effect starts to appear at temperatures less than $<0.4T_m$ (where *in situ* recovery effect is less) and at radiation damage of >0.1 dpa.

Figure 6.23a–c shows tensile stress–strain curves of irradiated (increasing fluence) BCC-based alloy (A533B – low-alloy ferritic steel), an FCC-based alloy (316-type stainless steel), and a HCP-based alloy (zircaloy-4), respectively. Generally, FCC- and HCP-based alloys do not show discontinuous behavior in unirradiated state. However, in irradiated state they exhibit yield point-like behavior, as shown in Figure 6.23b and c.

In BCC metals where yield points appear along with Luders strain before exposure to high-energy radiation, Luders strain increases following irradiation and at high neutron fluxes ($>10^{19}$ n cm^{-2}) fracture occurs during Luders strain itself. Figure 6.24 depicts a series of stress–strain curves in mild steel tested at ambient temperature following neutron radiation exposures.

An actual example of the effect of neutron irradiation on tensile strength and ductility properties is shown in Figure 6.25. Here, we describe one example from

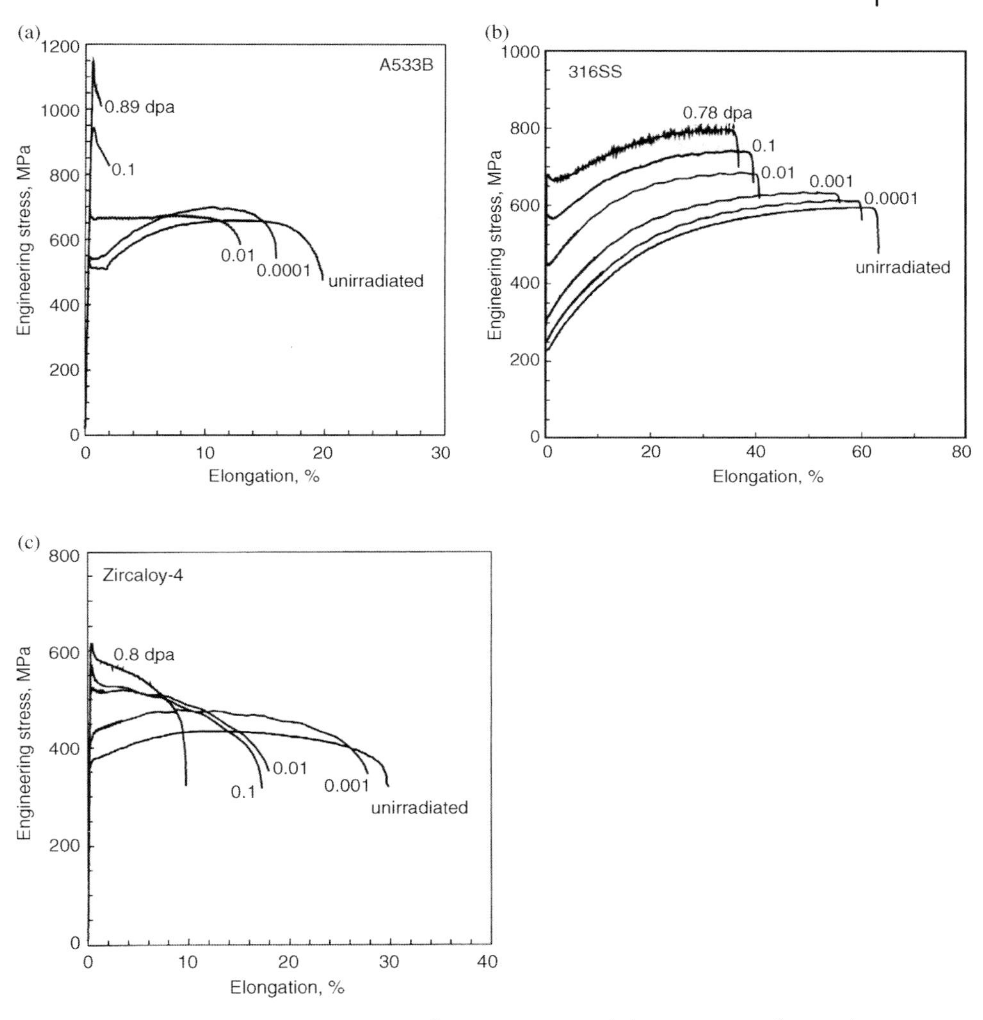

Figure 6.23 Engineering stress–strain curves of (a) A533B RPV steel, (b) 316-type stainless steel, and (c) zircaloy-4, irradiated at different radiation damage (dpa) levels [21].

9Cr–1MoVNb steel (T91) on the radiation effect in F/M steels. Irradiation exposure dose of 9 dpa resulted in appreciable radiation hardening due to the formation of a wide range of radiation-produced defects in a temperature range of 425–450 °C (Figure 6.25a). Figure 6.25b shows the corresponding ductility as a function of test temperature. Hardening causes a decrease in ductility at the lowest temperature. However, interestingly, the ductility of the irradiated alloy increases at 450 °C compared to that of the aged alloy. It is interesting to note that the aged alloy shows a maximum strength at a temperature where the irradiated alloy shows much higher

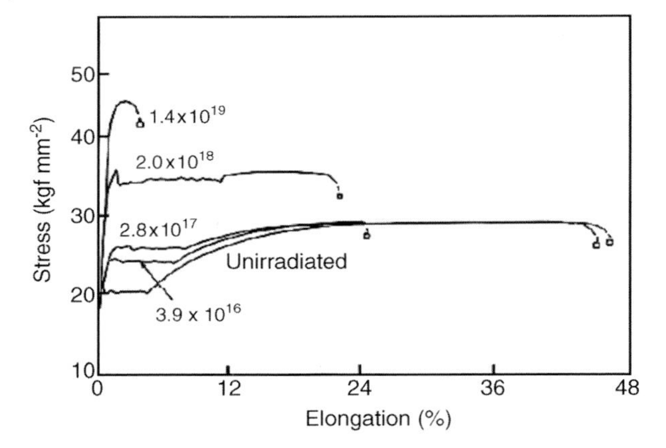

Figure 6.24 Effect of neutron fluence on engineering stress–strain curves for mild steel at ambient temperature [22] neutron fluences are given in neutrons per cm^2.

ductility. However, note that the ductility data of the unirradiated alloy are not available at 450 °C. Radiation hardening saturates by around 10 dpa. For irradiation above 425–450 °C, there may be enhanced softening due to increased recovery and coarsening.

Understanding radiation hardening would need our understanding of the dislocation theories and strengthening mechanisms. First, we will discuss the two major components of radiation hardening – source hardening and friction hardening. As discussed in Chapter 5, yield stress (τ_y) can be regarded as composed of source hardening (τ_s) and friction hardening (τ_i) terms, representing the hardening due to solute atoms locking dislocation sources and due to subsequent dislocation movement through the lattice.

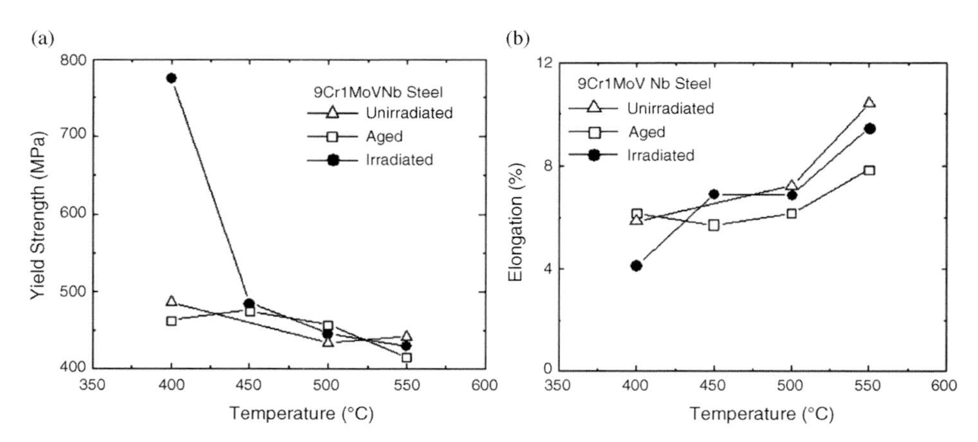

Figure 6.25 Variation of (a) yield strength and (b) total elongation as a function of test temperature in a T91 steel Ref. [16].

a) **Source Hardening (τ_s)** This hardening can increase the stress required to start a dislocation moving on its glide plane. This can be found in irradiated FCC metals, and in both unirradiated and irradiated BCC metals. In case of FCC metals/alloys and most HCP metals/alloys, the unirradiated materials do not show source hardening behavior. This is shown by continuous stress–strain curve without yield point phenomenon (note in Section 5.1). However, the unirradiated BCC metals (low-alloy ferritic steels) manifest a source hardening-like phenomenon that occurs due to the dislocation–interstitial impurity interaction as seen in the yield point phenomenon (Figure 6.23a). Source hardening observed in irradiated FCC metals is due to the formation of irradiation-produced defect clusters near the Frank–Read sources, consequently raising the stress ($\tau_{FR} = Gb/L$) required to activate the loop by decreasing the pinning point distance (L). However, once the loop starts forming, it sweeps away the defect clusters and the stress drops.

In a polycrystalline material, majority of the dislocation sources are on or near the grain boundaries and the dislocation pileups create stress concentration at the boundaries to activate dislocation sources and generate dislocations in the other grain. Basically, the Hall–Petch strengthening ($k_y d^{-1/2}$) effect, discussed in Section 5.1 contributes profusely to the source hardening term in polycrystalline materials.

b) **Friction Hardening (τ_i)** After being generated from the source, the dislocation encounters a number of obstacles that lie on the slip plane or near the slip plane while moving on it. This raises the stress needed to move dislocations on the slip plane and in aggregate is called friction hardening. Friction stress (τ_i) consists of two components: long-range stresses (τ_{LR}) and short-range stresses (τ_{SR}):

$$\tau_i = \tau_{LR} + \tau_{SR}, \tag{6.3}$$

The long-range stresses generally arise from the repulsive interaction between a moving dislocation and the dislocation network. This effect is termed as long range as it works over a distance from the gliding dislocation (Taylor Equation).

$$\tau_{LR} = \alpha G b \varrho_d^{1/2}, \tag{6.4}$$

where ϱ_d is the dislocation density.

The short-range stresses may have two origins – athermal and thermally activated. The short-range stresses may arise out of precipitates, such as precipitate hardening in terms of Orowan bowing and particle cutting, in the presence of voids/bubbles (void hardening) and dislocation loops. One general way of expressing the short-range stresses is the summation of the contributions of precipitation hardening (τ_p), void hardening (τ_v), and loop hardening (τ_l):

$$\tau_{SR} = \tau_p + \tau_v + \tau_l. \tag{6.5}$$

If it is assumed that these obstacles are dispersed in a random fashion, it can be shown that the average interparticle spacing (l) between the defects characteristics (number density N and average diameter d) can be described as Eq. (6.6):

$$l = (Nd)^{-1/2}. \tag{6.6}$$

Thus, the general form of τ_{SR} can be given by

$$\tau_{SR} = \alpha Gb(Nd)^{1/2}. \tag{6.7}$$

At a very low dose, the irradiated microstructure would contain defect clusters and small loops. With increasing dose, the loop microstructure saturates at a particular number density and size as the loops unfault and become part of the dislocation line network, thus increasing dislocation density. At higher temperatures, voids/bubbles would be present and irradiation-induced precipitation can also contribute to the radiation hardening effect.

6.2.1.1 Saturation Radiation Hardening

According to the discussion above, the yield strength increase should be proportional to $N^{1/2}$. In the absence of annihilation mechanisms of obstacles, N is proportional to fluence and thus radiation hardening should be proportional to $(\phi t)^{1/2}$. But this is not the case in reality. It has been shown that the radiation hardening does not increase with neutron fluence indefinitely, but seems to saturate as shown in Figure 6.26.

Makin and Minter [23] postulated a theory to explain the observation. According to their model, as the defect concentration increases, it becomes harder to form

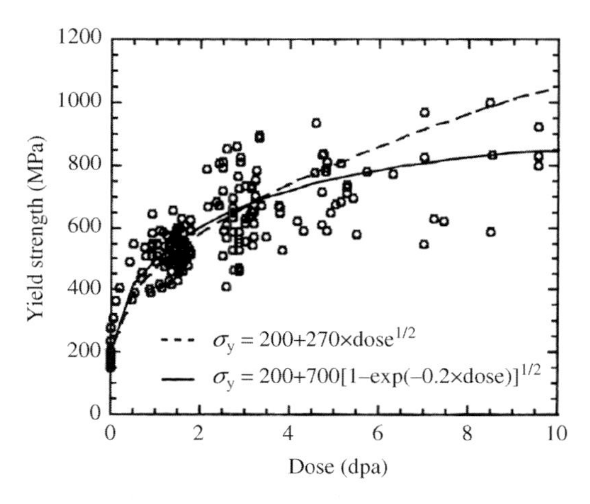

Figure 6.26 Evidence of saturation of radiation hardening for 300 series austenitic steels [24].

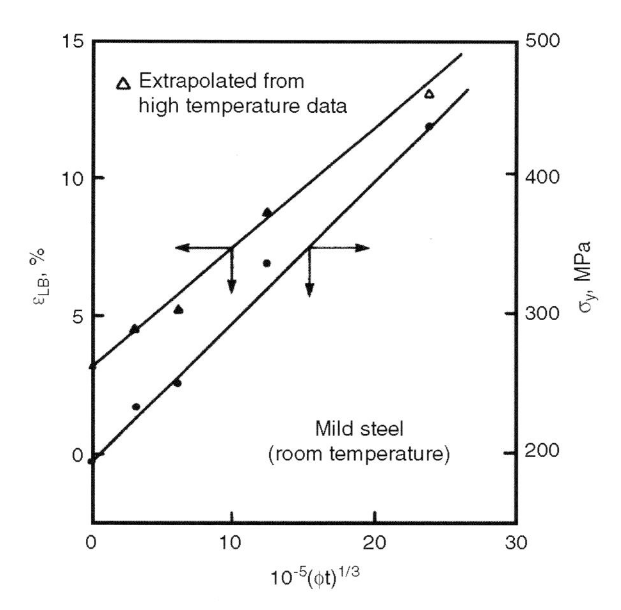

Figure 6.27 Effect of neutron fluence on yield stress and Luders strain in mild steel [25].

new zones because of the reduced volume available for new zone formation. The time rate change of the zone density (N) is given by

$$\frac{dN}{dt} = \varsigma \sum\nolimits_s \phi(1 - VN), \tag{6.8}$$

where ζ is the number of zones created per neutron collision ($\sim$1), Σ_s is the macroscopic scattering cross section, and ϕ is the fast neutron flux. The term in parentheses represents the fraction of solid volume available for creation of new zones. Radiation saturation takes place because of the dynamic balance reached between creation and annihilation of obstacles.

Studies on mild steel by Murty and Oh [25] revealed that the yield stress increases with fluence raised to 1/3 rather than 1/2; Luders strain also increased correspondingly as $(\phi t)^{1/3}$ as expected since in mild steel the Luders strain is proportional to the yield stress (Figure 6.27). They analyzed the results to determine the effect of neutron radiation fluence on friction and source hardening terms and demonstrated that while friction hardening (τ_i) increases as $(\phi t)^{1/2}$, the source hardening (τ_s) decreases (Figure 6.28), so the yield stress increases with fluence raised to a power less than 0.5.

6.2.1.2 **Radiation Anneal Hardening (RAH)**

An additional hardening effect occurs upon annealing of BCC metals after irradiation. This phenomenon is known as RAH. Note that hardening starts at 120 °C and increases to a maximum at 180 °C before decreasing. A second

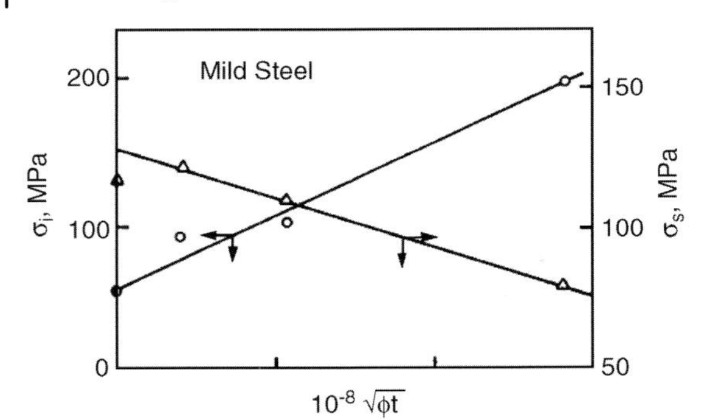

Figure 6.28 Effect of neutron fluence on friction and source hardening in mild steel [25].

hardening peak appears at 300 °C before the yield strength drops due to recovery effect. The first peak is due to migration of oxygen to defect clusters and the second peak is due to the migration of carbon (Figure 6.29). Formation of respective interstitial impurity and defect cluster complexes at the corresponding temperature leads to enhanced obstruction from the dislocation movement and manifests in the form of additional hardening peaks. This phenomenon has also been observed in molybdenum, vanadium, iron alloys where the interstitial impurities like carbon, nitrogen and oxygen are responsible. Irradiated FCC metals/alloys generally do not show this kind of effect upon annealing.

6.2.1.3 **Channeling: Plastic Instability**

In some highly irradiated metals, the onset of necking coincides with yielding with no uniform deformation. This kind of behavior has been shown in Figure 6.23a in the stress–strain curve of A533B irradiated to 0.89 dpa as well as in Figure 6.24 for

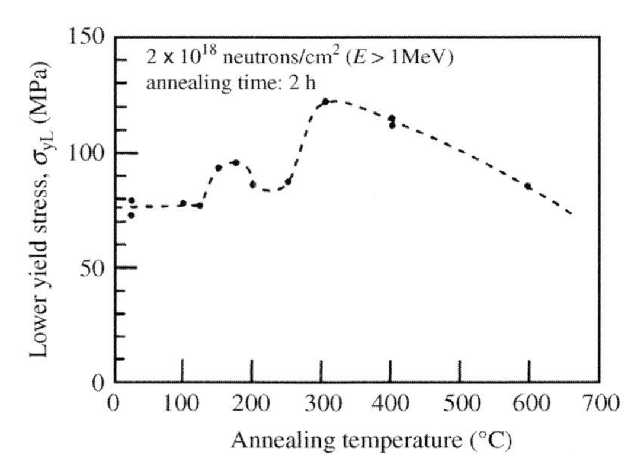

Figure 6.29 RAH in niobium containing 35 wppm C, 41 wppm O, and 5 wppm N following irradiation to 2×10^{18} n cm^{-2} and annealing for 2 h [24, 26].

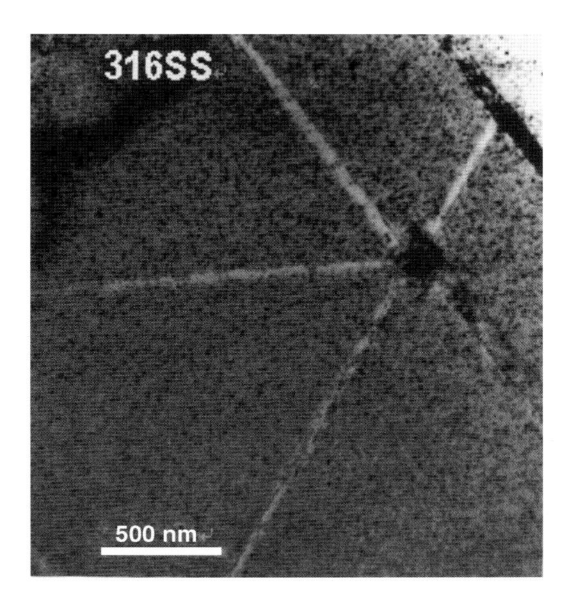

Figure 6.30 Channeling in 316SS tensile tested to 5% strain at room temperature (neutron irradiation: 0.78 dpa and 80 °C) [21].

mild steel at the highest fluence. This unusual macroscopic behavior is due to the microscopic phenomenon of dislocation channeling. However, this effect has nothing to do with the PKA channeling that occurs due to the crystallinity of materials – so avoid getting confused between the two terms. In dislocation channeling, an avalanche of dislocations can be released to move on particular slip planes along planar channels that have been cleared of obstacles. As dislocations see these paths as the path of least resistance, the dislocations generated move through these channels. Thus, the strain remains highly localized. In this way, eventually stress concentration sites are created where these dislocation channels intersect the grain boundaries. Figure 6.30 shows microstructural evidence of dislocation channels in a 316-type stainless steel irradiated to 0.78 dpa at 80 °C.

6.2.2 Radiation Embrittlement

As discussed in earlier chapters, ductility (or toughness) is an important property of any structural material or any other types of materials in load-bearing nuclear reactor components. Radiation hardening generally leads to radiation embrittlement and occurs in a wide range of materials. However, radiation embrittlement in BCC metals/alloys (such as ferritic and ferritic–martensitic steels) that exhibit ductile–brittle transition temperature (DBTT) refers to an increased DBTT along with decreased upper shelf energy and decreased slope

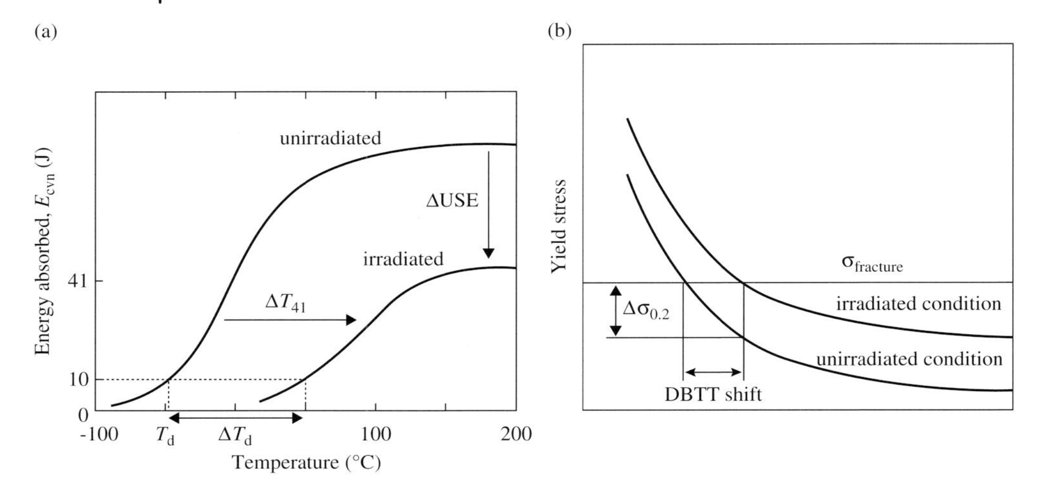

Figure 6.31 (a) A schematic illustration of the impact toughness as a function of temperature in both unirradiated and irradiated steels [24]. (b) Davidenkov's diagram [27].

in the ductile–brittle transition region, as shown in Figure 6.31a. Yield strength decreases with temperature, but fracture stress is roughly temperature independent. Irradiation causes an increase in the yield stress shifting the point at which fracture stress and flow stress curves intersect at higher temperature, thus raising the DBTT. Figure 6.31b is known as *Davidenkov's Diagram*. It is to be noted that in BCC metals the yield stress increases rapidly as temperature decreases, while in FCC and HCP metals the decrease is not that rapid, which results in DBTT occurring at very low temperatures and thus of less significance (Section 5.1). The effect of irradiation on the use is believed to be due to a reduction in strain hardening and increase in flow localization, leading to lower ductility (also dislocation channeling has an effect).

Earlier we have underscored the importance of having a decreased DBTT for structural applications. This is also applicable for reactor pressure vessel (RPV) steels, and this section is devoted to understanding the origin and implications of radiation embrittlement in RPV steels. Reactor pressure vessel is an integral part of a nuclear power reactor and it is considered a life-limiting reactor component, which means that the RPV is highly unlikely and extremely difficult (cost-prohibitive) to replace during the operational life of the reactor. Hence, the same RPV stays in place for the entire operational life of the reactor. Currently, many utilities are applying for relicensing their reactors for another 20–30 years as their original design life comes to an end. RPV surveillance program has facilitated the understanding of radiation embrittlement in RPV steels; however, understanding the behavior over very long period is still evolving. The relicensing of the commercial power reactors beyond their original design lives makes the understanding of radiation embrittlement in RPV steels even more important. Understanding the role of late blooming phases in the

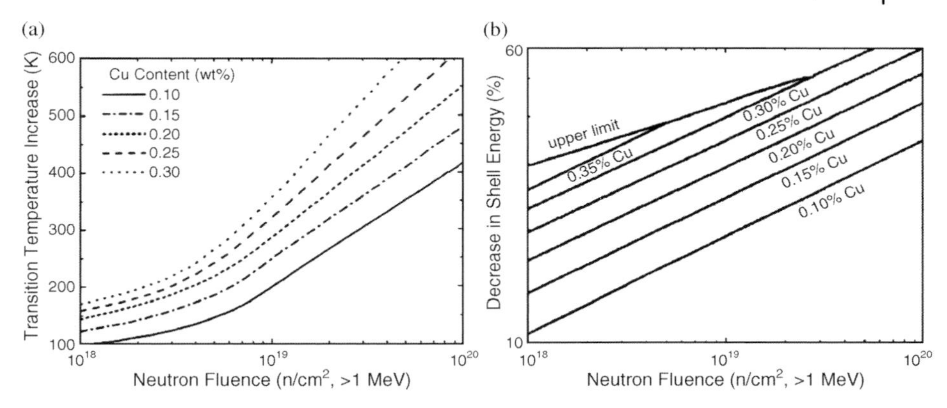

Figure 6.32 (a) The variation of the DBTT shift and (b) upper shelf energy as a function of neutron fluence at different copper contents.

radiation embrittlement of reactor pressure vessel steels will need more sustained studies in the wake of LWR sustainability efforts.

Generally, low-alloy ferritic steels (A508 and A533B grades) are used in the RPVs of the light water reactors. The RPV steel in a PWR would have to withstand irradiation temperature of 240–290 °C and fast neutron fluence of $<2 \times 10^{24}$ n m^{-2} over a time period of decades, leading to substantial changes in microstructure: (i) formation of radiation defects, (ii) phase transformations accompanied by the generation of various precipitate populations, (iii) formation of impurity–vacancy clusters (like copper–vacancy ones), (iv) intergranular segregation of phosphorus and other impurities, and (v) segregation of phosphorus to interfaces between secondary phases and matrix (i.e., intragranular segregation). These are the likely reasons for radiation embrittlement as observed in the RPV steels.

6.2.2.1 Effect of Composition and Fluence

This depends on the alloy composition (Cu, P, Ni, etc.), fluence, and irradiation temperature. Copper forms very fine (1–3 nm) coherent precipitates; nickel and manganese amplify the hardening effect of copper. Phosphorus segregates to the grain boundaries (contributes no more than 10–20% of the total effect) and the rest to the particle–matrix interfaces (this contribution is very important). Figure 6.32a and b shows the effect of copper content on the change of DBTT and upper shelf energy, respectively.

Recent advances in atom probe technique have positively identified the presence of these Cu- and P-enriched clusters. An atom probe map of a neutron-irradiated KS-01 weld is shown in Figure 6.33.

6.2.2.2 Effect of Irradiation Temperature

Higher irradiation temperature results in less embrittlement. This effect occurs due to the in-reactor annealing out of many radiation-produced defects before they can become stable defects.

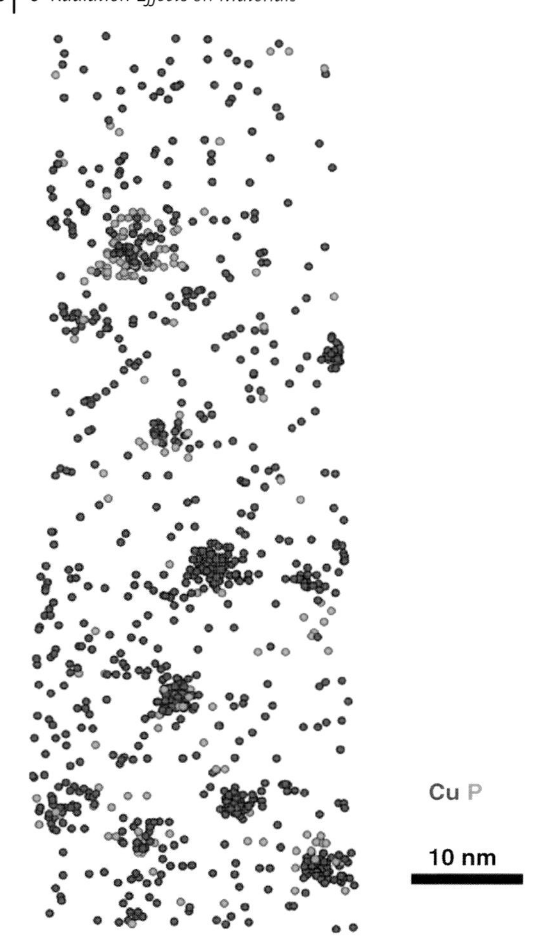

Figure 6.33 Atom map from the KS-01 weld that was neutron irradiated to a fluence of 0.8×10^{23} n m^{-2} ($E > 1$ MeV) at a temperature of 288 °C showing phosphorus segregation to the copper-enriched precipitates and a phosphorus cluster [28].

According to Cottrell–Petch theory of radiation embrittlement [11],

$$\Delta T_{\mathrm{DBTT}} = -\frac{\Delta\sigma_{\mathrm{i}}}{((\sigma_y/k_y)(\mathrm{d}k_y/\mathrm{d}T) + (\mathrm{d}\sigma_y/\mathrm{d}T))} \tag{6.9}$$

where $\Delta\sigma_i$ is the change in friction stress, σ_y is the yield stress, k_y is the unpinning parameter, $\mathrm{d}k_y/\mathrm{d}T$ is the change in the unpinning parameter as a function of temperature, and $\mathrm{d}\sigma_y/\mathrm{d}T$ is the yield stress as a function of temperature. Thus, the change in DBTT can be estimated by knowing these various parameters and their changes with temperature and fluence (Figure 6.34).

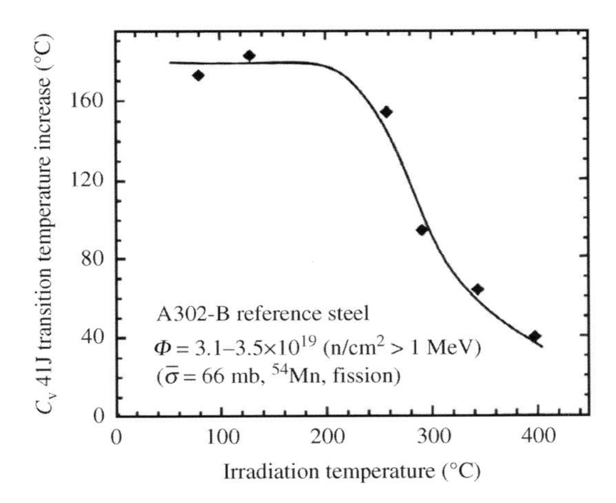

Figure 6.34 The variation of transition temperature increase as a function of irradiation temperature for an irradiated A302-B steel. From Ref. [29].

6.2.2.3 **Effect of Thermal Annealing**

Postirradiation annealing of irradiated RPV steels near 450–490 °C for several hours (70–150 h) could lead to complete recovery of the transition temperature, even after very high neutron fluences ($>10^{20}\,\text{n}\,\text{cm}^{-2}$). Thus, this is regarded as a potential technique to eliminate or minimize the radiation embrittlement effect. Figure 6.35 [27] illustrates the point in a ferritic steel containing 1.6 wt% Ni, 0.007 wt% P, and 0.06 wt% Cu. It is important to note

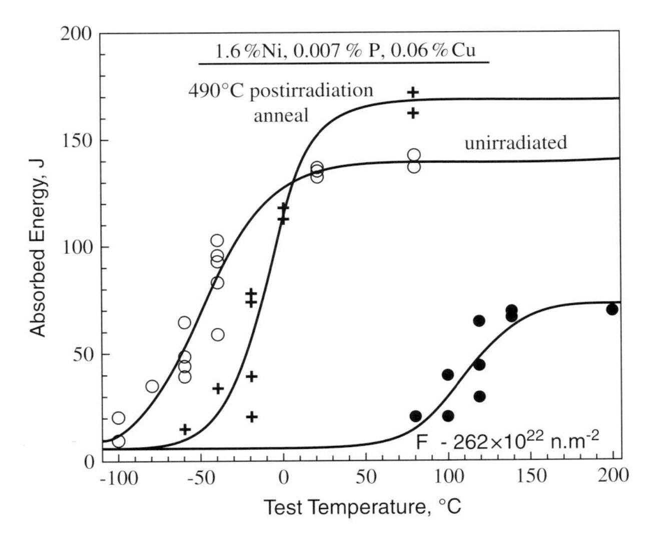

Figure 6.35 Effect of postirradiation annealing on the transition curve of a ferritic steel containing nickel, phosphorus, and copper [27].

that Ni, Cu, and P are known to promote radiation embrittlement. Even in this case, thermal annealing at 490 °C can recover the DBTT and the upper shelf energy. In this particular case, the upper shelf energy appears to go above that of the unirradiated alloys. It is thought that the phosphorus already was to some extent in segregated state before the irradiation and thermal annealing effect removes phosphorus segregation.

6.2.3 Helium Embrittlement

Helium gas is produced through transmutation of the component elements in austenitic (FCC) stainless steels and other materials. This can lead to embrittlement behavior that cannot be eliminated by high-temperature annealing. Helium is *practically insoluble* in metals and hence after generation, it tends to precipitate into bubbles particularly when the temperature is high enough ($>0.5T_m$) for helium atoms to migrate. Helium may produce severe embrittlement (*intergranular cracking*) to such an extent that at elevated temperatures even if the yield strength recovers in the irradiated alloy, the ductility is never regained. The extent of helium embrittlement depends on fast neutron fluence, alloy composition, and temperature.

The source of helium in steels is the component elements present in them. Helium is generated in threshold reactions due to the interaction of neutron with the specific isotopes comprising (n,α) reaction. Boron (B^{10}) and Ni^{58} are such elements important in generating alpha particle (i.e., helium nucleus or He^4) through the following reactions:

$$B^{10} + n^1 \rightarrow Li^7 + He^4. \tag{6.10}$$

$$Ni^{58} + n^1 \rightarrow Ni^{59}; \quad Ni^{59} + n^1 \rightarrow Fe^{56} + He^4. \tag{6.11}$$

These reactions occur in both thermal and fast neutron spectra. Similarly, helium (n,α) reactions between fast neutrons and Ni, Fe, Cr, and N atoms occur, although with different reaction cross sections. In fast reactors, helium embrittlement is more pronounced simply because the fast neutron flux in the fast reactor is about three to four orders of magnitude higher than the thermal flux, whereas the ratio is 1 : 1 in the thermal reactors. Helium embrittlement remains a widely studied topic. Olander [11] summarized various theories of helium embrittlement:

i) Woodford, Smith, and Moteff postulated that the increased strength of the matrix material due to the presence of helium bubbles in the grain interiors would lead to stress concentration at the grain boundary triple points during deformation. But as the stress concentration at the grain boundaries would not be able to relax itself, this would induce grain boundary triple point cracks and eventual propagation of cracks along the grain boundaries. So this theory espouses an indirect origin on helium embrittlement.
ii) Kramer and coworkers showed that helium bubbles can form on the grain boundary carbide particles ($M_{23}C_6$), thereby allowing cracks to form. Now

the question is why helium bubbles tend to nucleate on the carbide particles. One aspect of it could be that these carbide particle surfaces reduce the critical nucleation barriers of the helium bubbles. But it is not clear why it happens to be on these $M_{23}C_6$ particles. Boron was found to be associated with these structures as $[M_{23}(CB)_6]$. So, when boron transmutes through (n,α)-type reaction, helium is produced close to the carbide particles and forms bubbles on the particle itself. Thus, bubbles are formed on or very near to the grain boundary, thus promoting the possibility of helium embrittlement.

iii) Reiff showed that the presence of helium in triple-point cracks permits unstable growth of these cracks at stresses much lower than that required for a gas-free crack to propagate. The presence of helium deteriorates the grain boundary cohesion, thereby leading to weaker grain boundaries that cannot sustain larger loads.

iv) A majority of researchers believe that this phenomenon occurs due to the stress-induced growth of helium bubbles on the grain boundaries that eventually link up and cause intergranular failure. However, as you can see from above, all these theories are related and would explain the behavior depending on the situation. Perhaps all the above factors play a role in helium embrittlement. Figure 6.36 shows the various schematic locations of helium bubble formation. On the other hand, Figure 6.37 shows the formation of helium bubbles in different 82 series alloys with a base composition of Fe–25Ni–15Cr. It demonstrates the formation of helium bubbles at the secondary MX-type precipitates, primary MX precipitates, on the grain boundary $M_{23}C_6$ precipitates, and on the grain boundaries themselves.

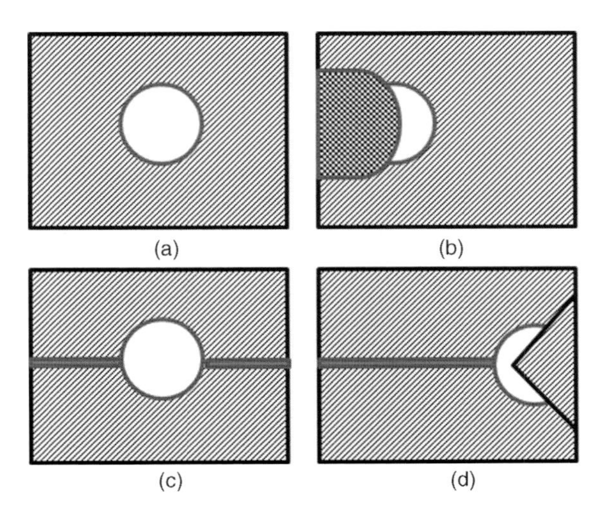

Figure 6.36 Helium bubble nucleation at various locations in the microstructure: (a) grain interior, (b) on a particle, (c) grain boundary, (d) grain boundary particle situated at the triple point.

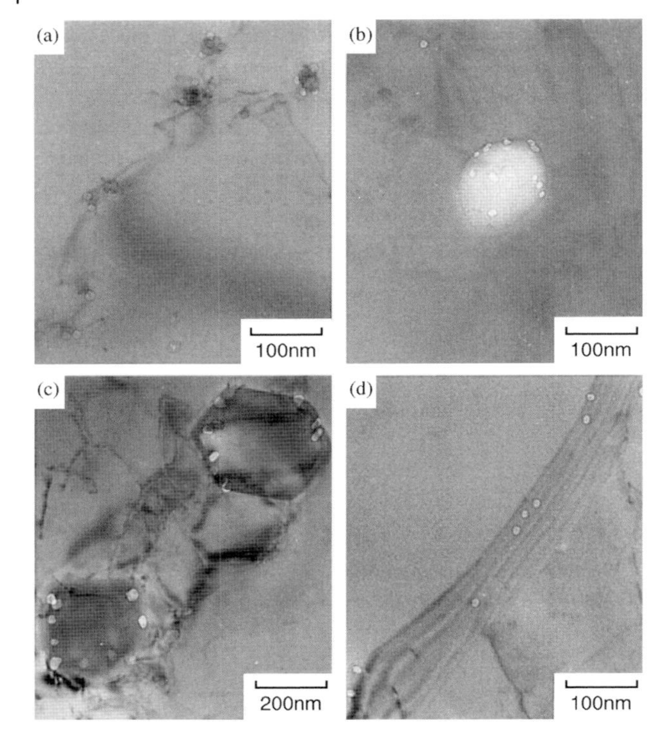

Figure 6.37 TEM micrographs showing various helium bubbles in 82 series alloy after helium implantation (He content: 50 appm) and creep rupture (stress of 100 MPa and temperature of 650 °C): (a) on secondary MX particles, (b) on primary MX particles, (c) on $M_{23}C_6$ particles, and (d) on the grain boundaries [20].

The effect of irradiation temperature on the ductility (percentage elongation to fracture) of the irradiated 304-type stainless steel is shown in Figure 6.38. The fast neutron fluence was kept at $>10^{22}\,\mathrm{n\,cm^{-2}\,s^{-1}}$ during irradiation. Tensile tests were conducted at 50 °C. Note the dip in ductility due to helium embrittlement near ~580 °C.

Interestingly, BCC metals/alloys are less vulnerable to helium embrittlement (i.e., no drastic loss in ductility). It is believed that the large diffusion coefficients in BCC materials due to their more open structure help in relaxing stress concentrations at the grain boundaries effectively and thus minimize the stress-enhanced helium bubble growth.

6.2.4 Irradiation Creep

Creep is time-dependent, thermally activated plastic deformation process, as described in Chapter 5 where we entirely dealt with the conventional thermal creep

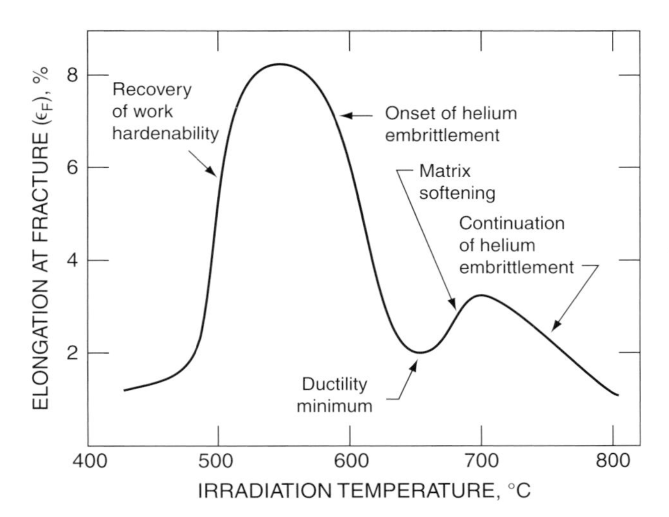

Figure 6.38 The effect of irradiation temperature on an irradiated austenitic stainless steel (fast neutron fluence of $>10^{22}$ n cm^{-2} s^{-1} ref. [11].

that generally occurs at elevated temperatures ($>0.4T_m$) under stress. Materials under stress could undergo creep effects (contributes to the dimensional instability of an irradiated material) under energetic particle flux (such as fast neutron exposure) even at much lower temperatures where thermal creep is essentially negligible in the absence of neutron irradiation. The generation of point defects is at the heart of the irradiation creep process. One way to understand the process is from the point of vacancy production in materials from two sources – thermal vacancies (C_ν^{th}) and neutron-induced (C_v^*). That is, the total vacancy concentration $C_\text{v} = C_\nu^{\text{th}} + C_\text{v}^*$. Thus, the total irradiation creep rate ($\dot{\varepsilon}^{\text{irr}}$) can be expressed in two components:

$$\dot{\varepsilon}^{\text{irr}} = \dot{\varepsilon}^* + \dot{\varepsilon}^{\text{th}} \tag{6.12}$$

where $\dot{\varepsilon}^*$ is affected by the radiation component and $\dot{\varepsilon}^{\text{th}}$ is affected by the thermal creep contribution. However, irradiation creep effect could be quite complex and research is ongoing to fully understand the effect in different material systems. Here, we lay out a general discussion on irradiation creep by categorizing it into two types: radiation-induced creep and radiation-enhanced creep. This is a rather simplistic way of describing irradiation creep, although it has substantial pedagogic advantages.

Radiation-induced creep occurs at lower homologous temperatures at which thermal creep is negligible ($\dot{\varepsilon}^{\text{th}}$), that is, not thermally activated. At these lower temperature regions, the vacancy concentration produced by atomic displacements due to irradiation (that are not in thermal equilibrium, but produced as a function of fluence) could be large enough to induce creep deformation under the application of stress. A simple relation used for describing radiation-induced creep rate ($\dot{\varepsilon}^{\text{irr}}$) is given by

$$\dot{\varepsilon}^{\text{irr}} \approx \dot{\varepsilon}^{*} = B\sigma\varphi, \tag{6.13}$$

where B is a constant relatively insensitive to the test temperature, σ is the applied stress, and φ is the neutron flux. From the above relation, it is clear that the radiation-induced creep rate is directly proportional to the stress and the neutron flux. In essence, it means that with increasing stress and neutron flux, the creep effect would accelerate. If one integrates the above equation over time, it can be seen that the creep strain would vary with the neutron fluence (the product of flux and time). In 1967, Lewthwaite and coworkers in Scotland [30] demonstrated irradiation creep in several metals and alloys at 100 °C and published their findings in *Nature* (a well-known journal). Later, the radiation-induced creep has been observed in a number of alloys, including ferritic–martensitic steels (HT9 and T91) and austenitic steels as well as zirconium-based alloys. One way to study the irradiation creep behavior has been to irradiate stressed specimens under neutron exposure and study the stress relaxation behavior. There have been several studies using this mode of testing [31]. Irradiated annealed 304-type austenitic stainless steel specimens at 30 °C in a reactor under a neutron flux of $10^{13}\,\text{n cm}^{-2}\,\text{s}^{-1}$ for ~127 days to a total fluence of $\sim 1.1 \times 10^{19}\,\text{n cm}^{-2}$. The specimens were subjected to torsional strain due to the application of ~30 MPa. It was found that the level of relaxation in the irradiated material was about a factor of 40 more compared to that in the unirradiated (control) material under comparable temperature and stress.

Radiation-enhanced creep, as the term suggests, is the creep process enhanced by irradiation. This occurs at homologous temperatures at which thermal creep can also operate. As we know, generation of defects like vacancies at higher temperatures increases the thermal vacancy concentration in the material. This translates into the increase of diffusivity. Thermal creep rate can be shown to be proportional to diffusivity (Section 5.1). Now the addition of more vacancies produced through fast neutron irradiation can augment the vacancy concentration further, enhancing the overall creep rate. In this case, total radiation enhanced creep rate is given by

$$\dot{\varepsilon}^{\text{irr}} = \dot{\varepsilon}^{*} + \dot{\varepsilon}^{\text{th}} = B\sigma\varphi + AD\sigma^{n}, \tag{6.14}$$

where $\dot{\varepsilon}^{\text{th}} = AD\sigma^{n}$. Here A is a constant, D is the diffusivity, σ is the applied stress, and n is the stress exponent (n could be 5 or some other number depending on the conditions) (see Eq. (5.40)). In Eq. (6.14), D is proportional to the vacancy concentration that comprises contributions from thermal vacancies ($e^{-E_f/kT}$) and atomic displacements due to radiation ($\propto$ dpa).

One example of radiation-enhanced creep is shown in Figure 6.39 from the work of J.R. Weir [32]. Weir determined in-reactor stress–rupture properties of hot-pressed beryllium. Figure 6.39 shows the stress versus rupture time for three beryllium materials under three conditions. The neutron flux was $9 \times 10^{13}\,\text{n cm}^{-2}\,\text{s}^{-1}$. The results of the unirradiated material are compared with those of two types of irradiated materials. Some specimens were loaded after placing it in-reactor;

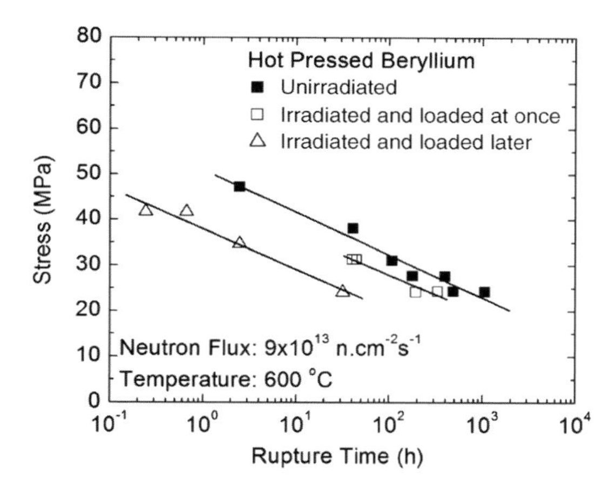

Figure 6.39 In-reactor stress–rupture properties of hot pressed beryllium at 600 °C.

however, a few specimens were loaded later after the specimens were irradiated for 800 h. The temperature was kept at ~600 °C (i.e., a homologous temperature of ~0.56). The stress–rupture life of the specimens got reduced when the specimen was loaded at once. But the specimen that was irradiated for 800 h accumulated more radiation damage leading to less rupture life. Similar reductions in stress–rupture life were found when stress–rupture tests were conducted on an irradiated 316-type austenitic stainless steel in the temperature range of 540, 600, 650, and 760 °C. Prior to stress–rupture tests, the steel specimens were irradiated up to a total neutron fluence of $1.2 \times 10^{22}\,\text{n cm}^{-2}$ at an irradiation temperature of 440 °C.

Irradiation creep sometimes operates at the same time as swelling and radiation growth (if applicable). In such situations, it becomes important to distinguish the contribution of irradiation creep to the total strain. Toloczko and Garner [33] have analyzed irradiation creep data from HT-9 and used the concept of creep compliance (B_0) to estimate the contribution of irradiation creep independent of swelling. Figure 6.40 shows the inclusion of an irradiation creep regime along with other thermal creep mechanisms on a deformation mechanism map of a 316-type stainless steel.

6.2.5
Radiation Effect on Fatigue Properties

Here, we will only briefly discuss the effect of radiation on fatigue properties. Recall the universal slopes method in Section 5.1, given by

$$\Delta\varepsilon = AN_f^{-0.6} + BN_f^{-0.12}, \tag{6.15}$$

where the first term represents the low cycle fatigue (LCF) that is controlled by ductility, while the second term represents high cycle fatigue (HCF) controlled

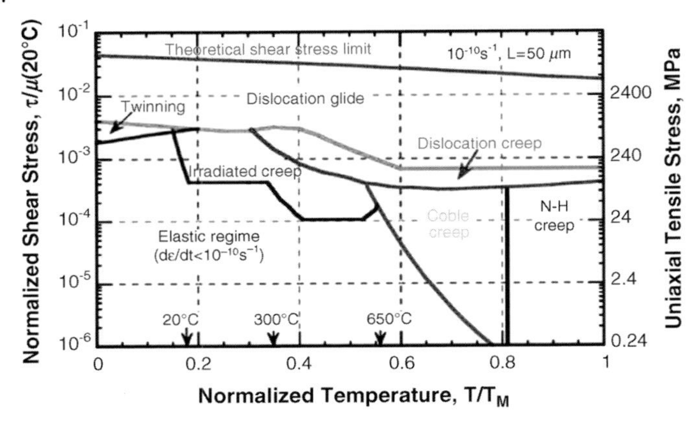

Figure 6.40 Deformation mechanism map of an irradiated 316-type stainless steel [34].

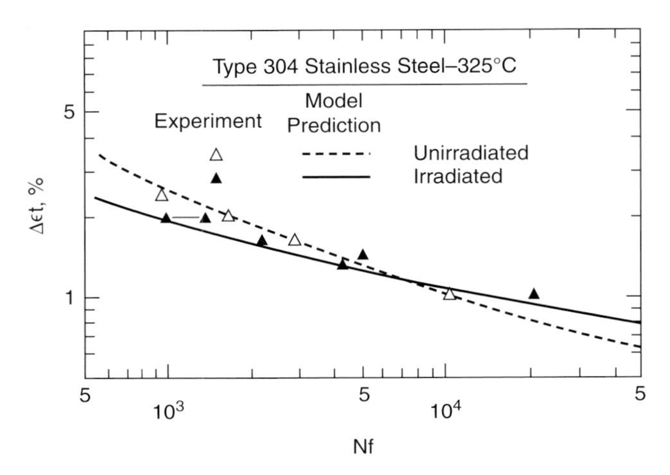

Figure 6.41 Effect of neutron radiation exposure on fatigue of 304-type stainless steel [35].

by strength. Now we know the constant *A* is directly proportional to ductility (i.e., reduction in area), while the constant *B* is proportional to strength (such as ultimate tensile strength). Since radiation exposure leads to hardening and embrittlement, fatigue life decreases in LCF and increases in HCF. Figure 6.41 illustrates the features clearly delineated in irradiated stainless steel tested at 325 °C, where it is noted that radiation exposure results in decreased fatigue life in LCF while improved life in HCF.

6.3
Radiation Effects on Physical Properties

Various physical properties such as thermal conductivity, thermal expansion coefficient, density, elastic constant, and so on are of interest for nuclear applications. Hence, it is important to understand the irradiation effects on the physical properties. Before we embark on discussing these, we must accept that irradiation can cause various changes in the structure of the materials and thus there may not be a general trend – but the effects will depend on particular situations. So, one needs to be prudent while analyzing these conditions and drawing inferences.

6.3.1 Density

Calculations indicate that vacancy–interstitial pairs should cause substantial changes in the density of the irradiated material as they would increase the volume ~1.5 times theoretically. However, experimental observations show very little or no change in the density (which should decrease due to the generation of Frenkel pairs), except in the radiation swelling regime where volume increase occurs through the creation of voids/bubbles (discussed in Section 6.1.2.2). It is thought that due to the greater mobility of interstitials and their clusters, they would diffuse even at homologous temperatures of 0.15–0.20 and get trapped or annihilated. This implies that we would expect to see little or no change in density of metals/alloys irradiated near or below room temperature. Some exceptions have been seen in very high melting metals such as refractory metals. In such cases, at lower temperatures, the interstitials are not that mobile, resulting in some significant changes in volume and in turn resulting in decreased density, as observed from lattice parameter measurements.

6.3.2 Elastic Constants

From various theoretical calculations, it has been noted that point defects created during irradiation may affect the elastic constants (elastic modulus and shear modulus). Interstitials are noted to have the greatest effect. Experimental results have not readily shown much changes. Also, point defect diffusion and annihilation lessen the effect. Higher fluences tend to show some effect. However, as defect insensitive properties, elastic constants do not change significantly with change in microstructure.

6.3.3 Thermal Conductivity

As noted in Chapter 1, thermal conductivity is a very important property of fuel cladding materials as it is needed to take the heat away from the fuel to the coolant. Theory indicates that one would expect a decrease in the thermal conductivity due to increased phonon scattering in the irradiated materials. However, in actual reactor operating conditions, the changes have been insignificant.

6.3.4
Thermal Expansion Coefficient

Point defects do not tend to change the thermal expansion coefficient, which have been confirmed experimentally.

6.4
Radiation Effects on Corrosion Properties

The chemical environment found in the nuclear reactors is quite harsh as many of the electrochemically active metals/alloys constitute the materials used in nuclear reactors. Furthermore, existence of crevices allowing chemicals to collect can lead to crevice corrosion. Sometimes "crud" formation also causes problems; the word "Crud" stands for Chalk River Unidentified Deposit. "Crud" is the corrosion product that is created in the steam generator, piping, and reactor pressure vessel walls, get transported via the core thus acquiring induced radioactivity, and get deposited in various locations of the reactor primary system. These result in problem with heat transfer and exacerbation of corrosion issues. Stobbs and Swallow [36] explained the effect of radiation in terms of metal, protective layer, and environment (corrodent).

6.4.1
Metal/Alloy

Radiation damage by generation of Frenkel defects, spikes, or transmutation can affect corrosion to some degree. Possible mechanisms are as follows:

a) **Increased chemical activity:** This effect becomes less important at higher temperatures.
b) **Irradiation-induced dimensional changes:** These may lead to enhanced corrosion by cracking the surface film so that the surface underneath becomes exposed to the corrodent. Thermal cycling is more important than density changes.
c) **Radiation-induced losses:** Such losses in ductility accompanied by stressing may result in stress corrosion. This has been observed in control rod alloys of boron in steel, titanium, or zirconium.
d) **Radiation-induced phase changes:** Such changes (e.g., precipitation) may affect the corrosion behavior.

6.4.2
Protective Layer

Lattice defects introduced into the protective oxide layer may affect the corrosion rate in the metallic substrate.

a) Diffusion of anions or cations through the oxide layer – irradiation would increase the number of defects.

b) Activity changes in the oxide layers are a potential source of enhanced corrosion.
c) Phase changes are a possible source of accelerated corrosion. If irradiation influences the transformation in a system such as ZrO_2 (monoclinic to tetragonal), an increase in rate may occur.

6.4.3 Corrodent

Various reactions can occur in the coolant exposed to irradiation. Short-lived radicals may form and secondary ionization produced; impurities generated, and secondary ionization occurred.

a) In gases, the effect is quite limited. Apparently, carbon dioxide can decompose to form oxygen in the reactor with increased attack on zirconium, Inconel, and 446, 310, and 316-type stainless steels.
b) Radiolysis of water can pose problem – such as radiolysis of water producing hydrogen may lead to hydriding of zirconium alloy cladding.
c) Liquid metals are generally stable under irradiation.
d) Fused salts are usually unaffected by irradiation.

6.4.3.1 LWR Environment

As noted before, radiation exposure generally leads to increased rates of corrosion and oxidation. Three basic effects on materials such as zircaloys are (i) oxidation/corrosion, (ii) hydriding, and (iii) stress corrosion cracking and corrosion fatigue.

Oxidation

In LWRs, radiolytic decomposition of water can lead to the creation of various free radicals (such as hydrogen, oxygen, hydroxyl radicals, and hydrogen peroxide), which may accelerate the corrosion effect. Some of the reactions occurring in water during radiolysis are given below:

$$H_2O \rightarrow H + OH \tag{6.16}$$

$$2H_2O \rightarrow 2H_2 + O_2 \tag{6.17}$$

$$OH + OH \rightarrow H_2O_2 \tag{6.18}$$

$$H + H \rightarrow H_2 \tag{6.19}$$

Figure 6.42 shows the schematic representation of zircaloy corrosion in the temperature range of 260–400 °C. In zircaloys, ZrO_2 forms as protective layer following a cubic rate law, meaning that oxidation rate decreases as ZrO_2 layer thickens. However, after a certain point (breakaway transition), the oxidation rate (as measured in terms of weight gain) becomes constant following a linear rate law, as shown in Figure 6.42. The breakaway transition occurs due to the destruction of the ZrO_2 film. It is thought that polymorphic phase transformation of ZrO_2 takes place as the stress develops in the ZrO_2 film as the thickening continues. This polymorphic phase transformation leads to sudden volume expansion, thus resulting in

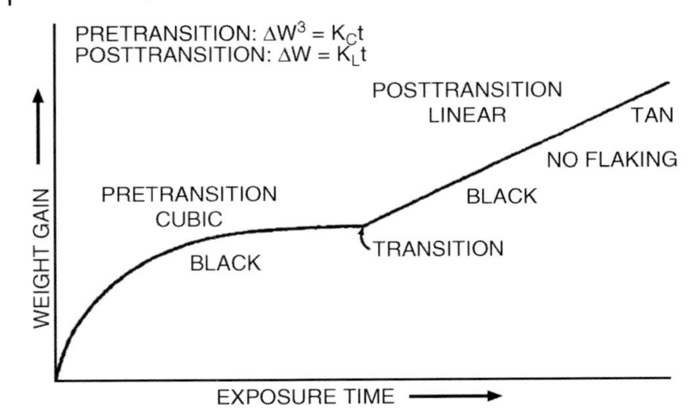

Figure 6.42 Schematic oxidation plot in terms of weight gain versus exposure time in zircaloys Ref. [37].

cracking/flaking of the protective film (Figure 6.43). This also means that the longer the breakaway time, the better would be the component sustaining the chemical environment. Recently, research has shown that Nb-containing zirconium alloys (like Zirlo) has greater longer term corrosion resistance in LWR environment.

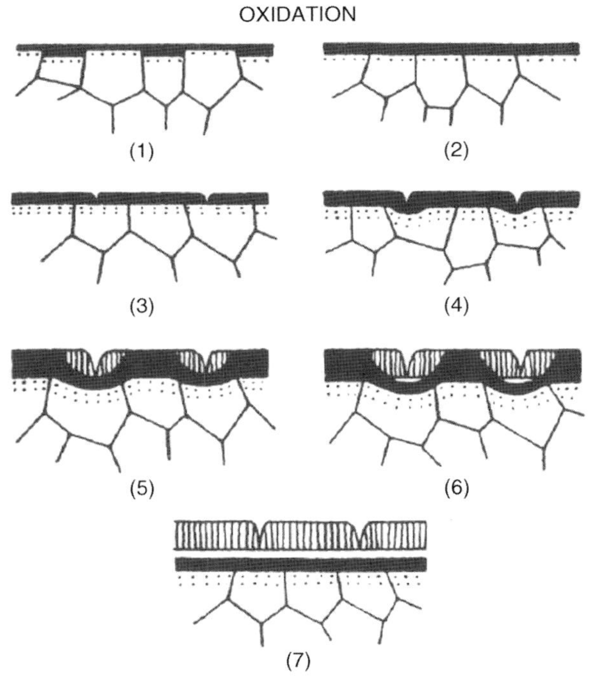

Figure 6.43 Schematic illustration of corrosion film breakaway Ref. [38].

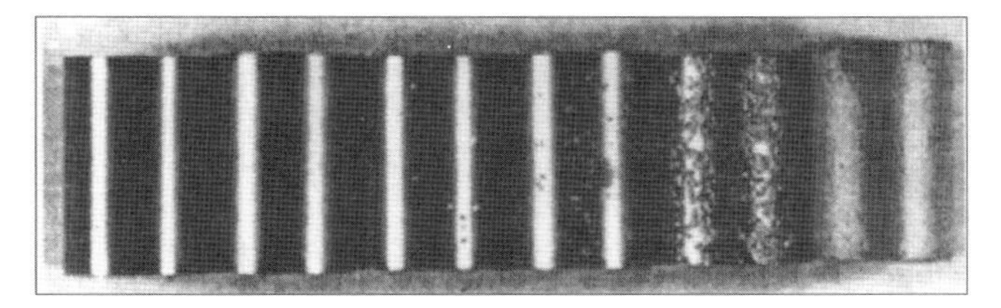

Figure 6.44 An example of nodular corrosion on the zircaloy fuel cladding surface.

Generally, materials chosen to serve in nuclear reactors should have the ability to form self-healing, passivating oxide films. But prior experience has shown that it is quite difficult to maintain the passivating oxide film in the long term in a reactor environment. Various flaws can be introduced as a result of stress and irradiation, which may turn into crack initiation sites. Hence, both uniform corrosion and nodular corrosion have been observed to be an issue. Evidence of nodular corrosion on zircaloy fuel cladding tube is shown in Figure 6.44.

Hydriding

This hydriding effect is particularly related to the use of zirconium alloy as fuel cladding. Hydrogen is produced due to reaction between water and zirconium in LWR environment. But moisture can also attack from within the fuel pellets if the pellets are not dried properly before insertion into the reactor. Hydrogen can then be absorbed and diffuse in the interior of zirconium cladding. The low-temperature α-zirconium (HCP) phase has a very low solubility of hydrogen, leading to any excess hydrogen-forming zirconium hydride precipitates. This leads to embrittlement, delayed hydride cracking, and hydride blistering, all of which diminish the lifetime of fuel rods and cause safety concerns in spent nuclear fuel rod repositories. It has been known that crystallographic texture of the zirconium alloy fuel cladding tube can be tailored to minimize the effect of hydride formation. With a suitable texture, the hydrides form along the hoop direction in place of radial direction, making it less susceptible to fracture. Figure 6.45 shows a microscopic cross section of hydrides oriented along the hoop direction in a zircaloy tube (the usual

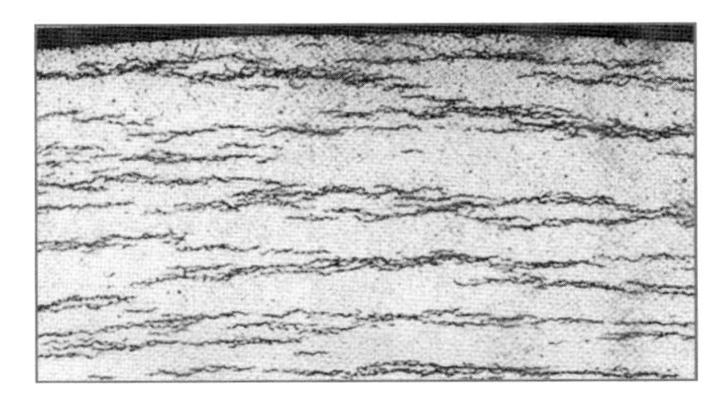

Figure 6.45 A microscopic cross section of a hydride zircaloy, showing zirconium hydride platelets oriented along the hoop direction.

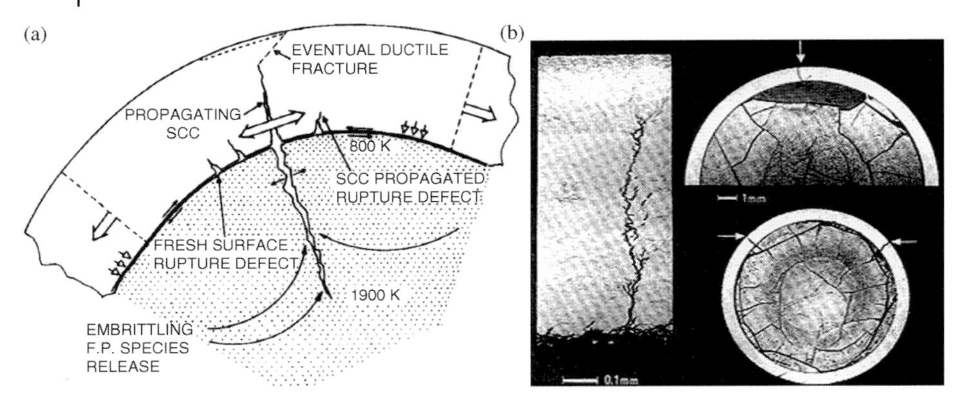

Figure 6.46 (a) A schematic of pellet–cladding interaction effect in zircaloy fuel cladding. (b) An actual case of PCI failure [39].

texture consists of majority of grains with *c*-axis close to the radial direction: $+/-30°$ from the radial direction toward the hoop).

Stress Corrosion Cracking

In Section 5.3, we have learned about corrosion basics, including stress corrosion cracking. Inside the reactor, these effects manifest themselves in various ways with respect to different components. Here, we introduce an important effect that occurs in the fuel rod known as pellet–cladding interaction (PCI). Figure 6.46a shows a schematic of the PCI effect. PCI is associated with local power ramps at start-up or during operation and occurs due to the effect of fission products like I, Cs, Cd, and so on, resulting in stress corrosion cracking. The crack generally initiates in the inner wall of the cladding and then progresses outward. Figure 6.46b shows an actual example of PCI failure in a fuel cladding system in a commercial power reactor. PCI minimization/elimination can be achieved by the following: (i) reduced ramp rates (flux variation or thermal gradient), but it is not an easy solution, (ii) coated (barrier) fuel, that is, surface coated with proper lubricant, and (iii) barrier cladding, that is, inner surface coated with graphite, copper, or pure zirconium to minimize stresses at the ID surface of the cladding; in BWRs, crystal bar Zr and zircaloy-2 are coextruded to form a thin Zr liner on the ID. The latter is discussed in detail in the following.

A modified Zr-lined barrier cladding known as TRICLAD™ has been developed by GE by adding a thin layer of corrosion-resistant zircaloy-2 bonded to the inner surface of the Zr barrier [40, 41]. In addition, the outer surface is made resistant to nodular corrosion by heat treatment that results in small second-phase particles, while majority of the parent zircaloy-2 material contained characteristic large SPP size distribution for improved crack growth resistance (toughness) [42, 43]. Figure 6.47 is a schematic of the Triclad for BWR service with the following four layers from ID to OD: (i) an inner layer of corrosion-resistant zircaloy-2 to slow oxidation and hydrogen generation, and to delay local hydride formation in the case of rod perforation, (ii) a Zr barrier for PCI resistance to blunt cracks nucleated at the inner surface, (iii) bulk

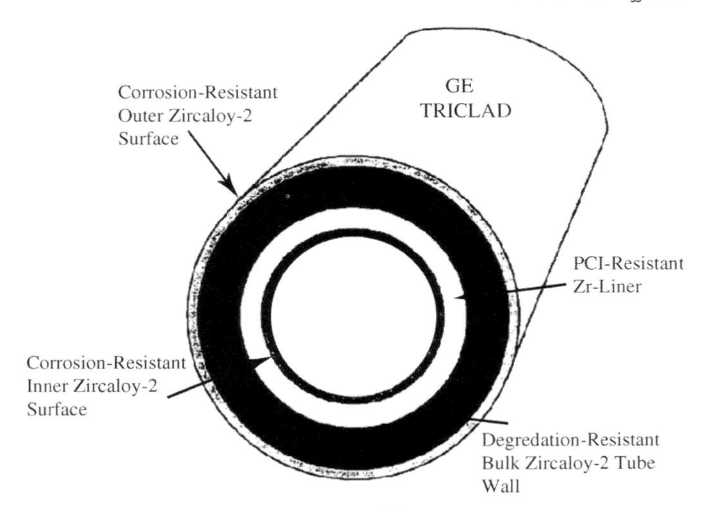

Figure 6.47 Cross section of GEs Triclad™ with zircaloy-2 lined Zr barrier tubing [44].

parent zircaloy-2 tubing with good toughness during irradiation, and (iv) an outer layer of zircaloy-2 processed for high resistance to nodular corrosion. An important feature of the final product is the integrity of the metallurgical bonding between the Zr barriers with both the inner zircaloy-2 liner and the bulk zircaloy-2 tubing.

6.4.3.2 Liquid Metal Embrittlement

In Chapter 1, we have discussed liquid metal fast breeder reactor (LMFBR). Both EBR-1 and EBR-2 used liquid metal coolants. Hence, it is important to know the effect of liquid metals on the nuclear components. Due to thermodynamic imbalance between structural metal and liquid metal (Na, K, Li, NaK, etc.), the structural metals/alloys exhibit liquid metal embrittlement as per the following:

i) Due to dissolution and precipitation (because of ΔT, . . .), depending on the solubility of liquid in the solid.
ii) Particle migration/diffusion of solid metal into liquid and vice versa.
iii) Penetration of the liquid metal atoms into solid metal (structural) mainly through GBs leading to rupture.

Radiation has minor effect on liquid metal embrittlement, but in general enhances due to decomposition of liquid metal or transmutation of atoms/nuclei.

Generally, austenitic stainless steels 304 and 316 SS are excellent candidates, but swelling could be a major limiting factor. Refractory metals tend to have high degree of resistance against liquid metals.

6.4.4 Irradiation-Assisted Stress Corrosion Cracking (IASCC)

Although the mechanism by which irradiation affects stress corrosion cracking is not precisely known, existing theories fall in the following categories: (a) radiation-induced grain boundary chromium depletion (radiation-induced segregation), (b) radiation hardening, (c) localized deformation, (d) selective internal oxidation, and (e) irradiation creep.

Corrosion by liquid metal and molten salts (especially fluorides) is seldom affected by radiation. Liquid metals appear to be unaffected by radiation (except to develop radioactivity). Reader is referred to various other references for further details.

6.5 Summary

Here, some general aspects of radiation effects are summarized.

- **Mechanical Properties**
  a) Hardness and strength increase (radiation hardening) due to increased defects mainly dislocations, precipitates, and so on.
  b) Ductility decreases (radiation embrittlement).
  c) Strain hardening exponent decreases (i.e., uniform ductility decreases).
  d) Ductile–brittle transition temperature increases (radiation embrittlement).
  e) Fracture toughness decreases (i.e., upper shelf energy decreases).
  f) Creep enhancement occurs (radiation-induced and radiation-enhanced) because of increased defect concentration and diffusivity.
  g) Low cycle fatigue life decreases and high cycle fatigue life increases due to embrittlement and hardening, respectively.
- **Physical Properties**
  a) Density decreases, that is, volume increases (radiation swelling) due to the formation of voids, bubbles, and depleted zones.
  b) Electrical resistivity increases (or conductivity decreases).
  c) Magnetic susceptibility decreases.
  d) Thermal conductivity decreases due to the increased defect concentration.
- **Corrosion Properties**
  a) Corrosion is enhanced by radiolytic dissociation of the environment.

Detailed discussion on radiation effects on materials is impossible in a single chapter. To have more detailed information on this vast topic, readers can refer to the excellent texts mentioned in the reference list [11, 24] or numerous papers published in nuclear materials specific journals.

Problems

6.1 A pressure vessel steel specimens (A533B, low-alloy ferritic steel) of diameter 1 mm and grain size 0.2 mm exhibited a yield point with the lower yield stress equal to 520 MPa. The source hardening term was found to be

150 MPa. For this steel, k_y decreases with temperature by $2\,\text{MN m}^{-3/2}$ for 50 °C increment. Assume that this linear dependence is valid and also the yield stress is temperature insensitive. The surveillance capsule tests revealed that the DBTT increased from −20 to +90 °C following radiation exposure in an LWR environment.

a) Estimate the increase in friction hardening (σ_i) required for this change.
b) Assuming that all this increase in friction hardening is due to dislocations, determine the dislocation density of the irradiated steel.
c) During radiation exposure in the reactor, the steel specimens were at around 425 °C that resulted in an increase in grain size from 0.2 to 0.4 mm. Determine the yield stress of the irradiated steel.

6.2 What are the effects of radiation on the following: (a) dislocation density, (b) vacancies, (c) diffusion, (d) corrosion, (e) strength (hardness), (f) ductility, (g) toughness, (h) DBTT, (i) upper shelf energy, (j) strain hardening, (k) creep (low temperature versus high temperature), (l) low and high cycle fatigue, (m) burnup, (n) density, and (o) thermal conductivity.

6.3 Describe the source and friction hardening in BCC and FCC alloys and the effects of neutron radiation exposure on them.

6.4 Describe the pellet–cladding interaction in LWR fuels and possible solutions to mitigate PCI.

6.5 In a fusion reactor blanket coolant channel, a copper (fcc) coolant tube (exposed to 14 MeV neutron radiation to a fluence of $5 \times 10^{22}\,\text{n cm}^{-2}$) exhibited radiation hardening and embrittlement with twice the tensile strength accompanied by reduced elongation to fracture (by 1/2 of that before irradiation) with *no necking*. The following properties were reported on the unirradiated copper: Young's modulus = 30×10^6 psi, nominal tensile strength = 50 ksi, fracture strain = 35%, and uniform strain = 25%.

a) Estimate the change in the toughness of the material following radiation exposure.
   This material is known to follow the universal slopes equation relating the fatigue life to the applied strain: $\Delta\varepsilon = (S_f/E)(N_f)^{-0.12} + \varepsilon_f^{-0.6}$, where S_f is the fracture strength, ε_f is the fracture strain, and N_f is the number of cycles to fatigue failure.
b) What are the effects of radiation exposure on fatigue life in LCF ($N_f \leq 50\,000$) and HCF ($N_f \geq 10^6$)? Does neutron irradiation *always* decrease fatigue life (explain your answers)?
c) Calculate the endurance limits before and after radiation exposure?
d) The radiation exposure resulted in 0.15 dpa. If one vacancy survived per million atomic displacements due to recombination and so on, calculate the density of Frenkel defects following radiation exposure? What is the probability (%) that a unit cell contains a vacant lattice site?
e) How does radiation exposure influence self-diffusion?
f) Show the effect on an Arrhenius plot indicating the relevant parameters.

6.6 Both zircaloys and stainless steels exhibit dimensional changes in-service (in-reactor) known as radiation growth and radiation swelling, respectively.

Describe these two phenomena with emphasis on distinctions between them.

6.7 A copper specimen is exposed to 2 MeV monoenergetic neutrons of flux, ϕ of $2 \times 10^{16}\,\text{n cm}^{-2}\,\text{s}^{-1}$ for 10 h.
   a) Determine the number of atomic displacements per atom (dpa).
   b) If PKAs were produced with 80 keV, calculate the number of atoms displaced by a PKA (assume Kinchin–Pease model)
   c) The steady-state creep rate ($\dot{\varepsilon}$) of Cu follows a power law: $\dot{\varepsilon} = AD_L\sigma^3$, with $A = 5 \times 10^{-7}$ ($\dot{\varepsilon}$ in h^{-1}, D in $cm^2\,s^{-1}$, and σ in MPa). If the creep rupture life of the unirradiated Cu at 400 °C at 20 MPa is 100 days, estimate the rupture life of the irradiated material at 300 °C under the same stress (assume that the same equation is valid for the irradiated material also).

Bibliography

1 Bacon, D.J., GAo F., and Ostesky, Yu N (2003) J. *Nucl. Mater.*, **276**, 152–162.
2 Li, M., Byun, T.S., Snead, L.L., Zinkle, S.J., (2007) Defect cluster and radiation hardening in molybdenum neutron-irradiated at 80 °C, *Journal of Nuclear Materials*, 367–370, 817–822.
3 Jenkins, M.L. and Kirk, M.A. (2001) Characterization of Radiation Damage by Transmission Electron Microscopy, Institute of Physics, Philadelphia, PA.
4 Yoshiie, T., Kiritani, M. (1999) Destination of point defects and microstructural evolution under collision cascade damage, 271–272, 296–300.
5 Russell, K.C. and Powell R.W. (1973) Dislocation loop nucleation in irradiated metals, *Acta Metallurgica*, **21**, 187–193.
6 Matsui, T., Muto, S., and Tanabe, T. (2000) TEM study on deuterium-induced defects in tungsten and molybdenum, *Journal of Nuclear Materials*, 283–287, 1139–1143.
7 Katoh, Y., Hashimoto, N., Kondo, S., Snead, L.L., Kohyama, A. (2006) Microstructural development in cubic silicon carbide during irradiation at elevated temperature, *Journal of Nuclear Materials*, **351**, 228–240.
8 M. Kiritani, (1994) *J. Nucl Mater*, **216**, 200–264.
9 Murase, Y., Nagakawa, J., Yamamoto, N., and Shiraishi, H. (1998) *Journal of Nuclear Materials*, **258–263**, 1639–1643.
10 Straalsund, J.L., Powell, R.W., and Chin, B.A. (1982) An overview of neutron irradiation effects in LMFBR materials. *Journal of Nuclear Materials*, **108–109**, 299–305.
11 Olander, D.R. (1985) *Fundamental Aspects of Nuclear Reactor Fuel Elements*, Technical Information Center, Springfield, VA.
12 Harkness, S.D. and Li, C.-Y. (1971) A study of void formation in fast neutron irradiated metals. *Metallurgical and Materials Transactions B*, **2**, 1457.
13 Corbett, J.W. and Ianniello, L.C. (1972) Radiation Induced Voids in Metals, AEC Symposium Series No. 26, p. 125.
14 Guthrie, G.L. *et al.* (1970) Absence of voids in neutron irradiated sputtered nickel. *Journal of Nuclear Materials*, **37**, 343.
15 Holmes, J.J. (1969) Transactions of American Nuclear Society, 12, 117.
16 Klueh, R.L. (2005) Elevated temperature ferritic and martensitic steels for future nuclear reactors, *International Materials Review*, **50**, 287–310.
17 Kimura, A., Cho, H., Toda, N., Casada, R., Kishimoto, H., Iwata, N., Ukai, S., Ohtsuka, S., and Fujiwara, M. (2006) Paper 6456, Proceedings of the International Congress on Advances in Nuclear Power Plants (ICAPP'06), p. 2229.
18 Cawthorne, C. and Fulton, E. (1967) Voids in irradiated stainless steel. *Nature*, **216**, 576.
19 Bruemmer, S.M., Simonen, E.P., Scott, P.M., Andresen, P.L., Was, G.S., and

Nelson, J.L. (1999) Radiation-induced material changes and susceptibility to intergranular failure of light-water-reactor core internals. *Journal of Nuclear Materials*, **274**, 299–314.

20 Nemoto, Y., Hasegawa, A., Satou, M., Katsunori, A., and Hiraoka, Y. (2004) Microstructural development and radiation hardening of neutron irradiated Mo-Re alloys. *Journal of Nuclear Materials*, **324**, 62.

21 Farrell, K., Byun, T.S., and Hashimoto, N. (2004) Deformation mode maps for tensile deformation of neutron irradiated structural alloys. *Journal of Nuclear Materials*, **335**, 471–486.

22 Murty, K.L., Charit, I. (2008) Structural materials for Gen-IV nuclear reactors: Challenges and opportunities. *Journal of Nuclear Materials*, **383**, pp. 189–195.

23 Makin, M.J. and Minter, F.J. (1960) Irradiation hardening in copper and nickel, **8**, 691–699.

24 Was, G.S. (2007) *Fundamentals of Radiation Materials Science: Metals and Alloys*, Springer.

25 Murty, K.L. and Oh, D.J. (1983) Friction and source hardening in irradiated mild steel. *Scripta Metallurgica*, **17**, 317.

26 Ohr, S.M., Tucker, R.P., Wechsler, M.S. (1970) Radiation anneal hardening in niobium—an effect of post-irradiation annealing on the yield stress, *Physica Status Solidi*, **2**, 559–569.

27 Gurovich, B.A., Kuleshova, E.A., Nikolaev, Y.A., Shtrombakh, Y.I. (1997) *Journal of Nuclear Materials*, **246**, 91–120.

28 Miller, M.K. and Russell, K.F. (2007) Embrittlement of RPV steels: an atom probe tomography perspective. *Journal of Nuclear Materials*, **371**, 145–160.

29 Steele, L.E. (1975) Neutron Irradiation Embrittlement of Reactor Pressure Vessel Steels, IAEA, Vienna.

30 Lewthwaite, G.W., Mosedale, D. (1967) Irradiation creep in several metals and alloys at 100 °C, *Nature*, **216**, 472–473.

31 Hesketh, R.V. (1963) A transient irradiation creep in non-fissile materials, *Philosophical Magazine*, **8**, 1321–1333.

32 Weir, J.R. (1963) Work at the Oak Ridge National Laboratory in.

33 Toloczko, M.B., Garner, F.A. (1999) Variability of irradiation creep and swelling of HT-9 irradiated to high neutron fluence at 400–600 °C, Effects of Radiation on Materials: 18th International Symposium, ASTM STP 1325, 765.

34 Zinkle, S.J. and Ghoniem, N.M. (2000) Operating temperature windows for fusion reactor structural materials. *Fusion Engineering and Design*, **51–52**, 55.

35 Murty, K.L. and Holland, J.R. (1982) Low-cycle fatigue characteristics of irradiated 304SS. *Nuclear Technology*, **58**, 530.

36 Stobbs, J.J. and Swallow, A.J. (1962) Effects of radiation on metallic corrosion. *Metallurgical Review*, **7**, 95.

37 Hillner, E. (1977) Corrosion in Zirconium-Base Alloys — An Overview, Zirconium in the Nuclear Industry, A. Lowe and G. Parry (Eds.) ASTM STP 633.

38 Douglas, D.L. (1971) The Metallurgy of Zirconium, IAEA, Vienna.

39 Garzarolli, F., Jan, R.V., and Stehle, H. (1979) The main causes of fuel element failure in water cooled power reactor. *Atomic Energy Review*, **17**, 31–128.

40 Edsinger, K. and Murty, K.L. (2001) LWR pellet–cladding interactions: materials solutions to SCC. *Journal of Metals*. July, pp. 9–13.

41 Armijo, J.S., Coffin, L.F., and Rosenbaum, H.S. (1995) Method for making fuel cladding having zirconium barrier layers and inner liners, US Patent # 5,383,228.

42 Cheng, N. and Adamson, R.B. (1985) Mechanistic study of zircaloy nodular corrosion. *Zirconium in the Nuclear Industry*, ASTM STP 939, p. 387.

43 Marlowe, M.O., William, C.D., Adamson, R.B., and Armijo, J.S. (1995) Degradation resistant fuel cladding. Proceedings of Jahrestagung Kerntechnik, 1995, Nurnberg, Germany, p. 329.

44 Murty, K.L. and Charit, I. (2006) Texture development and anisotropic deformation of zircaloys. *Progress in Nuclear Energy*, **48**, 325–359.

7
Nuclear Fuels

"We believe the substance we have extracted from pitch-blende contains a metal not yet observed, related to bismuth by its analytical properties. If the existence of this new metal is confirmed we propose to call it *polonium*, from the name of the original country of one of us."

—*Marie Curie*

7.1
Introduction

The heart of a nuclear reactor is the "reactor core" that contains nuclear fuels among other components/materials. Nuclear fuel forms consist of radioactive materials that may create the fission chain reaction under suitable conditions creating a large amount of heat that is then utilized for producing the electrical power. The following are the basic requirements of a nuclear fuel:

a) The capital installation costs for nuclear power plants are substantial. In order to maintain profitability in the power production, the fuel costs must be minimal.
b) Adequate thermal conductivity of nuclear fuels is necessary to ensure that they can withstand the thermal gradients generated between the fuel center and periphery.
c) The fuel should be able to resist repeated thermal cycling due to the reactor shutdowns and start-ups.
d) It should have adequate corrosion resistance against the reactor fluids.
e) It should transmit heat quickly out of the fuel center.
f) The fuel should be relatively free from the constituent elements or impurities with high neutron capture cross section in order to maintain adequate neutron economy.
g) It must be able to sustain mechanical stresses.
h) The fuels should be amenable for reprocessing or disposal.

Nuclear fuel materials developed over decades include metals/alloys (uranium, plutonium, and thorium) and ceramics (oxides, carbides, nitrides, and silicide compounds containing the former radioactive elements). Nuclear fuels are fabricated in a wide

An Introduction to Nuclear Materials: Fundamentals and Applications, First Edition.
K. Linga Murty and Indrajit Charit.
© 2013 Wiley-VCH Verlag GmbH & Co. KGaA. Published 2013 by Wiley-VCH Verlag GmbH & Co. KGaA.

variety of configurations, such as cylindrical pellets, long extruded rods (for metal/alloy fuels only), spherical particles, coated particles, dispersion fuels (such as cermets), and fluid forms (for aqueous homogeneous reactors and molten salt cooled reactors).

Some Basic Terms Regarding Nuclear Fuels

Burnup: Fuel burnup is an important property of nuclear fuels. Burnup is generally defined as the amount of heavy metal (in the form of uranium and higher actinides) in the fuel that has been fissioned. This term can be expressed either as a percentage of heavy metal atoms that have fissioned (atom%) or in units of fission energy (gigawatt-day or GWd; 1 GWD $= 8.64 \times 10^{13}$ MWd) produced per metric ton of the heavy metal (MTHM), that is, GWd/MTHM or MWd/kgHM. One atom% burnup corresponds to approximately 9.4 GWd per MTHM. However, the fuel burnup is often limited by the fuel cladding performance. Superior cladding performance allows higher burnups in fuels.

Blanket fuel: Nuclear reactor fuel that contains the fertile isotopes that are bred into fissile isotopes

Driver fuel: Nuclear reactor fuel that contains the fissile isotopes along with fertile isotopes that are bred into fissile isotopes

Reproduction factor: It is generally represented by η, which is the number of neutrons created per neutron absorbed in fuel. If ν neutrons are produced per fission reaction, the number ratio of fission to absorption in fuel is σ_f/σ_a and the number of neutrons per absorption is

$$\eta = \frac{\sigma_f}{\sigma_a}\nu. \tag{7.1}$$

The value of η is higher in fast reactor compared to that in thermal reactors.

Conversion ratio: The ability to convert fertile isotopes into fissile isotopes can be measured by the conversion ratio (CR) defined as

$$\text{CR} = \text{Fissile atoms produced/fissile atoms consumed.} \tag{7.2}$$

If CR is >1 such as in a fast reactor, it is called the breeding ratio (BR). The breeding gain (BG) is given by (BR − 1), which represents the additional plutonium produced per atom burned.

Fission products (Fp): According to Kleycamp [1], the fission products can be classified as follows – (1) fission gases (fg) and other volatile elements – Br, Kr, Rb, I, Xe, Cs, and Te; (2) fission product forming precipitates – Mo, Tc, Ru, Rh, Pd, Ag, Cd, In, Sn, Sb, Se, Te; (3) Fp forming oxide precipitates – Rb, Sr, Zr, Nb, Mo, Se, Te, Cs, and Ba; and (4) Fp dissolved as oxides in fuel matrix – Rb, Sr, Zr, Nb, La, Ce, Pr, Nd, Pm, Sm, and Eu.

Fissium (Fs) or fizzium (Fz): Fissium is nominally 2.4 wt% Mo, 1.9 wt% Ru, 0.3 wt% Rh, 0.2 wt% Pd, 0.1 wt% Zr, and 0.01 wt% Nb and is designed in such a way that it can mimic noble metal fission products remaining after a simple reprocessing technique based on melt refinement.

7.2 Metallic Fuels

The history of metallic nuclear fuels dates back to the first developmental stages of nuclear reactors. U- and Pu-based fuels were used in the Experimental Breeder Reactor-1 (i.e., EBR-1) that produced useful electricity for the first time in December 1951. In addition, EBR-2, the first-generation Magnox reactors (such as Calder Hall in the United Kingdom), and many other subsequent fast reactors have used metallic fuels. When water-cooled reactors were being developed, the metallic fuels were not chosen mainly because of the compatibility issues between water and metallic fuel at elevated temperatures arising during the event of a cladding breach resulting in the formation of metal hydrides or oxides. However, in the mid-1960s, when the fast reactor development was gaining ground, designers chose oxide fuels in place of metallic fuels as they envisioned that the metallic fuels would have only limited burnups because of the presumed swelling problems and anticipated creation of liquid phases in fuels during the higher temperature operations. During that time, oxide fuels were recommended for power reactors even though limited information was available on them. Later simple design changes for the fuel elements, widening the gap between the fuels and cladding materials and providing a plenum volume for accumulating fission gases, have shown marked improvement (1% versus 20% burnup) over the earlier designs where there was no gap or very little gap between the fuel and the cladding materials. However, it is important to note that research and test reactors have traditionally used metallic fuels mainly because of the lower temperature operations involved.

Here we highlight three main metallic nuclear fuel materials. Metallic fuels have a number of advantages as well as disadvantages often specific to the fuel types. However, the metallic fuels generally have higher thermal conductivity, high fissile atom density (improved neutron economy), and fabricability as their prime advantages, whereas lower melting points, various irradiation instabilities, poor corrosion resistance in reactor fluids, and various compatibility issues with the fuel cladding materials are some prominent disadvantages. Metallic fuels can also be used in alloy forms to improve corrosion resistance and irradiation performance among others.

7.2.1 Metallic Uranium

Uranium (atomic number 92) takes up about 4 ppm (parts per million) of the earth crust. Uranium is more abundant than the common elements such as silver, mercury, cadmium, and so on. The amount of economically recoverable uranium in the world has been estimated to be about 5.5 million tons (2007 estimate). Uranium mined from the earth crust is known as "natural uranium" that contains two main isotopes U^{235} (0.71%) and U^{238} (99.28%) along with a very minor presence (0.006%) of U^{234}. Uranium in general is found in a variety of minerals, such as pitchblende (U_3O_8), Uraninite (UO_2) followed by carnotite ($K_2O \cdot 2UO_3 \cdot V_2O_5 \cdot nH_2O$). High-grade uranium ores are mined in Kazakhstan, Canada, Australia, Namibia, South Africa, Niger, the Rocky Mountain region of the United States, and many other nations. Kazakhstan leads the world production of uranium (27%) followed by Canada (20%) and Australia (20%), as per 2010 estimate. However, it can be noted that a sizable portion of uranium is also produced by reprocessing spent nuclear reactor fuels. Note that existing laws presently forbid reprocessing of spent nuclear fuels in the United States. So, almost all commercial uranium fuels in the United States need to be made following extraction from the ores. On the other hand, France uses reprocessing of spent nuclear fuels as one of the major methods for meeting its fuel needs.

7.2.1.1 Extraction of Uranium

Over several decades, many uranium extraction methods have been pursued. Here, only some of the very common extraction techniques are discussed. Almost all uranium minerals are present in the ore with a variety of gangue (impurity) materials. Hence, it is essential to separate the gangue minerals from the mineral with a metallic value. The metal extraction process thus contains mineral beneficiation or ore dressing techniques as the first steps that increase the metal value of the ore by removing the gangue materials. The uranium extraction is generally achieved by using chemical methods, as entirely physical beneficiation methods are not effective enough to liberate ore minerals of value. For recovery of the most metal content, it becomes necessary to pursue leaching and precipitation reactions. One of the most common processes is to leach finely ground uranium ores by dilute acids. Sulfuric acid (H_2SO_4), nitric acid (HNO_3), and hydrochloric acid (HCl) can be used for leaching. However, the latter two acids adversely affect the process economics. Also, they present corrosion problems for the process equipment. However, if the solvent extraction process is to be used after leaching, nitric acid must be used. Under certain conditions, sodium carbonate (Na_2CO_3) is also used as suitable leaching agent.

Leaching is nothing but a dissolution reaction in which uranium forms a soluble compound that remains dissolved in the solution. It is known that uranium can go into solution only when it is in a hexavalent state. This is a common requirement for both acid and alkali leaching. Thus, if uranium is present in the tetravalent state, it needs to be oxidized to the higher oxidation state. This is obtained by the presence of trivalent iron or pentavalent vanadium, which happens to coexist with uranium in most uranium ores. The reactions for acid and alkali leaching are given below:

$$2U_3O_8 + 6H_2SO_4 + (O_2) = 6UO_2SO_4 + 6H_2O \qquad (7.3)$$

$$2U_3O_8 + 18Na_2CO_3 + 6H_2O + (O_2) = 6Na_4UO_2(CO_3)_3 + 12NaOH \qquad (7.4)$$

Acid leaching generally leads to greater recovery of uranium than carbonate leaching. However, there are some limitations of acid leaching. This cannot be used for ores that contain magnesium and/or calcium carbonates as these compounds tend to react with acid leachants wasting an excessive amount of acid. Furthermore, because of corrosion problems, the equipment and procedure used in acid leaching are far more expensive. Conversely, corrosion problems are not so severe (thus reducing the process costs) in alkali leaching. It also allows reagent recovery. However, alkali leaching is not suitable for leaching ores with contents of high gypsum or sulfide or refractory constituent.

After leaching, uranium is to be recovered from the leach solution. This can be done by following one of the methods – chemical precipitation, ion exchange, and solvent extraction. Here, we discuss the ion exchange method. This method is used for recovering uranium after acid leaching by sulfuric acid. The method is based on the principle that uranyl sulfate ions can be selectively removed by allowing them to be adsorbed by an ion exchange resin, and subsequently the adsorbed ions could be separated from the loaded resin with a solution concentrated with nitrate or chloride ions. It can be noted that the ion exchange process is also applicable to alkali leaching process. From the processes discussed above, eventually a yellow powder containing 80–85% of UO_2 can be obtained. Nitric acid is used as a dissolving agent for this powder subsequent filtration. This reaction produces uranyl nitrate that is extracted with the help of ether. Then water is used to wash the filtrate to produce an aqueous solution from which pure ammonium diuranate [$(NH_4)_2U_2O_7$] is precipitated by passing ammonia gas. Following the separation of diuranate, it is dried and reduced to uranium dioxide in hydrogen at 650 °C. In another process (Dryway process), ammonium diuranate can be converted to uranium trioxide (UO_3) by heating it to remove ammonia and steam. Uranium trioxide is then treated with hydrogen at ~600 °C to produce uranium dioxide and then this dioxide is reacted with anhydrous hydrogen fluoride to obtain the uranium tetrafluoride (UF_4) that can then be reduced with either calcium or magnesium to produce metallic uranium. Some of the reactions involved in the Dryway process are given below:

$$UO_3 + H_2 \rightarrow UO_2 + H_2O \tag{7.5}$$

$$UO_2 + 4HF \rightarrow UF_4 + 2H_2O \tag{7.6}$$

$$UF_4 + 2Mg \rightarrow U + 2MgF_2 \tag{7.7a}$$

$$UF_4 + 2Ca \rightarrow U + 2CaF_2 \tag{7.7b}$$

For enrichment of uranium (i.e., to increase the amount of fissile atom U^{235} density), there are specific methods like membrane separation, centrifuging, and so on, using UF_6. Depending on the condition of uranium (unenriched, slightly enriched, or highly enriched), specific details of the process may change to avoid any possibility of obtaining critical mass for fission chain reaction.

7.2.1.2 **Nuclear Properties**

The nuclear properties of uranium are summarized in Table 7.1. U^{235} and U^{233} nuclides have substantial fission cross sections at the thermal neutron energy level (average energy of 0.025 eV), but U^{238} has negligible fission cross section as it is not fissile (rather fertile). Natural uranium contains only 0.7% U^{235}. So, as the enrichment of U^{235} is increased, the fission cross section increases. The absorption cross section in Table 7.1 is the sum of fission and capture cross sections.

Table 7.1 Thermal nuclear cross sections and other parameters for uranium (for thermal neutron of average energy 0.025 eV).

	U^{233}	U^{235}	U^{238}	Natural U
Fission cross section	531.1	582.2	<0.0005	4.18
Capture cross section	47.7	98.6	2.71	3.50
Absorption cross section	578.8	680.8	~2.71	7.68
ν (Neutrons/fission)	2.52	2.47	—	2.46
η, Fast neutrons/thermal neutron absorbed	2.28	2.07	—	1.34

7.2.1.3 **Uranium Crystal Structure and Physical Properties**

Despite being a metal, uranium has chemical bonding characteristics of metalloids like arsenic, antimony, or bismuth.

Up to 666 °C, uranium assumes an orthorhombic crystal structure (α-U), with 4 atoms per unit cell: density – 19.04 g cm^{-3} and lattice constants ($a = 2.8541 \pm 0.003$ Å, $b = 5.8692 \pm 0.0015$ Å, and $c = 4.9563 \pm 0.0004$ Å) (at 25 °C). The structure is somewhat unique in that it can be thought of as stacks of "corrugated" sheets with atoms parallel to a–c plane with ~2.8 Å distance between atoms in the sheets and ~3.3 Å distance between the sheets. It can also be described as a distorted HCP crystal structure!

In the temperature range of 666–771 °C, uranium has a complex tetragonal crystal structure (30 atoms in a unit cell) and is called β-U. Density of β-U is 18.11 g cm^{-3}, and lattice constants $a = 10.759 \pm 0.001$ Å and $c = 5.656 \pm 0.001$ Å (at 720 °C).

In the temperature range of 771–1130 °C, uranium assumes a simple body-centered cubic crystal structure (γ-U), that is, with 2 atoms per unit cell. Density is 18.06 g cm^{-3}, and lattice constant $a = 3.524 \pm 0.002$ Å (at 805 °C).

Because of the anisotropic nature of the crystal structure of alpha-uranium, thermal expansion coefficients are anomalous along the crystallographic directions determined by lattice parameter measurements and shown in Figure 7.1. That is, the linear thermal expansion coefficient (both linear and volume) increases in the direction of [100] and [001], and decreases along [010] with increasing temperature. However, the volumetric thermal expansion coefficient (i.e., the overall thermal expansion effect due to combination of linear expansion and contraction) does increase with increasing temperature. The dilatometry has also been used to measure thermal expansion coefficients and they have shown comparable trend. As noted before, uranium shows allotropic transformation and thus shows increased volumetric thermal expansion coefficients, as the phase transformation occurs as a function of temperature.

Thermal conductivity is a important property with respect to heat removal from the fuel through cladding (by conduction) to the coolant (by convection) in a nuclear reactor. The linear power rating of a reactor fuel element is generally limited by the thermal conductivity of the fuel to avoid center melt. Figure 7.2 shows thermal conductivity of a well-annealed high purity polycrystalline uranium as a

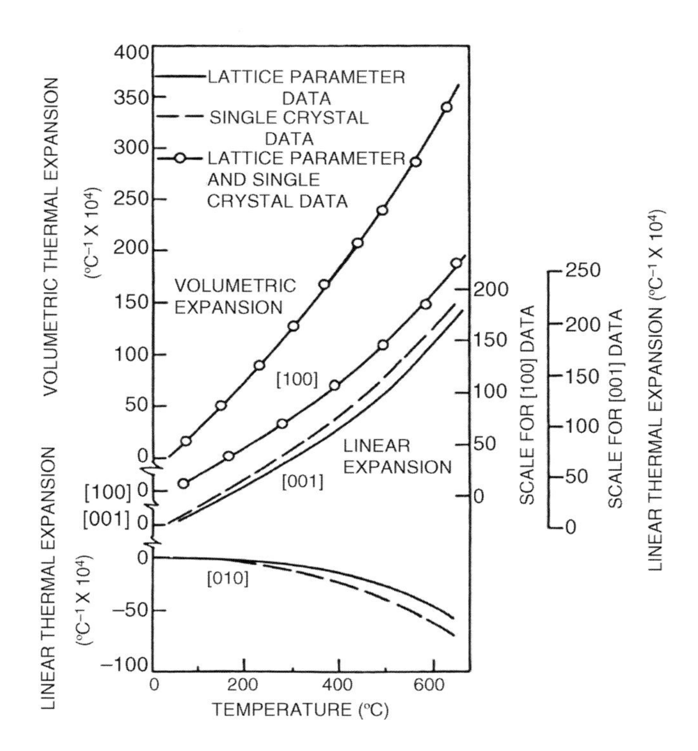

Figure 7.1 Thermal expansion coefficient of α-U is anisotropic as a function of temperature [2].

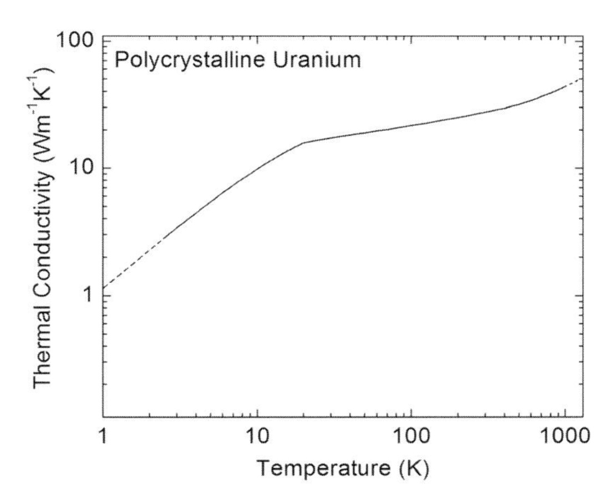

Figure 7.2 Thermal conductivity of a well-annealed high purity polycrystalline uranium as a function of temperature [3].

function of temperature. Interestingly, the thermal conductivity of uranium keeps on rising as the temperature increases, thus offering the advantage of having better heat conduction at elevated temperatures! However, depending on various factors, thermal conductivity may vary and fall in a data-band.

Heat capacity of uranium in the range of 20–669 °C (293–942 K) is calculated by expression given by Rahn *et al.* [4]:

$$C_p\ [\mathrm{J\ kg^{-1}\ K^{-1}}] = 104.82 + (5.3686 \times 10^{-3})T + (10.1823 \times 10^{-5})T^2, \qquad (7.8)$$

where T is in K.

The average C_p in the temperature regime of 669–776 °C (beta-phase regime) is 176.4 J kg^{-1} K^{-1}, whereas the average C_p is 156.8 J kg^{-1} K^{-1} in the temperature regime of 776–1132 °C (gamma-phase regime).

7.2.1.4 **Mechanical Properties**

Pure uranium is a moderately ductile material. However, the mechanical properties depend on crystallographic texture (i.e., preferred orientation of grains) and, thus are in alpha-uranium. The texture is affected by the fabrication history and heat treatment. Grain size and shape are also important parameters affecting the mechanical properties. The tensile properties are sensitive to impurities like carbon or fission products or alloying elements. A typical stress-strain curve of uranium is shown in Figure 7.3. The strength decreases precipitously with increasing temperature, as shown in Table 7.2.

The plastic deformation of uranium generally involves the following mechanisms: (i) slip in the {010}⟨100⟩ system, (ii) {130} twinning, (iii) {172} twinning, and (iv) kinking, cross-slip, {176} twinning, and {011} slip under special conditions. Overall, twinning appears to be the major deformation mode at room temperature. However, the contribution of slip to plastic deformation increases as the temperature is increased, and above ~450 °C, slip becomes the predominant plastic deformation mechanism.

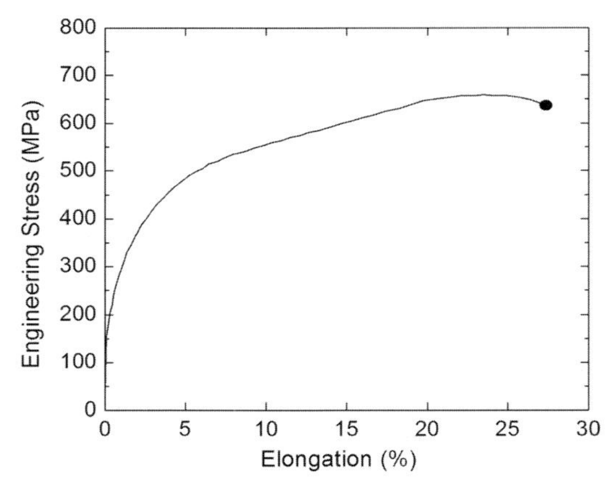

Figure 7.3 A typical engineering stress-strain curve of uranium.

Table 7.2 Summary of tensile properties of uranium [2].

Condition	Test temperature (°C)	Yield strength (MPa)	Ultimate tensile strength (MPa)	Elongation (%)
Rolled at 300 °C				
Alpha annealed[a)]	Room temperature	296	765	6.8
	300	121	241	49.0
	500	35	77	61.0
Beta annealed[b)]	Room temperature	169	427	8.5
	500	49	72	44.0

a) Heated at 600 °C for 12 h, slowly cooled.
b) Heated at 700 °C for 12 h, slowly cooled.

7.2.1.5 **Corrosion Properties**

Uranium is very active chemically and rapidly reacts with most environments (air, oxygen, hydrogen, water, water vapor, and others). Freshly polished uranium has a dull silvery color. However, when exposed to air for a few minutes, the surface shows straw-like color and darkens to a blue-black color within a few days. The oxide films formed are not quite protective. At elevated temperatures as the film thickens with time, the characteristic black color of UO_2 develops, and it starts cracking and crumbling exposing fresh uranium metal from underneath the oxide film to be attacked.

Unalloyed uranium reacts with water almost readily. Figure 7.4 shows the corrosion behavior of uranium in aerated distilled water. At 50–70 °C, an initially formed UO_2 film provides corrosion protection to the metal with considerable incubation period. However, beyond this temperature regime, corrosion rate picks up as the surface oxide film becomes porous and the protection of the oxide film is lost. Also, no incubation period is noted at these

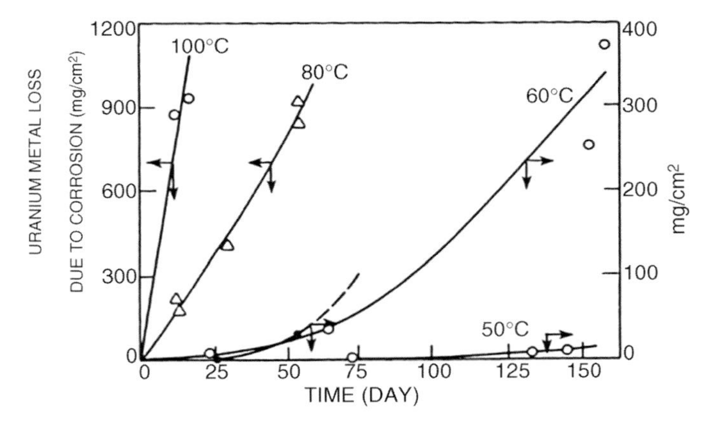

Figure 7.4 Corrosion of unalloyed uranium in aerated distilled water Ref. [2].

temperatures. Conversely, in hydrogen-saturated or degassed water, the corrosion rates remain linear with respect to time in the moderate temperature regime. It is postulated that hydrogen diffusion takes place through the thin oxide film to form uranium hydride (UH_3) between the oxide and the metal and also into the grain boundaries of uranium.

7.2.1.6 **Alloying of Uranium**

Alloying of uranium is done to improve the mechanical properties, dimensional stability and corrosion resistance of uranium. However, selection of alloying elements should not adversely affect the neutron economy; hence, a lot of emphasis was placed on the alloying elements like Al, Be, and Zr. The alloying elements like Ti, Zr, Nb, and Mo have extensive solid solubility in uranium at higher temperatures, V and Cr have moderate solubility, and Ta and W are further less soluble in γ-U. Figure 7.5 shows the equilibrium-phase diagram of U–Mo system. U–Zr and U–Pu–Zr fuels in EBR-II were used as the alloying raised the alloy solidus

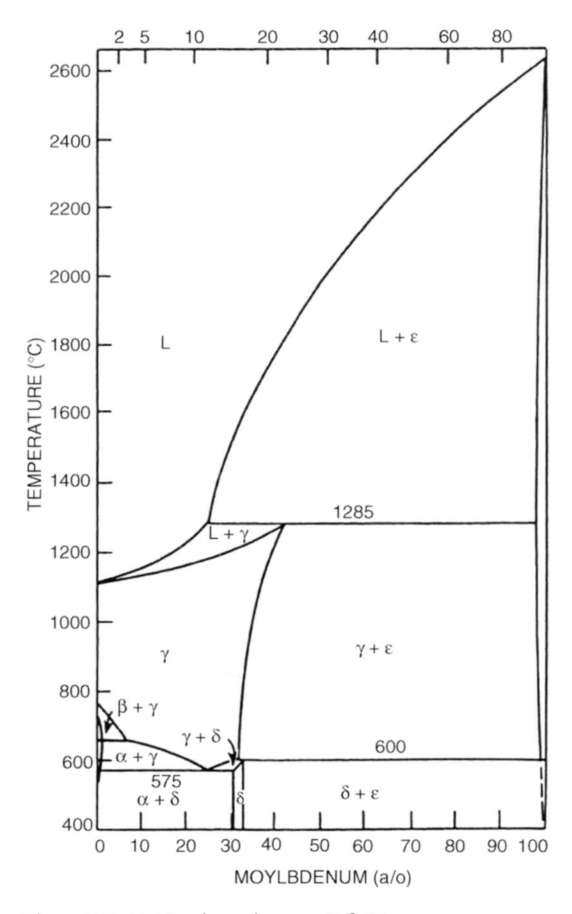

Figure 7.5 U–Mo phase diagram Ref. [2].

temperature, enhanced dimensional stability under irradiation, and reduced fuel–cladding material chemical interaction. Furthermore, uranium–fissium/fizzium (U–Fs or U–Fz) alloys are being utilized in LMFBRs. U–Fs alloys can be developed during the reprocessing of spent fuels in which part of the fission products such as Mo, Nb, Zr, Rh, Ru etc. are left in the uranium matrix. These types of alloys (e.g., U–15 wt% Pu–10 wt% Fs or U–5 wt% Fs) show better irradiation stability.

Addition of alloying elements to small concentrations in uranium can improve the high-temperature strength of the alloy. This is beneficial since the strength of uranium falls drastically at elevated temperatures. For example, addition of Cr to the tune of 0.5 wt% or Zr to 2.0 wt% can increase the yield strength by four–five times. Addition of Si and Al may also improve strength when added in small amounts. However, addition of larger amounts may result in the formation of brittle intermetallics, adversely affecting the ductility and fabricability of the alloy. Martensitic transformation is another way of hardening the uranium alloys. The addition of Zr to the tune of about 5–10 wt% can be water quenched from the gamma-phase regime to produce supersaturated metastable alpha-phase (alpha-prime) regime. The as-quenched U–5 wt% Zr alloy (900 °C at 1 h and quenched) is very hard (~535 VHN). Upon tempering at 650 °C for 2 h, it loses its hardness (~315 HVN). However, a range of microstructure can be developed by manipulating the tempering parameters. Similar martensitic transformations also occur in U–Mo, U–Ti, and U–Nb alloy systems.

Uranium alloys exhibit better corrosion resistance by forming and retaining a protective oxide film up to 350 °C if the alloy is in the form of (i) metastable gamma-phase, (ii) supersaturated alpha-phase, and (iii) intermetallic compounds.

The first type of alloys contains 7 wt% or more Mo or Nb. The alloying elements remain dissolved in the gamma-matrix (BCC) by cooling at moderate or rapid rates from the gamma-phase regime. As long as the gamma phase is retained in the alloy, the corrosion rate remains low. It can be noted here that U-Mo alloys are being developed under the 'Reduced Enrichment for Research and Test Reactors (RERTR)' program. The RERTR program was started by the US Department of Energy in 1978 to develop technologies essential for enabling the conversion of civilian nuclear facilities using high enriched uranium (HEU; > 20 wt.% U^{235}) to low enriched uranium (LEU; <20 wt.% U^{235}). By the end of 2011, over 40 research reactors have been converted from HEU to LEU.

Supersaturated alpha-phase alloy is formed by adding a small amount of niobium (up to 3 wt%) and letting it cool rapidly leading to martensitic transformation. As long as the martensitic structure is maintained, the corrosion resistance property is retained. Further improvement in corrosion resistance can be achieved by adding zirconium. For example, a ternary uranium alloy with 1.5 wt% Nb and 5 wt% Zr has good corrosion resistance. However, the alloy is susceptible to embrittling hydrogen attack.

Uranium-based intermetallic compounds may provide better corrosion resistance as typified by uranium silicide (U_3Si). This class of uranium-based materials includes a range of intermetallics such as UAl_2, UAl_3, U_6Ni, U_6Fe, and so on. The main advantage of these compounds is that they provide corrosion resistance at elevated temperatures at which the first two types of alloys (metastable gamma and supersaturated alpha) cannot.

7.2.1.7 Fabrication of Uranium

Fabrication of uranium is guided by factors like its reaction tendency with air and hydrogen and anisotropic properties. Nevertheless, a host of fabrication techniques including rolling, forging, casting, extrusion, swaging, wire and tube drawing, machining, and powder metallurgy can be used. Here, we discuss some aspects of rolling and powder metallurgy.

In theory, uranium can be rolled in any of the three phase regimes (alpha, beta, or gamma). However, since the beta phase is relatively hard, it needs almost two–three times more force when rolling in the temperature range of 660–770 °C compared to rolling at 650 °C. On the other hand, the gamma phase has adequate ductility and is easily susceptible to warping and sagging effects. Hence, most rolling is done in the alpha-phase regime. The recrystallization temperature for heavily cold worked uranium is about 450 °C. For hot rolling, special protection sheaths like nickel along with graphite barrier or other barrier materials used in between the sheath and uranium sheet are used to prevent oxidation of the surface during rolling.

Uranium powder is generally obtained via hydriding method. Uranium powders are highly pyrophoric. That is why special cautions should be taken while using powder metallurgy technique. Paraffin wax or oil is used to protect the powder surfaces during powder processing. As uranium is considered somewhat toxic, uranium powders should be handled under controlled atmosphere like in a "glove box." Powder products are generally made by the following techniques:

i) Cold compaction with sintering in the high-temperature gamma-phase regime (1095–1120 °C).
ii) Cold compaction, sintering, repressing, and final annealing either in the alpha- or gamma-phase regime.
iii) Hot pressing in the high-alpha-phase regime.

7.2.1.8 Thermal Cycling Growth in Uranium

Polycrystalline uranium undergoes a kind of dimensional instability when subjected to repeated heating and cooling (i.e., thermal cycling) in the alpha-phase regime. This phenomenon is known as thermal cycling growth and results in (i) growth (change in length, that is, increase or decrease) and (ii) surface roughening arising out of wrinkling effect. The growth effect comes from a thermal ratcheting mechanism involving (i) relative movement between two neighboring grains with different thermal expansion coefficient due to the basic anisotropy of alpha-uranium, and (ii) stress relaxation in one of the grains by plastic deformation or creep. An interesting example of thermal cycling growth is shown in Figure 7.6, where an alpha-uranium rod has elongated over several times due to thermal cycling between 50 and 500 °C. In superplasticity literature, it is known as thermal cycling superplasticity! Thermal cycling growth coefficient (G_t) is expressed by

$$G_t = \frac{\text{Percentage length increase}}{\text{Total number of cycles}}. \tag{7.9}$$

As thermal cycling is an inherent feature of nuclear reactor kinetics, it can thus have important influence on the thermal stability of the fuel. The gamma phase of

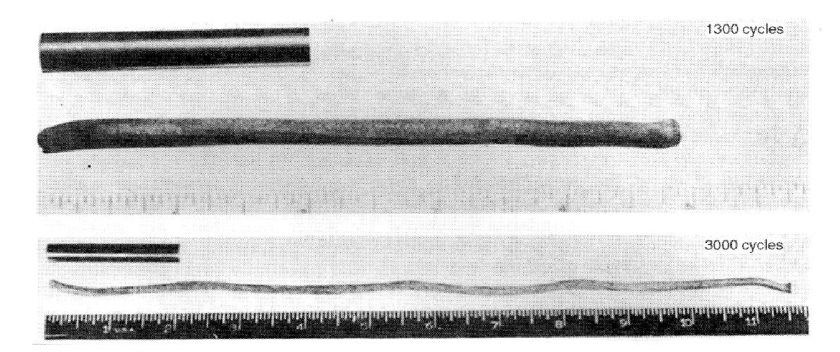

Figure 7.6 Effect of thermal cycling in highly oriented fine-grained uranium (rod rolled at 300 °C) between 50 °C and 500 °C for 1300 cycles (top) and 3000 cycles (bottom). Taken from Ref. [5].

uranium does not show thermal cycling growth phenomenon and may thus be desirable. Thus, suitable amounts of gamma stabilizing alloying additions (Al, Mo, and Mg) help in avoiding thermal cycling growth effect as illustrated with an example from U–Mo system (Figure 7.7). Note that U–Mo alloys generally contain at least 6 wt% Mo in order to avoid thermal cycling growth.

7.2.1.9 **Irradiation Properties of Metallic Uranium**

Radiation Growth

Radiation growth of uranium is another form of dimensional instability that occurs under irradiation without the need of any stress in a lower temperature (i.e.,

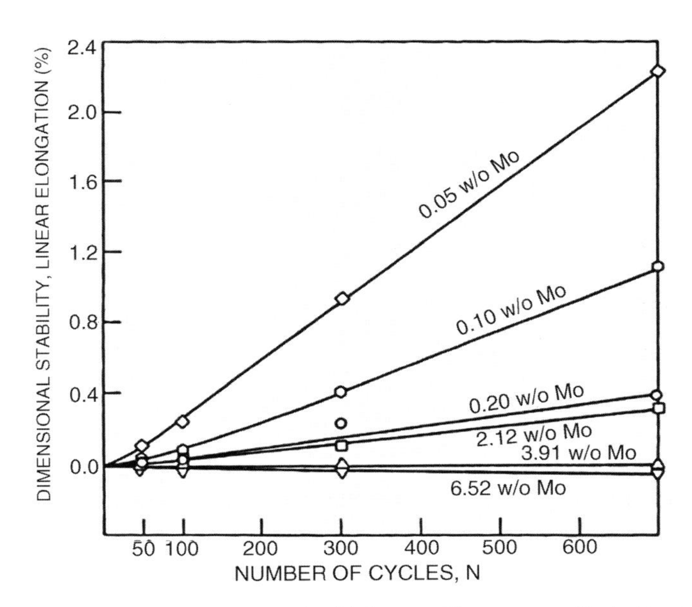

Figure 7.7 Suitable amount of molybdenum addition can modify the kinetics of alpha-or gamma-phase uranium and remove the thermal cycling growth effect. Taken from Ref. [2].

around 300 °C) regime. Since it does not require stress to occur, it is not considered radiation-induced creep. Also, since the volume of the material remains constant during radiation growth, it is not considered as irradiation swelling. Under radiation exposure, a single crystal of alpha-uranium expands in the [010] direction, contracts along the [100] direction, and remains more or less unaltered in the [001] direction. The result of this characteristic expansion/contraction is that the volume remains essentially constant. However, in order for radiation growth to take place, single crystal of uranium is not essential but polycrystalline uranium strong crystallographic texture can also exhibit the effect. Deformation processing and heat treatment is important to minimizing or eliminating the radiation growth effect by suitable texture engineering. Minimization of radiation growth can be achieved by processing the material to produce a fine-grained microstructure with randomly oriented grains. Also, suitable alloying additions to stabilize isotropic phases can help.

The radiation growth coefficient (G_t) is given by

$$G_t = \frac{\text{Percentage length increase}}{\text{Percentage atom burnup}}. \tag{7.10}$$

There have been a number of studies to understand radiation growth, but it remains elusive. A leading hypothesis of radiation growth by Buckley [ ] is based on differential directional rates of interstitial atoms and vacancies. The interstitials have a tendency to migrate along the [010] direction and to the vacancies in the [100] or [001] directions, leading to basically removal of mass from one side and plating them on the other side. Figure 7.8 shows the irradiation growth effect in a uranium fuel against the fuel burnup.

Radiation Swelling

Irradiation swelling is a type of dimensional instability (in the form of volume expansion) that is encountered through the formation of voids/bubbles and fission

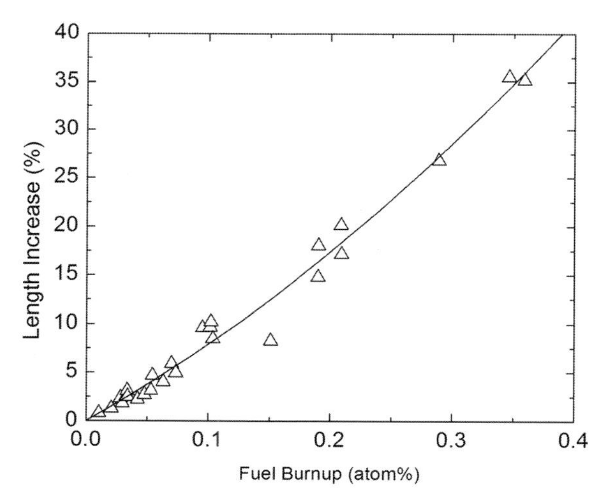

Figure 7.8 Radiation growth effect (length increase) in an irradiated uranium fuel specimens rolled at 600 °C as a function of fuel burnup. Adopted from Ref. [2].

Note

It is important to understand the similarities and differences between the thermal cycling and radiation growth phenomena [2, 5].

Similarities

i) Irradiation growth and thermal cycling growth can create significant dimensional and structural instability in uranium.
ii) Both produce growth in the ⟨010⟩ direction.
iii) Maximum growth for both requires ⟨010⟩ texture (preferred orientation of grains) of the polycrystalline uranium.
iv) Mechanical deformation within grains and at grain boundaries exist for both, but differ in behavior.
v) Basic anisotropy of the uranium crystal structure is essential for both to happen and thus alpha-uranium (orthorhombic lattice structure) is prone to these effects.
vi) Very little change in orientation occurs in both cases.

Differences

i) Thermal cycling growth requires polycrystalline uranium (i.e., with grain boundaries) for thermal ratcheting effect to take place. On the other hand, radiation growth can operate just as well in single crystals as it can in polycrystals.
ii) Thermal cycling can occur at all temperatures, whereas radiation growth occurs at relatively low temperatures.
iii) Radiation growth is generally accompanied by radiation hardening and embrittlement, but thermal cycling is not.
iv) Microstructures undergoing radiation growth are often twinned and deformed, but thermal cycling can produce polygonization with little twinning.

product retention (Xe^{133}, He^4, Kr^{85}, etc.) in fuels. The phenomenon generally occurs in the temperature regime of 0.3–0.6T_m. One question that may arise immediately relates to the difference between radiation growth and radiation swelling. They are in the following ways:

i) Radiation swelling occurs at a higher temperature regime compared to radiation swelling.
ii) Radiation swelling leads to volumetric growth in the fuel, while radiation growth changes the shape of the uranium fuel keeping the volume essentially constant.
iii) Radiation swelling can occur in all forms of uranium under suitable conditions as its origin is attributed to generation of voids/bubbles and fission gases. On the other hand, in order for radiation growth to occur, basic anisotropy is important. That is why radiation growth can be observed in alpha-uranium but

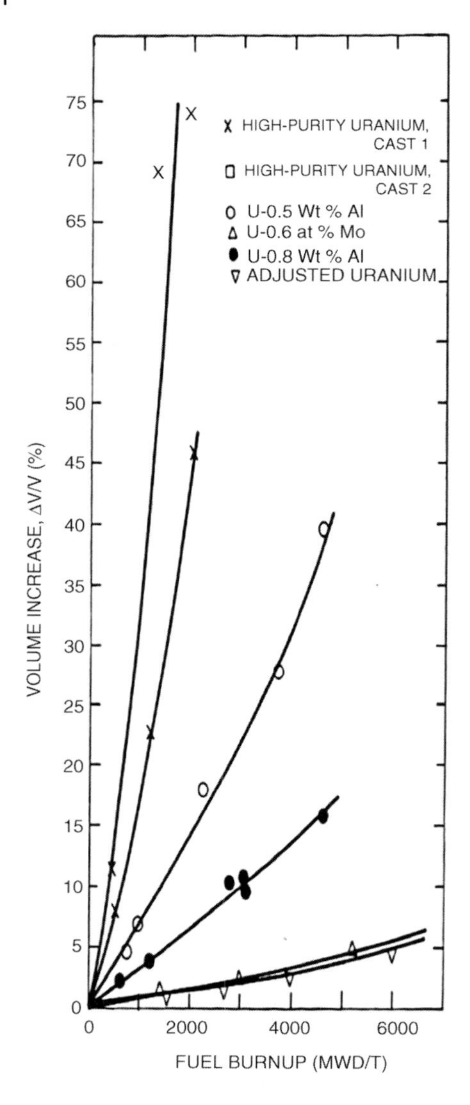

Figure 7.9 The extent of swelling in various uranium-based materials as a function of fuel burnup. Taken from Ref. [2].

not in gamma-uranium, while both phases can exhibit radiation swelling even though alloying may minimize the radiation swelling effect.

Figure 7.9 shows the volume change as a function of fuel burnup for a number of uranium-based materials. This does show the effect of alloying in suppressing radiation swelling. Note that the *adjusted uranium* shown in the plot is known as the British Standard Fuel Produce (contains 400–1200 ppm, 300–600 ppm C, and small amounts of Mo, Nb, and Fe) and it shows better radiation swelling performance.

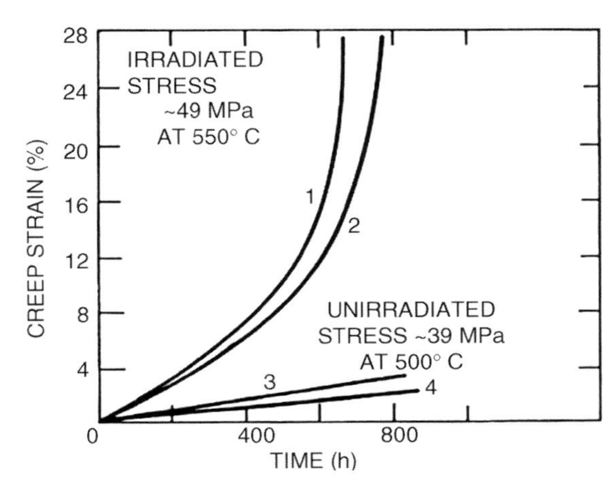

Figure 7.10 Thermal creep and irradiation creep curves of hot rolled uranium under differently processed conditions - 1: Beta-cooled in air, 2: beta-quenched in water, 3: gamma-cooled in air, 4: gamma-quenched in water; after Ref. [2].

Irradiation Creep

Creep is an important time-dependent mechanical property for high-temperature application. In Chapter 6, we have learned how thermal creep is different from irradiation creep. So, here we are not going to repeat the fundamental principles. Figure 7.10 shows the effect of irradiation on the creep behavior of hot rolled uranium.

7.2.2 Metallic Plutonium

Plutonium (atomic number 94) and its alloys can be used as nuclear fuels in nuclear reactors and space batteries. Notably Pu^{239} is the major fissile isotope of plutonium. Among them, plutonium (Pu^{239}) serves as a fissile fuel as its fission cross section is high (742.5 b) with thermal neutrons. Pu^{241} isotope also has significant fission cross section in the thermal spectrum (1009 b), whereas Pu^{240} can act as burnable poison allowing reactor to have constant reactivity throughout reactor lifetime. Plutonium is found in natural uranium only in trace quantity. It is mainly produced artificially by the transmutation reaction of U^{238} isotope (fertile fuel) with a neutron as described in Eq. (1.2). Radioisotopic thermoelectric generators also use plutonium (Pu^{238}) to power spacecrafts. Plutonium appears originally as a bright silvery-white metal, but soon loses its bright color when oxidized in air. Smaller critical mass of plutonium (almost one-third of that of uranium), its high toxicity, and pyrophoricity warrant safe handling of this metal. Plutonium can be recovered from the spent fuel of a thermal reactor through chemical treatment. However, depleted uranium can be kept together with plutonium for fuels used in fast breeder reactors such as LMFBRs. In other cases, separated plutonium can be used in plutonium-burning reactors.

Hecker [5] noted the following in his paper about plutonium:

> "PLUTONIUM is a notoriously unstable metal – with little provocation, it can change its density by as much as 25 pct; it can be as brittle as glass or as malleable as aluminum; it expands when it solidifies; and its silvery freshly machined surface will tarnish in minutes, producing nearly every color in the rainbow. In addition, plutonium's continuous radioactive decay causes self-irradiation damage that can fundamentally change its properties over time."

As we present in the next several sections various characteristics of plutonium, we must bear in mind this *fickle* nature of plutonium.

7.2.2.1 **Crystal Structure and Physical Properties of Plutonium**

There are six allotropes of plutonium. Given the melting temperature of just ~640 °C, it means the various allotropes are stable only in limited temperature ranges. In plutonium, the narrow conduction bands and high density of states of 5f electrons make it energetically favorable for the ground-state crystal structure to distort to a low-symmetry monoclinic lattice at lower temperatures. Plutonium adopts more typical symmetric structures only at elevated temperatures or with suitable alloying.

Up to ~122 °C, plutonium is known as alpha phase (α-Pu) and has a simple monoclinic crystal structure. The density of this phase at 21 °C is 19.816 g cm^{-3}, and the lattice parameters are $a = 6.1835$ Å, $b = 4.8244$ Å, $c = 10.973$ Å, and $\beta = 101.81°$. Figure 7.11a illustrates the (020) plane of the α-plutonium monoclinic lattice, which resembles an HCP plane. In Figure 7.11b, two stacked (020) planes of the crystal contain four unit cells. The bond lengths, however, can be categorized into two groups (long bonds ~3.19–3.71 Å, and short bonds ~2.57–2.78 Å); each of the eight numbered sites is crystallographically unique. Lawson *et al.* [7] showed that there are 16 atoms per unit cell with 8 distinct lattice positions.

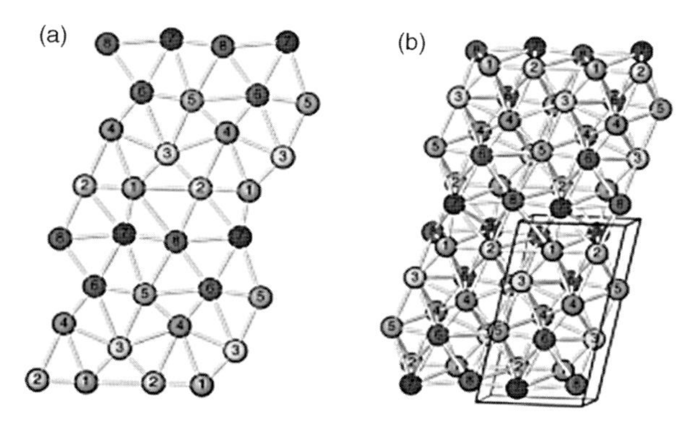

Figure 7.11 (a) The (020) plane configuration in alpha-plutonium. (b) Unit cell configuration of alpha-plutonium Ref. [6].

From ~122 °C to ~206 °C, plutonium assumes a body-centered monoclinic lattice structure (34 atoms per unit cell) and is known as β-Pu. At 190 °C, the density of β-Pu is 17.70 g cm^{-3} and the lattice parameters are $a = 9.284$ Å, $b = 10.463$ Å, $c = 7.859$ Å, and $\beta = 92.13°$.

From ~206 °C to ~319 °C, plutonium assumes a face-centered orthorhombic (8 atoms per unit cell) crystal structure. This phase is also known as gamma-plutonium (γ-Pu). At 235 °C, the density of γ-Pu is 17.14 g cm^{-3} and the lattice parameters are $a = 3.1587$ Å, $b = 5.7682$ Å, and $c = 10.162$ Å.

From ~319 °C to ~451 °C, plutonium takes up a face-centered cubic (FCC) crystal structure with usual 4 atoms per unit cell, and is known as delta-plutonium (δ-Pu). At 320 °C, the density of δ-Pu is 15.92 g cm^{-3} and the lattice constant is $a = 4.6871$ Å.

From ~451 °C to ~476 °C, plutonium assumes a body-centered tetragonal lattice structure with usual 2 atoms per unit cell. This phase is known as delta-prime (δ′-Pu). At 465 °C, the density of δ′-Pu is 16.00 g cm^{-3} and the lattice constants are $a = 3.327$ Å and $c = 4.482$ Å.

From ~476 °C to just before the melting point (639.5 °C), plutonium maintains a body-centered cubic structure (2 atoms per unit cell) and is called "epsilon" phase. The lattice constant and the physical density of this phase are 3.6361 Å and 16.51 g cm^{-3}, respectively.

The allotropic (or polymorphic) transformation kinetics is very sensitive to prior processing history and presence of impurities. In case of pure metal, significant hysteresis occurs on cooling. Plutonium has many anomalous characteristics. For instance, the transformation from β-Pu to α-Pu is very slow and generally other high temperature phases are found to be retained even below the transformation temperature except under application of high pressures. So, generally high strain is retained at room temperature in α-Pu when cooled at atmospheric pressure. So, martensitic type of transformation modes can also occur in plutonium, such as transformations from the delta phase to the gamma phase and from the gamma phase to the beta phase. Linear thermal expansion coefficient of plutonium as a function of temperature follows the trend as shown in Figure 7.12 from dilatometry

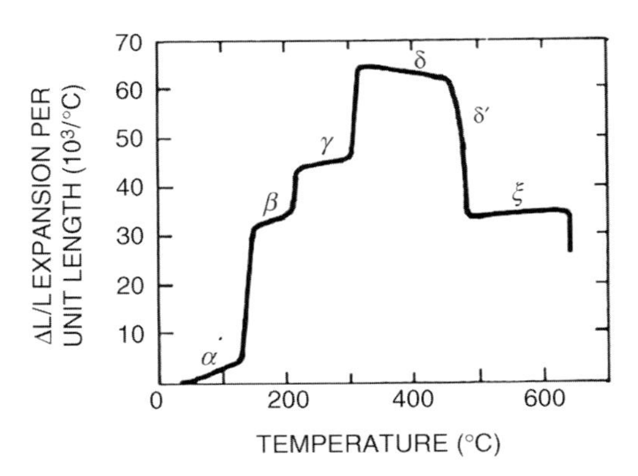

Figure 7.12 Linear thermal expansion coefficient of plutonium against temperature Ref. [2].

experiment. The temperature is generated due to its alpha-emitting characteristics with about 1.923 $W g^{-1}$. A specimen of plutonium bit larger than a foil or a wire can be used as an adiabatic self-heater in the thermal expansion measurement so that no external heat needs to be provided for experimentation. The thermal expansion coefficient increases in the alpha-, beta-, and gamma-phase regimes. However, the delta phase contracts as the temperature rises. It also has positive temperature coefficient of resistivity. The delta-prime phase shows anomalous thermal expansion behavior. The electrical resistivity of plutonium is one of the highest in all metals with much like semiconductors. Thermal conductivity and specific heat of plutonium increase with increasing temperature. Interestingly, plutonium has the largest low-temperature specific heat of any pure element [8].

7.2.2.2 Fabrication of Plutonium

Plutonium is mainly produced by reprocessing of spent fuels since it is not a naturally occurring element. Recycling of dismantled nuclear weapons could also provide some plutonium! Here, we will not discuss those details, rather we will focus on the fabrication of plutonium. Fabrication of plutonium needs to be carried out with extreme caution as plutonium may cause extreme health hazard. A host of techniques, including casting, rolling, extrusion, drawing, and machining can be used. The characteristics of plutonium, such as low melting point, high fluidity, small-volume change, and high density, all are favorable for casting. However, the inherent differences between different forms of allotropes make precision casting of plutonium practically implausible. Generally, plutonium is melted and cast under controlled environment (under vacuum or inert gas) in a resistively heated or induction furnace. A number of materials can be used for the crucibles (for melting) and molds (for casting).

Alpha plutonium is relatively brittle and can be fabricated using machining or press forging. The beta and gamma phases also show brittle behavior, but can be plastically worked with due care. On the contrary, delta phase is relatively ductile and can be formed by traditional mechanical working techniques. Although elevated temperature working gives better deformation characteristics, oxidation is a problem. So, working at temperatures that has limited oxidation is desired. However, the large volume changes associated with allotropic transformation may introduce distortions in the components. The delta phase can be extruded in the temperature range of 320–400 °C with care.

7.2.2.3 Mechanical Properties of Plutonium

Plutonium is considered a weak material compared to most other structural metals. Also, because of its low melting temperature, even room temperature may show up effects of high homologous temperatures. The mechanical properties are very sensitive to impurities, temperature, crystal defects, anisotropy, and phase transformation. Thus, high-temperature application of pure plutonium is not possible. The properties vary with the allotropes. The elastic constants for alpha-plutonium are Young's modulus: ~82.7–97 GPa, shear modulus: ~37.2–43.4 GPa, and Poisson's ratio (0.15). Figure 7.13a shows a comparison of stress–strain curves of alpha-

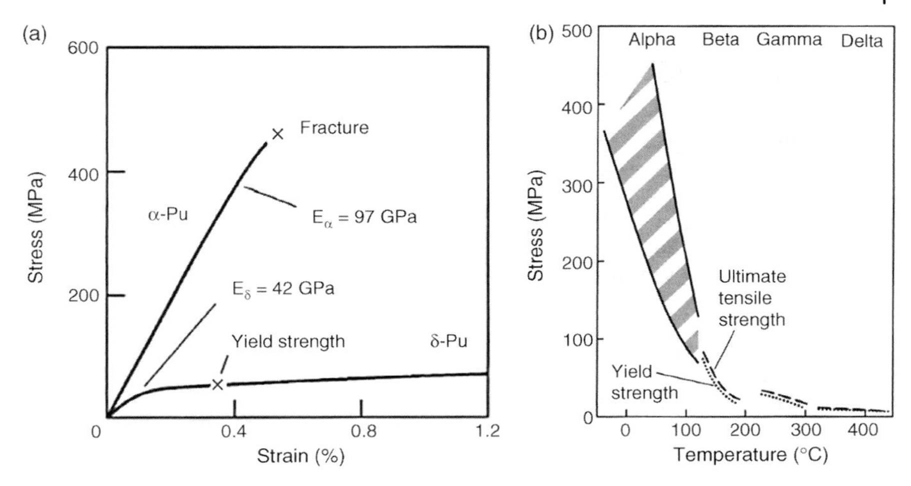

Figure 7.13 (a) Stress–strain curves in alpha- and delta-plutonium. (b) Strength versus temperature plots in various phase regimes of plutonium. Hecker [6].

plutonium and Pu–1.1 at% Ga (delta-phase) alloy. Tensile yield strength and tensile strength vary between ~310 and ~380 MPa. Compressive yield strengths are approximately in the range of 345–517 MPa. Elongation to failure and reduction in area are less than 1% at room temperature. The mechanical strength of polycrystalline plutonium (both α- and β-phases) is quite sensitive to temperature, as illustrated in Figure 7.13b. The upper range shows the trend in ultimate tensile strength and the lower boundary shows the trend in yield strength. There is a considerable scatter in the data available for alpha-plutonium, and thus the data are shown in the form of a band. The delta phase has very low strength. Also, the strength of epsilon phase (BCC lattice structure) is very low as the diffusivity is fast through the BCC crystal lattice (more open structure).

Merz and Nelson [9] demonstrated that the tensile behavior of polycrystalline alpha-plutonium is much more sensitive to strain rate at the proximity of ambient temperature than the first thought. This finding is illustrated in Figure 7.14a. Fine-grained (grain size of 1–3 μm) extruded alpha-plutonium at room temperature and with a strain rate of $7 \times 10^{-4}\,s^{-1}$ shows good ductility (see Figure 7.14b). This provided the first evidence that deformation mechanisms that can operate at the higher end of the alpha-phase temperature range and into the beta-phase range (i.e., at lower homologous temperatures) could be grain boundary sliding in addition to dislocation glide. For example, grain boundary sliding was shown to play an important role in the deformation of fine-grained alpha phase. At 108 °C, it exhibited superplastic elongation of ~218%, as shown in Figure 7.14c.

7.2.2.4 Corrosion Properties

The corrosion properties of plutonium are not good. It corrodes much like uranium, although at a greater rate. Freshly cleaned plutonium has a bright luster on the surface, but loses it as soon as it is exposed to air forming olive green

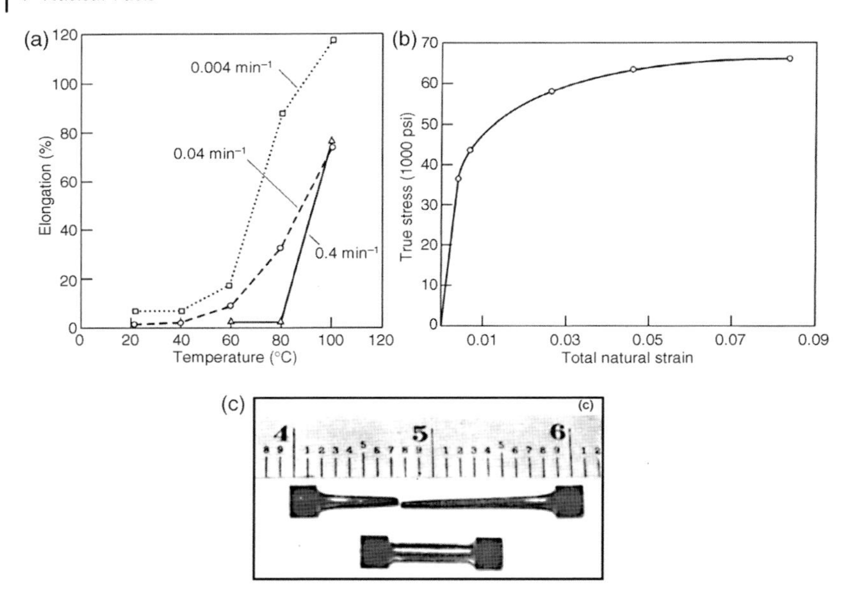

Figure 7.14 (a) Percentage elongation of failure versus temperature data in an extruded alpha-plutonium. (b) True stress–true strain curve of a fine-grained alpha-plutonium at room temperature showing elongation of ~8%. (c) The fine-grained material tested at 108 °C exhibits ~218% elongation. Merz and Nelson [9].

plutonium dioxide (PuO_2), which forms a powdery surface on the metal. Recent research has noted that hyperstoichiometric version of plutonium oxide (PuO_{2+x}) is actually thermodynamically stable in air at 25–350 °C. The metal does not react with nitrogen even at elevated temperatures. Atmospheric corrosion gets aggravated if the metal is exposed to humid air (i.e., containing moisture). Experiments have shown that a plutonium sample in dry air for 200 h has a weight loss (a measure of the extent of corrosion) of ~0.015 mg cm^{-2}, whereas the weight loss increases to ~1 mg cm^{-2} in air with 5% relative humidity. It has been observed that at 100 °C, this difference between the corrosion rates in dry air and moist air can reach to the order of 10^5. The chemical reactions involved with molecular oxygen and moisture can be shown by Eqs (7.11a) and (7.11b), respectively:

$$\mathrm{Pu\ (s) + O_2\ (g) \rightarrow PuO_2\ (s)} \tag{7.11a}$$

$$\mathrm{Pu\ (s) + 2H_2O\ (g) \rightarrow PuO_2\ (s) + 2H_2\ (g)} \tag{7.11b}$$

It is important to note that the loose oxide formed becomes easily airborne and can cause more health hazard than the metal itself. The oxidation rate decreases as plutonium is heated beyond the beta-phase regime into the gamma-phase region. Interestingly, the oxide formed in the delta phase is more tenacious and protective, resulting in less oxidation rate compared to that at lower temperature (i.e., 50 °C). However, plutonium becomes susceptible to formation of a brown powdery oxide and ignition may take place at temperatures above 450 °C. Aqueous corrosion can lead to the formation of oxides and/or hydrides via diffusion of oxide or hydroxyl ions. Pinning corrosion is an important corrosion issue for clad plutonium fuel. However, there are still many things not known well about the corrosion/oxidation behavior of plutonium and need sustained studies.

7.2.2.5 Alloying of Plutonium

As noted before, plutonium is a highly concentrated fissile material. Thus, it must be diluted before it can be used. Furthermore, its physical, chemical, and mechanical properties do not allow it to be used in unalloyed metallic form. Plutonium has a stronger tendency to form intermetallic compounds than uranium. However, they have similar behavior of alloy formation. There are a few elements that can form alloy with plutonium. A plutonium alloy with intended application must have the following characteristics: (i) plutonium required for criticality is kept to a minimum, (ii) should have good fabricability features, (iii) good thermal and irradiation stability, (iv) high corrosion resistance, and (v) available alloying elements.

Alloying elements like Al, Ga, Mo, Th, and Zr can stabilize the delta phase, even though it could make it metastable much like the stabilization of gamma-uranium. Gschneider *et al.* [10] reported that the negative thermal expansion coefficient of delta-plutonium becomes less negative and eventually becomes positive with increased concentrations of alloying additions of Al, Zn, Zr, In, and Ce because of increased electron concentration.

One of the well-known alloys is Pu–3.5 at% Ga alloy. This alloy was developed during the days of the Manhattan project and was used as a fuel in the erstwhile Los Alamos fast reactor. In this alloy, delta phase is stabilized in a wider temperature range. It improves the corrosion resistance of plutonium by several times; for example, the weight loss was only 0.1 mg cm^{-2} during the test of exposure of 27 000 h in moist air.

There has been a deep interest in developing metallic alloy fuels involving fissile–fertile combinations for a breeder reactor like LMFBR as the metallic fuels have the inherent advantages of high fissile atom density leading to higher breeding ratio along with shorter doubling time compared to ceramic fuels. The phase diagram of U–Pu is shown in Figure 7.15. Pu–U is a complex system with complicated phase relations. Plutonium content more than the solubility limit in gamma-uranium leads to an extremely brittle phase. This makes casting of this alloy very difficult and also results in alloys that are quite susceptible to thermal cycling and radiation damage. However, addition of molybdenum has some beneficial effects in that it suppresses the creation of the embrittlement phase. There are many different compositions that have been studied. A single-phase alloy with better corrosion resistance has been found to be U–21Pu–16Mo (in atom%). Other ternary alloy systems like U–Pu–Th, U–Pu–Al, U–Pu–Fe, and so on have also been studied.

7.2.3 Metallic Thorium Fuel

Thorium (atomic number 90) is a soft, silvery-white metal when present in pure form. It retains its luster for a long period of time. However, given a chance to oxidize to thorium oxide (ThO_2), it quickly loses its luster turning into gray and

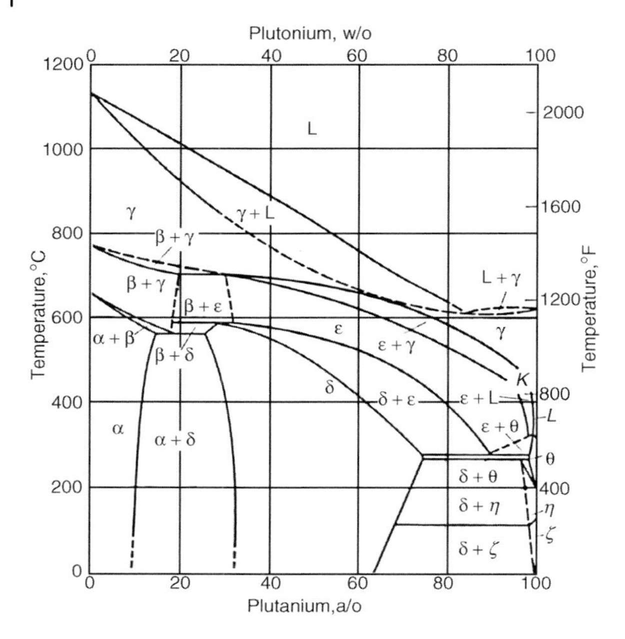

Figure 7.15 Uranium–plutonium phase diagram Ref. [5].

finally black in color. It is another nuclear fuel that has not been tapped to its full potential to date. As we have already discussed in Chapter 1, Th^{232} is a fertile isotope that could produce fissile U^{233} isotope upon capturing a neutron and then fission to produce energy needed for electric power. The relevant reaction has been shown in Eq. (1.3). Thus, thorium is an important breeder material. Thorium exists in only one isotope form (Th^{232}) in nature and decays very slowly (its half-life is 14.05 billion years, that is, thrice the age of the earth). Other isotopic forms of thorium (Th^{228}, Th^{230}, and Th^{234}) occur as decay products of thorium and uranium at some point, but they are present in trace amounts only.

Thorium is far more abundant than uranium in nature. Most rocks and sands on earth surface contain minute amount of thorium. Monazite sand that is rare earth phosphate mineral is an important source of thorium, and almost two-third of this high-quality deposit (thorium content of 6–7 wt% on average) is found in southern and eastern coasts of India. Other thorium resource also includes thorite or thorium silicate ($ThSiO_4$). A large vein deposit of thorium is present in the state of Idaho of the United States.

7.2.3.1 **Extraction of Thorium and Fabrication**

Thorium is mainly extracted from the monazite ore via a multistage process. The first stage is the process of digestion that involves dissolving monazite sands in concentrated sulfuric acid (93–98%) at 120–150 °C for several hours [11]. As an alternative, alkaline digestion process can also be followed. In the digestion process, thorium, uranium, and rare earth metals pass into solution to form sulfates in phosphoric acid. Following the digestion process, the resulting solution is diluted to pH of 1 using ammonium hydroxide, and all the thorium from solution gets precipitated out of the solution along with some rare earths. But subsequently increasing the pH to ~2.5, the rest of the rare earth metals and uranium also get precipitated. The precipitate residue is then collected and treated with nitric acid (solvent extraction process) and thorium compound is separated.

There are several methods to obtain thorium from thorium compounds. Metallic thorium can be obtained by reduction in a sealed container or *bomb*-reacting thorium tetrachloride ($ThCl_4$) or tetrafluoride (ThF_4) with calcium, sodium, or magnesium. Because thorium has a high melting point, zinc is often added to create a low-melting eutectic from which Zn is later distilled off under vacuum to obtain the so-called "bomb-reduced" thorium. A relevant chemical reaction is shown below:

$$ThF_4 + Zn + 2Ca \rightarrow Th\text{-}Zn + 2CaF_2 \qquad (7.12)$$

Highly pure thorium (>99.9%) can be obtained by using an iodide treatment (DeBoer process).

The morphology of the *bomb reduced thorium* is sponge-like and iodide thorium is loosely packed crystals of highly pure thorium. That is why they need to be consolidated through ingot or powder metallurgy process. In ingot metallurgy, two types of methods are generally employed: induction melting/casting under vacuum and arc melting/casting. If thorium is low in oxygen, silicon, nitrogen, and aluminum impurities, it can be fabricated by various deformation processing techniques like extrusion, hot and cold rolling, hot forging, and swaging. However, wire drawing presents challenges since thorium has great tendency to stick with the drawing dies. In the powder metallurgy, thorium can be fabricated into cold compacts (with 95% of theoretical density) from powders produced by hydride method. Then, hot pressing under vacuum at 650 °C at a nominal pressure can produce almost full density metal. The machining of thorium has been found to be easier, especially with greater tool feed rate and low spindle speeds.

7.2.3.2 Crystal Structure and Physical Properties of Metallic Thorium

Thorium has only two allotropes, both of which are of cubic type. The alpha phase of thorium is FCC (4 atoms per unit cell) and retained up to ~1400 °C. The alpha phase has a lattice constant (a) of 5.086 Å and density of 11.72 g cm^{-3}, at ambient temperature (25 °C). Above ~1400 °C, thorium assumes BCC lattice structure (beta phase) and is stable up to its melting point of ~1750 °C. The lattice constant (a) of beta-thorium is 4.11 Å at 1450 °C and density is 11.1 g cm^{-3}. Hence, it can be noted that thorium has a higher melting point and lower density than uranium. Thermal conductivity of thorium is about 30% higher than that of uranium at 100 °C and about 8% greater at 650 °C. Figure 7.16 shows the variation of specific heat at constant pressure (C_P) for (highly pure) thorium produced by iodide process.

7.2.3.3 Mechanical Properties

The mechanical properties of thorium are sensitive to the impurities present as well as to the crystallographic texture and amount of cold work. The purest form of

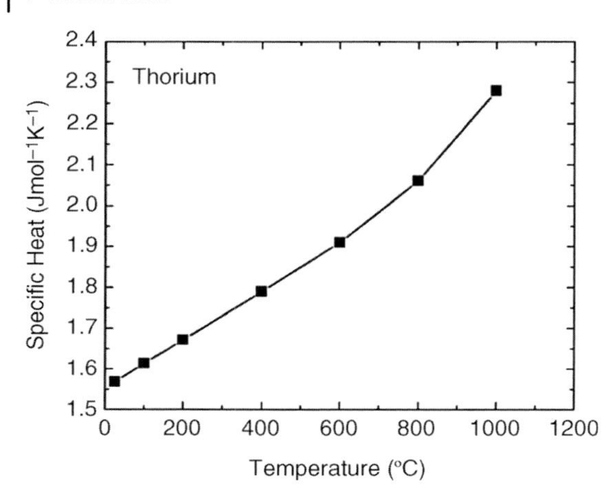

Figure 7.16 The variation of specific heat at constant pressure as a function of temperature in iodide thorium Ref. [12].

thorium produced by DeBoer iodide process in annealed condition can have yield strength of ~34–124 MPa, ultimate tensile strength of ~110–138 MPa, and adequate ductility (percentage elongation of 28–51%), depending on various other factors. However, commercial thorium produced via bomb reduction of ThF_4 contains higher amounts of carbon, oxygen, and other impurities than that found in iodide thorium. These impurities harden thorium. In the cast and/or wrought and annealed condition, this type of metal shows yield and tensile strength of up to ~124 and ~172 MPa, respectively. Thorium work hardens readily resulting in the yield strength increase by 100% and tensile strength increase by 50% with cold reductions (means plate/sheet thickness reduction) of 50% or less. Table 7.3 summarizes room temperature tensile data of wrought annealed and cold worked thorium.

7.2.3.4 **Corrosion Properties of Thorium**

As mentioned before, freshly cut thorium is bright with silvery luster, but darkens quickly on exposure to air. The oxidation creates protective oxide film of thorium oxide (ThO_2) up to a temperature of ~350 °C. At further higher temperatures, the breakaway transition occurs when the oxide film cracks and the oxidation proceeds almost linearly. At about 1100 °C, the oxidation rate becomes parabolic again. Thorium corrodes at a slow rate at around 100 °C in high-purity water with the formation of an adherent oxide film. In the temperature range of 178–200°C in water, the oxide growth rate becomes rapid and eventually starts to spall (i.e., break up). The reaction becomes very rapid at 315 °C. Thorium has good resistance against most metals barring aluminum below 900 °C.

7.2.3.5 **Alloying of Thorium**

Many alloying additions have been attempted for improving mechanical properties and corrosion resistance of thorium. Only few elements (like zirconium and

Table 7.3 Effect of processing on the tensile properties of wrought-annealed and cold worked thorium at room temperature Ref. [13].

Condition[a)]	Yield strength (MPa)	Ultimate tensile strength (MPa)	Elongation (%)	Gage length (mm)	Reduction in area (%)
Iodide					
Wrought-annealed	47.5	119	36	50.8	62
Wrought-annealed	77.0	136	44	25.4	60
Bomb reduced					
Extruded-annealed rod	149.5	207	51	50.8	74
Extruded-annealed rod	190	237	51	50.8	73
Wrought-annealed rod	181	232	55	35.6	69
Wrought annealed rod	219.8	265	48	50.8	69
Wrought annealed sheet	209	273	—	—	52
Cold rolled 37.5% rod	313	337	20	35.6	61
Cold rolled 25% sheet	378	404	11	25.4	39
Cold rolled 50% sheet	424	451	5	25.4	16

a) Note that the materials were in variously processed initial conditions before tensile testing at room temperature.

hafnium) allow extensive solid solubility. However, many reactive elements form intermetallic compounds instead of forming solid solutions. Addition of two known alloying elements, namely, uranium and indium, to thorium improves the mechanical strength of thorium, whereas three elements, namely, zirconium, titanium, and niobium, improve corrosion resistance. Thorium–uranium and thorium–plutonium alloys provide opportunities for combining fertile and fissile materials to develop potential thorium-based fuel cycles. Table 7.4 summarizes tensile properties of Th–U alloys as a function of uranium content. It clearly shows that uranium addition to thorium increases the yield strength and tensile strength; however, it decreases ductility. Thorium–uranium alloys in excess of 50 wt% U can be readily melted and cast. Powder metallurgy techniques can also be used.

Table 7.4 Tensile properties of (bomb-reduced, annealed) Th–U alloys after [2, 5].

Uranium content (%)	Yield strength (MPa)	Tensile strength (MPa)	Elongation (%)	Reduction in area (%)	Poisson's ratio
Unalloyed	134	218	46	50	0.25
1.0	176	265	38	49	0.25
5.1	189	291	37	47	0.24
10.2	207	310	35	44	0.24
20.6	212	328	32	41	0.24
30.9	249	384	28	36	0.23
40.6	266	430	24	34	0.23
51.2	276	445	17	26	0.22
59.1	300	458	11	23	0.22

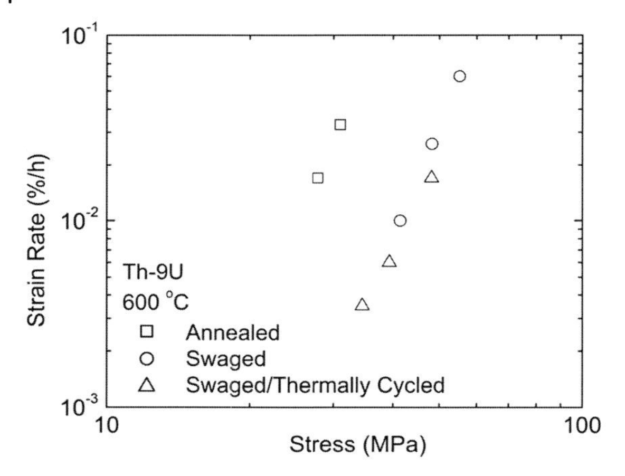

Figure 7.17 Strain rate versus stress from the creep tests at 600 °C of differently processed Th–9 wt% U alloys Ref. [13].

The creep data of Th–9 wt% U alloy are shown in Figure 7.17. The plot shows retention of strengthening due to the mechanical effect at 600 °C.

7.2.3.6 **Radiation Effects**

Thorium has cubic crystal structure and is thus isotropic. It does not show radiation growth effect and thus has better dimensional stability than α-U under irradiation.

7.2.3.7 **Pros and Cons of Thorium-Based Fuel Cycles**

Thorium is more abundant in nature compared to uranium and naturally there is interest in having an economic fuel cycle based on thorium. Thorium-based fuel cycles offer attractive features that are low level of waste generation along with a less amount of transuranics in the waste and provide a robust diversification option for nuclear fuel supply. Also, the use of thorium in majority of reactors leads to significant additional safety margins. However, the full commercial exploitation of thorium fuels has some significant obstacles in terms of building an economic case to undertake the necessary developmental work. A great deal of R&D, testing, and qualification work is required before any thorium fuel can be considered for routine commercial application. Other obstacles to the development of thorium fuel cycle are the greater fuel fabrication and reprocessing costs to include the fissile plutonium as a driver material. The high cost of fuel fabrication is partly because of the high level of radioactivity that is involved in the presence of U-233, chemically separated from the irradiated thorium fuel. But the U-233 gets contaminated with traces of U-232 that decays (69-year half-life) to daughter nuclides such as thallium-208 that are high-energy gamma-emitters [14]. Even though this improves the proliferation resistance of the fuel cycle, it also makes U-233 hard to handle and easy to detect. Notwithstanding, thorium fuel cycle provides hope for long-term energy security benefits without the need for fast reactors.

7.3 Ceramic Fuels

Metallic fuels with their high neutron economy, good thermal conductivity, and thermal shock resistance should be the natural choice for fuels. However, they are not adequate for high-temperature reactors due to low strength at high temperatures, phase transformations, and so on. The other ceramics have superior strength at higher temperatures, low thermal expansion, good corrosion resistance, and good radiation stability. A wide range of compounds are considered as ceramics, which include oxides, carbides, nitrides, borides, silicides, sulfides, selenides, and so forth. But ceramics also suffer from brittleness especially at lower temperatures. Here, we will discuss few ceramic nuclear fuels and discuss their salient features.

7.3.1 Ceramic Uranium Fuels

Three main ceramic uranium fuels are UO_2, UC, and UN (to a smaller extent U_3Si and US), i.e. uranium sulfide. The operating experience with UO_2 is the greatest even though UN and UC remain the best potential fuels for higher performance in the long term. Improvement in fuel performance and enhanced thermal efficiency require the fuel element temperature to be as high as possible. However, with metallic fuels, two main problems may occur: (a) central fuel melting and (b) excessive irradiation swelling and creep deformation due to irradiation instability at higher temperatures. In this regard, ceramic fuels have certain advantages over metallic uranium fuels: (a) higher permissible fuel and plant operating temperatures due to higher melting point, (b) good irradiation stability due to the absence of polymorphic phase transformation, and (c) high corrosion resistance to the environmental attack as a result of its chemical inertness and compatibility with cladding. The basic nuclear properties of competitive ceramic fuels are as follows: (a) large number of fissile uranium (U^{235}) atoms per unit volume of the fuel in order to avoid necessity for high enrichment, and (b) small neutron absorption cross section of the nonfissile components of the compound for preserving the neutron economy. The following sections discuss various aspects of uranium dioxide, which is the mainstay of nuclear fuels used in current generation of power reactors. We will briefly discuss UN and UC.

7.3.1.1 Uranium Dioxide (Urania)

Fabrication

Uranium dioxide production can follow the same methods described in Section 7.1.1.1, except the steps involved to produce metallic uranium. UO_2 can also be processed into bulk shapes such as pellets, tubes, rods, etc. by usual ceramic processing methods, including powder metallurgy. Sintering of UO_2 must be done

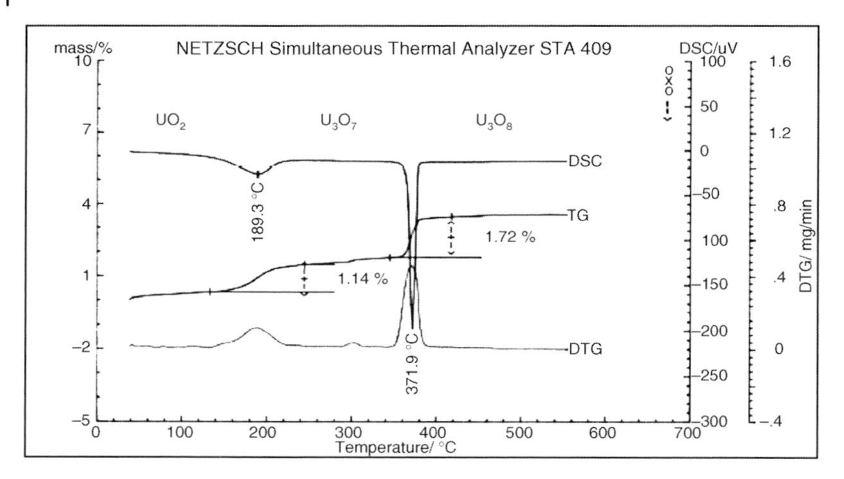

Figure 7.18 DSC (differential scanning calorimetry) and TGA (thermogravimetric) curves of UO_2. *Courtesy*: Dr. Jack Henderson, Netzsch.

under an atmosphere either inert or reducing since sintering in air has consequences. The UO_2 can exist in the form of a wide range of variable compounds depending on temperature and environment. U_3O_7 ($2UO_2 + UO_3$) in an unstable state of mixture forms at ~150–190 °C and U_3O_8 at about 375 °C. Figure 7.18 shows thermal analysis curves of UO_2 as a function of temperature, showing the evolution of U_3O_8 is unstable above ~500 °C and converts back to UO_2 at higher temperatures (1100–1300 °C).

$$UO_2 + 2UO_3 \rightarrow U_3O_8 \rightarrow 3UO_2 + O_2 \tag{7.13}$$

The change in density with all the phase changes causes disruption during sintering in air. That is why sintering in hydrogen at 1700–1725 °C for 8–10 h produces bulk UO_2 with 93–95% of the theoretical density. Minor additions of titanium dioxide (TiO_2) or cerium dioxide (CeO_2) can act as sintering additives and help to reduce the sintering temperature. However, there are now different kinds of sintering process that may allow better processing characteristics.

If the oxygen to uranium atom ratio is 2.0, the UO_2 is stoichiometric. If an oxygen-deficient or excessive uranium exists (i.e., $O/U < 2.0$), the fuel is called superstoichiometric fuel (UO_{2-x}). If $O/U > 2.0$, UO_{2+x} is called hypostoichiometric fuel (x is a small fraction). The departures from stoichiometry influence self-diffusion behavior in fuel itself and interdiffusion between fuel and cladding materials to form hyperstoichiometric or hypostoichiometric fuel during the reactor operation complicating the chemical composition of UO_2. It can also affect fuel density, melting point, and other physical and temperature-dependent properties.

Crystal Structure and Physical Properties

The crystal structure of uranium dioxide is of fluorite type, as described in Section 2.1. Theoretical density (TD) of UO_2 is 10.96 g cm^{-3} at room temperature.

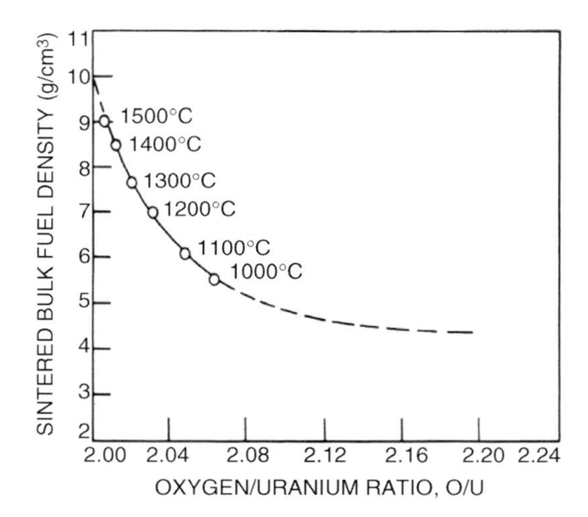

Figure 7.19 Bulk fuel density varies with oxygen to uranium content in moist hydrogen atmosphere at different temperatures [2].

The melting point of UO_2 is about 2850 °C (the literature gives a variety of values). However, it has appreciable vapor pressure at lower temperatures leading to weight losses during sintering. Fuel pellets of UO_2 are generally produced using powder metallurgy techniques. The properties of UO_2 often depend on the processing. The actual density (AD) of UO_2 fuel may vary from 80% to 95% of the TD depending on the size/crystallite shape of the powder particles and the actual fuel fabrication process. However, there are more fabrication techniques that have come to improve fuel fabrication. High density of UO_2 fuel fabricated has the following advantages: (a) high uranium density, (b) higher thermal conductivity, (c) high capability to contain and retain fission product gases in the fuel, and (d) large linear power rating of the fuel element. Figure 7.19 shows the bulk density of UO_2 fuels with respect to oxygen/uranium ratio. In general, UO_2 has a lower thermal conductivity compared to other uranium-based metallic and ceramic fuels, and its thermal expansion coefficient is relatively high. However, it got a smaller specific heat.

The thermal conductivity of UO_2 has been measured repetitively since the late 1940s; however, modern measurement techniques have produced a significant insight into the transport mechanisms within the UO_2 fuel. At temperatures ranging from room temperature to about 1800 K (1527 °C), the transport of energy within UO_2 is controlled by lattice vibrations that cause a temperature-based decrease in the thermal conductivity trend. However, above 1800 K and up to the melting temperature, a small ambipolar polaron contribution reverses the trend and begins to increase the thermal conductivity. UO_2 is a ceramic that is dominated by phonon–phonon interactions and as such the thermal conductivities are low at all temperatures. Figure 7.20 shows the thermal conductivity as a function of temperature.

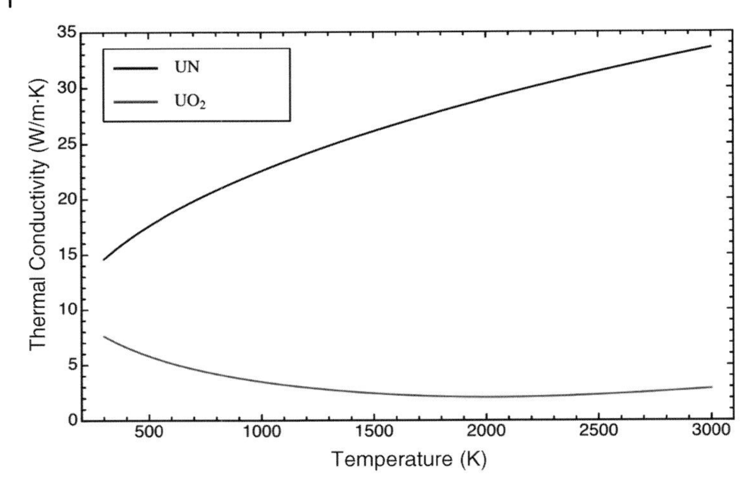

Figure 7.20 A plot of temperature-dependent thermal conductivity for UO_2 (gray). For comparison, thermal conductivity trend for UN is also shown (in black) Ref. [15].

Both UO_2 and UN exhibit very similar temperature-dependent specific heats (C_p) that rise much more rapidly at elevated temperatures than typical ceramic materials (Figure 7.21). Both fuels exhibit a behavior that presents specific heat values nearly twice the Dulong–Petit value near their melting points. Significant research has shown that at low temperatures, the specific heat of both UO_2 and UN is governed by lattice vibrations that can be predicted based on the Debye model. Over the range of 1000 (727 °C)–1500 K (1227 °C), the specific heat is governed by the harmonic lattice vibrations and above this temperature the specific heat is governed by crystal defects such as Frenkel pairs up to 2670 K (2397 °C). Above 2670 K, Schottky

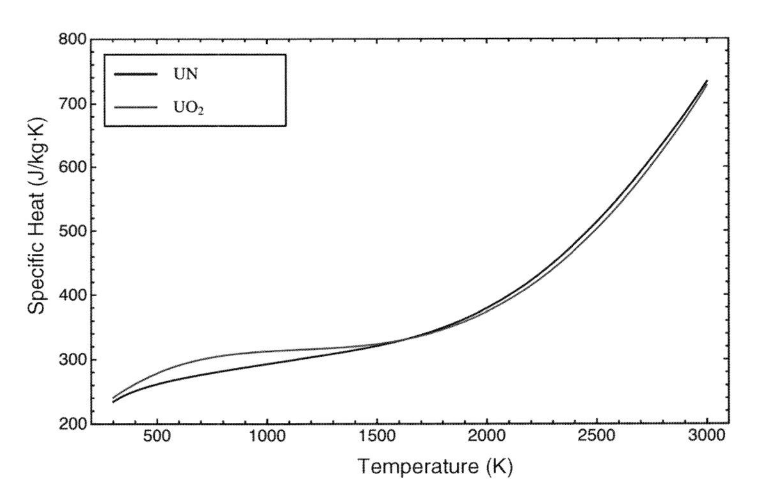

Figure 7.21 Plot of temperature-dependent specific heat at constant pressure for UO_2 (gray) and UN (black) Ref. [16].

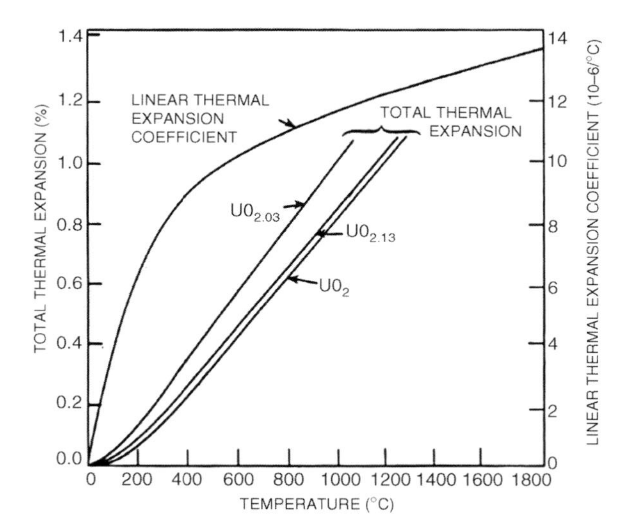

Figure 7.22 The variation of total thermal expansion and linear thermal expansion coefficients of UO_2 as a function of temperature Ref. [2].

defects become a dominant component in the specific heat for both ceramic materials.

The thermal expansion coefficients of both stoichiometric and nonstoichiometric uranium dioxide vary with temperature, as shown in Figure 7.22.

Mechanical Properties

Uranium dioxide has a tensile strength of just ~35 MPa and Young's modulus of ~172 GPa. Strength increases with temperature most probably due to closure of pores (sintering effect). Above 1400 °C, it rapidly loses strength and can be subject to plastic deformation.

Irradiation Effects

UO_2 is dimensionally stable to high-radiation exposure ($>10^{20}$ n cm^{-2}) and relatively high burnup. Under neutron irradiation of sufficient flux, the fuel pellets may fragment by radial cracking. Furthermore, axial and circumferential cracking can also be observed. It is noted that the pellet cracks only in the initial period into few pieces and then stay that way for prolonged duration, implying that the cracking is related to thermal stresses and is not due to mechanical degradation. However, displacement damage caused by fission fragments may enhance the cracking effect.

Uranium dioxide also tends to release volatile fission products from free surfaces. Fission gases could include Br, I, Te, Xe, Kr, and other related nuclides [17]. The amount of fission gas a depends on many factors, such as porosity, other microstructural characteristics, irradiation time, and irradiation temperature.

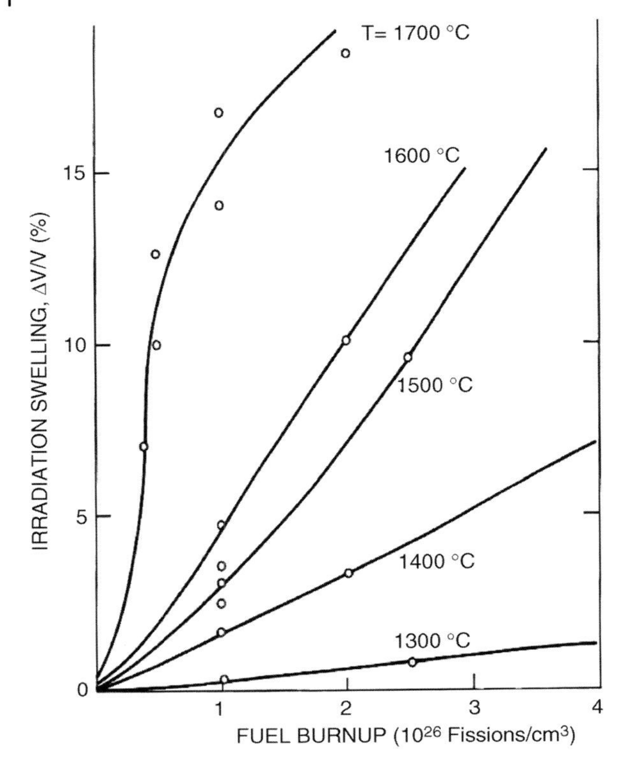

Figure 7.23 Comparison of predicted and observed irradiation swelling in UO_2 fuel as a function of fuel burnup and irradiation temperature Ref. [2].

Figure 7.23 shows the radiation swelling as a function of fuel burnup at different irradiation temperatures.

The creep rates of a stoichiometric UO_2 fuel as a function of stress in both unirradiated and irradiated conditions are shown in Figure 7.24. The in-reactor creep rate in the temperature regime of 800–900 °C does not depend much on temperature, but it depends on neutron flux and stress. So, radiation-induced creep operates in this temperature regime. However, in the higher temperature regime (>1200 °C), the creep rate becomes strongly temperature dependent.

7.3.2
Uranium Carbide

There are three uranium carbide compounds (UC, U_2C_3, and U_2C) with the greatest interest in UC. UC is often considered the ideal nuclear fuel compared to the metallic uranium and UO_2 fuels. UC does not undergo phase change until its melting point (2350 °C) and has a high uranium density and also a higher thermal conductivity than UO_2. UC also shows good thermal and irradiation stability (fission gas release and swelling are moderate and little cracking is observed). The higher

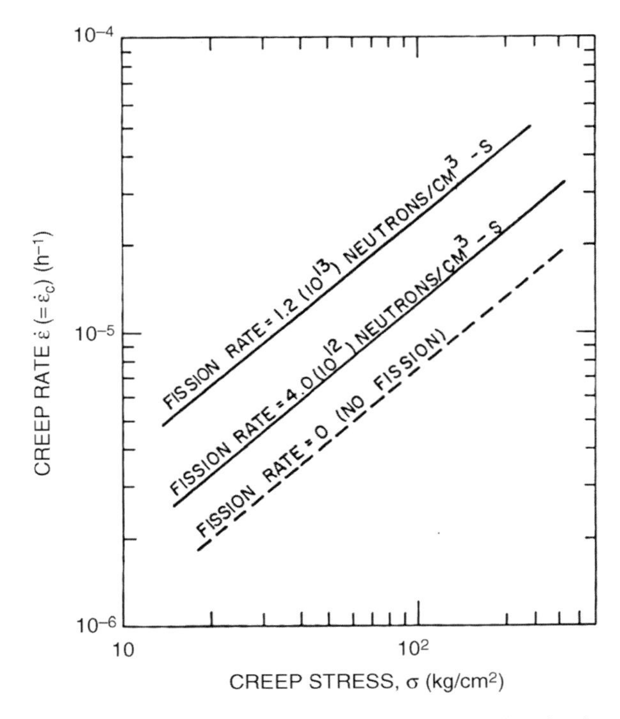

Figure 7.24 Variation of the UO_2 creep rate under irradiated and unirradiated conditions Ref. [2].

uranium atom density and higher thermal conductivity can result into the following:

i) Larger size of fuel elements desired for an economic fuel fabrication.
ii) Higher power density or specific power attainable.
iii) Smaller primary components such as pressure vessel, piping system, and so on.

7.3.3 Uranium Nitride

In uranium–nitrogen system, ceramic compounds such as UN, U_2N_3, and UN_3, with uranium mononitride (UN), are the most stable and the only compound with properties of a nuclear reactor fuel. UN has a NaCl-type (interpenetrating FCC) crystal structure. UN has a theoretical density of 14.32 g cm^{-3} under normal conditions. UN maintains its stoichiometry up to high temperatures, and becomes nonstoichiometric at >1500 °C. Melting temperature of UN is about 2650 °C. UN does not melt congruently and begins to dissociate into free uranium and gaseous nitrogen at a temperature that is a function of the system nitrogen overpressure, which can be analytically described by Eq. (7.14), where T_m represents the melting

Table 7.5 A comparative summary of a few characteristics of UO_2, UC, and UN Ref. [2, 5].

Fuel	Lattice structure	Dimensions (Å)	Melting point (°C)	U content (wt%)	Macroscopic cross section (fission) (cm^{-1})	Absorption cross section (cm^{-1})	Fast fission neutrons/thermal neutrons (η)
UO_2	Fluorite	$a = 5.469$	2760 ± 30	88.15	0.102	0.187	1.34
UC	Rock salt	$a = 4.961$	~2300	95.19	0.137	0.252	1.34
UN	Rock salt	$a = 4.880$	~2650	94.44	0.143	0.327	1.08

temperature in K and P_{N_2} represents the nitrogen partial pressure in the unit of atm (1 atm $\cong$ 101.3 kPa) Ref. [18].

$$T_m = 3035(P_{N_2})^{0.02832}. \tag{7.14}$$

The thermal conductivity and specific heat of UN have already been shown in Figures 7.20 and 7.21.

Here we summarize various features of different uranium ceramic fuels in a tabular form (Table 7.5).

7.3.4 Plutonium-Bearing Ceramic Fuels

Plutonium-based ceramic compounds are mainly plutonium dioxide (PuO_2), plutonium monocarbide (PuC), and plutonium nitride (PuN). But they are generally used with UO_2, UC, and UN, respectively, in the form of mixed oxide fuels. These are considered fast breeder reactor fuels. Mixed oxide fuels have drawn the most interest because of the long experience with them. The physical, mechanical, and chemical properties of UO_2 and PuO_2 are quite comparable. The nuclear properties of ceramic plutonium fuels are considered better than ceramic uranium fuels. However, plutonium fuels must be diluted. Mixed ceramic uranium–plutonium fuels have certain benefits: (i) high melting temperatures, (ii) well-developed fuel fabrication technology and operation experience (obtained from UO_2 fuel), and (iii) good thermal and irradiation stability. Given the lack of their near-term application as nuclear fuels, we will not discuss this further. Readers should refer to appropriate literature to gain information in this area.

7.3.5 Thorium-Bearing Ceramic Fuels

Thorium dioxide (ThO_2) is undoubtedly the best-characterized ceramic compound of thorium. Although this partly stems from its study for nuclear purposes, the majority of information exists because of the non-nuclear

Table 7.6 A comparison between a few characteristics of ThO_2 and UO_2 Ref. [13].

Characteristics	ThO_2	UO_2
Crystal structure	CaF_2 type	CaF_2 type
Lattice constant (Å)	5.5974 (at 26 °C)	5.4704 (at 20 °C)
	5.6448 (at 942 °C)	5.5246 (at 946 °C)
Theoretical density (g cm^{-3})	~10.0	~10.96
Melting point (°C)	3300 ± 100	2760 ± 30
Thermal conductivity (W m^{-1} K^{-1}):	10.3 (at 100 °C)	10.5 (at 100 °C)
	8.6 (at 200 °C)	8.15 (at 200 °C)
	6.0 (at 400 °C)	5.9 (at 400 °C)
	4.4 (at 600 °C)	4.52 (at 600 °C)
	3.4 (at 800 °C)	3.76 (at 800 °C)
	3.1 (at 1000 °C)	3.51 (at 1000 °C)
	2.5 (at 1200 °C)	

usefulness of the material. Since thoria has the highest melting point (~3300 °C) and is the most stable to reduction of all the refractory oxides, it is a superior crucible material for the melting of reactive metals. Thoria is generally prepared in powder form by the thermal decomposition of a purified salt, generally the oxalate. This powder can be consolidated by usual ceramic fabrication techniques, such as slipcasting, pressing, and sintering, or hot pressing. The fabricability and ceramic properties can often be related to conditions of preparation of the starting salt and firing.

Thorium dioxide exists up to its melting point as a single cubic phase with the fluorite (CaF_2 type) crystal structure and is isomorphous and completely miscible with UO_2 to a measurable extent. Therefore, it is stable to high temperatures in oxidizing environments. In vacuum, it darkens with loss of oxygen, even though the loss is insufficient to be reflected in chemical analysis or lattice constant measurement. Unlike UO_2, thoria does not dissolve oxygen even on prolonged heating to 1800–1900 °C. By reheating in air to 1200 or 1300 °C, the white color can be brought back.

When uranium dioxide is incorporated in thoria, the lattice can take up extra oxygen in proportion to the uranium content. Table 7.6 summarizes some important physical and mechanical properties of thoria, along with analogous properties of uranium dioxide taken from the compilation by Belle.

Thorium carbide and thorium mononitride fuels may have potential for use as nuclear fuels, but have not been thoroughly studied.

Note

Fuel fabrication is a multistep process. Figure 7.25 gives a flowchart showing the practical details of the fabrication process of metallic fuels for fast reactors.

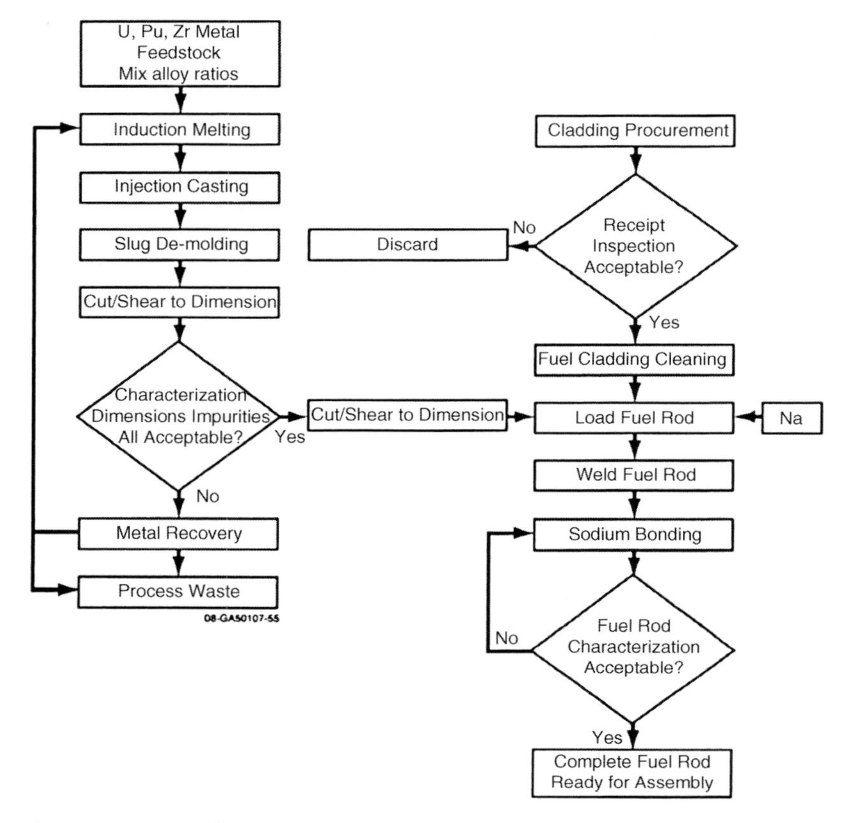

Figure 7.25 Process flowchart for fabrication of metallic fuels for fast reactors. Burkes et al., Ref. [19].

7.4 Summary

The topic of nuclear fuels is vast and is not easy to cover in a single chapter. Nonetheless, here a succinct review of both metallic and ceramic nuclear fuels is made and their various properties are discussed. Among metallic fuels, uranium, plutonium, and thorium are discussed. Among ceramic fuels, uranium dioxide, uranium nitride, and uranium carbide as well as plutonium-based oxide fuels and thorium oxide are covered. The metallic and ceramic fuels are found to have both advantages and disadvantages of their own.

Problems

7.1 What are the advantages and disadvantages of metallic nuclear fuels?
7.2 Describe various requirements imposed on a nuclear reactor fuel.

7.3 How many allotropic forms does uranium have? Discuss the effect of allotropic transformation on the properties of uranium.
7.4 Describe the plastic deformation mechanisms of alpha-uranium.
7.5 Discuss the beneficial effects of alloying on the properties of uranium as a nuclear fuel.
7.6 Distinguish between thermal cycling growth and radiation growth of alpha-uranium.
7.7 Martensitic transformation may occur in uranium system. Describe one example and its advantages and disadvantages.
7.8 How is plutonium produced?
7.9 Why is plutonium said to have *fickle* nature?
7.10 Discuss the origin of self-irradiation behavior of plutonium and its effects.
7.11 Discuss the origin of superplasticity in plutonium.
7.12 Can plutonium be alloyed with other metals? If so, compare it with uranium-based systems.
7.13 What are the main mineral sources of thorium?
7.14 What are the main impediments to realizing thorium cycle on a wider scale?
7.15 State the advantages and disadvantages of ceramic nuclear fuels.
7.16 Describe the effects of hypostoichiometry, stoichiometry, and hyperstoichiometry on the properties of uranium dioxide.
7.17 Compare and contrast between UO_2, UN, and UC fuels.

Bibliography

1 Kleycamp, H. (1985) The chemical state of the fission products in oxide fuels. *Journal of Nuclear Materials*, **131**, 221–246.
2 Ma, B.M. (1983) *Nuclear Reactor Materials and Applications*, Van Nostrand Reinhold, New York.
3 Ho, C.Y., Powell, R.W., and Liley, P.E. (1972) J. Phys. Chem. Ref. Data, **1** (2), 279–421.
4 Rahn, F.J., Adamantiades, A.G., Kenton, J. E., and Braun, C. (1984) *A Guide to Nuclear Power Technology*, John Wiley & Sons, Inc., New York.
5 Smith, C.O. (1967) *Nuclear Reactor Materials*, Addison-Wesley Massachusetts.
6 Hecker, S.S. and Stevens, M.F. (2000) Mechanical behavior of plutonium and its alloys. *Los Alamos Science*, **26**, 336–355.
7 Lawson, A.C., Goldstone, J.A., Cort, B., Martinez, R.J., Vigil, F.A., and Zocco, T.G. (1996) Structure of ζ-phase plutonium-uranium. *Acta Crystallographica*, **B52**, 32.
8 Hecker, S.S. (2004) The magic of plutonium: 5f electrons and phase instability. *Metallurgical and Materials Transactions A*, **35** (8), 2207–2222.
9 Merz, M.D. and Nelson, R.D. (1970) *Proceedings of the 4th International Conference on Plutonium and Other Actinides 1970* (ed. W.N. Miner), The Metallurgical Society of AIME, New York, p. 387.
10 Gschneider, K.A., Jr., Elliott, R.O., and Waber, J.T. (1963) *Acta Metallurgica*, **11** (8), 947–955.
11 Ray, H.S., Sridhar, R., and Abraham, K.P. (1985) *Extraction of Nonferrous Metals*, East-West Press, New Delhi, India.
12 IAEA (1997) *Thermophysical Properties of Materials for Water Cooled Reactors/IAEA-TECDOC- 949*, IAEA, Vienna, Austria.
13 Peterson, S., Adams, R.E., and Douglas, D.A., Jr. (1965) *Properties of Thorium, Its Alloys and Its Compounds*, ORNL Report ORNL-TM-1144.

14 World Nuclear Association, http://www.world-nuclear.org.
15 Webb, J.A. and Charit, I. (2012) Analytical determination of thermal conductivity of W-UO_2 and W-UN cermet nuclear fuels, *Journal of Nuclear Materials*, **427**, 87–94.
16 Webb, J.A. (2012) Ph.D. Analysis and Fabrication of Tungsten CERMET materials for Ultra-High Temperature Reactor Applications via Pulsed Electric Current Sintering, University of Idaho.
17 Rondinella, V. and Wiss, T. (2010) The high burn-up structure in nuclear fuel. *Materials Today*, **13**, 24–32.
18 Hayes, S. and Peddicord, T. (1990) Material properties correlations for uranium mononitride IV. *Journal of Nuclear Materials*, **171**, 300–318.
19 Burkes, D.E., Fielding, R.S., Porter, D.L., Crawford, D.C., and Meyer, M.K. (2009) A US perspective on fast reactor fuel fabrication technology and experience. Part I: metal fuels and assembly design. Journal of Nuclear Materials, **389**, 458–489.

Additional Reading

Allen, T., Busby, J., Meyer, M., and Petti, D. (2010) *Materials challenges for nuclear systems. Materials Today*, **13** (12), 14–23.

Buckley, S.N. (1961) Irradiation growth, Atomic Energy Research Establishment, Harwel ARE-R 3674, UK.

Appendix A
Stress and Strain Tensors

Stress is defined as the force acting on a plane, and in the case of a complex geometry under external force, it is convenient to define stress as

$$\sigma = \lim_{A \to 0} \frac{F}{A}. \tag{A.1}$$

We thus note that to designate stress (σ) we need to define the direction along which the force (F) acts on an area (A) designated with a normal to the area. Force F is a vector with magnitude and direction and is known as a first-rank tensor usually referred to as F_i (with i being x, y, or z). Similarly, an area A_i refers to the area whose normal is along the i-direction. On the other hand, a stress needs to be designated by two directions, one each for the force and area, respectively, and is referred to as a second-rank tensor (σ_{ij} acting on an area whose normal is along the "i" axis with a force along the "j" direction). When the plane normal and the force direction are same, they are referred to as "normal" stresses such as σ_{xx}, σ_{yy}, and σ_{zz}. Tensile stresses are regarded as positive and compressive stresses are regarded as negative. $i \neq j$ results in a shear stress such as σ_{ij} with force acting on a plane whose normal is "i" with "j" being the force direction; in this case, the shear stresses are considered to be "negative" if one of the indices (i or j) is negative (opposite to one another) and "positive" if both i and j are along positive or negative direction.

In general, stress tensor (σ_{ij}) may have nine different components:

$$\sigma = \begin{pmatrix} \sigma_{xx} & \sigma_{xy} & \sigma_{xz} \\ \sigma_{yx} & \sigma_{yy} & \sigma_{yz} \\ \sigma_{zx} & \sigma_{zy} & \sigma_{zz} \end{pmatrix}. \tag{A.2}$$

Since a body under force is regarded as rigid or with "no net moment," $\sigma_{ij} = \sigma_{ji}$ or stress is a "symmetrical" tensor, thereby having six terms with three normal (diagonal) and three shear (nondiagonal) terms,

$$\sigma = \begin{pmatrix} \sigma_{xx} & \sigma_{xy} & \sigma_{xz} \\ \sigma_{xy} & \sigma_{yy} & \sigma_{yz} \\ \sigma_{xz} & \sigma_{yz} & \sigma_{zz} \end{pmatrix}, \tag{A.3a}$$

An Introduction to Nuclear Materials: Fundamentals and Applications, First Edition.
K. Linga Murty and Indrajit Charit.
© 2013 Wiley-VCH Verlag GmbH & Co. KGaA. Published 2013 by Wiley-VCH Verlag GmbH & Co. KGaA.

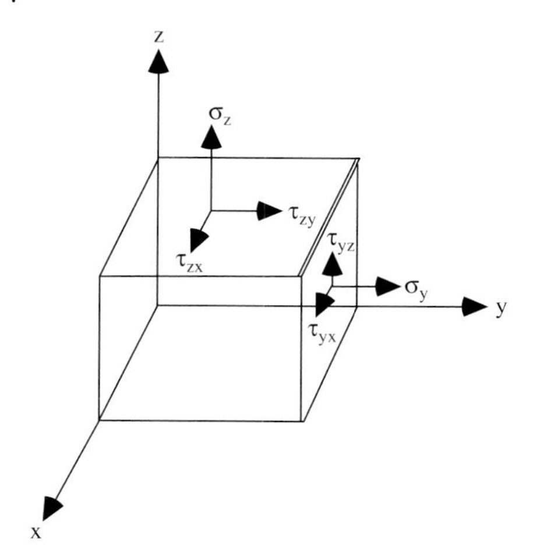

Figure A.1 Stress tensor designations.

which is often written as

$$\sigma = \begin{pmatrix} \sigma_x & \tau_{xy} & \tau_{xz} \\ & \sigma_y & \tau_{yz} \\ & & \sigma_z \end{pmatrix}. \tag{A.3b}$$

In Eq. (A.3b), the diagonal terms are referred by single index such as σ_y, which is the same as σ_{yy}, while shear stresses are usually referred to as "τ". These are illustrated in Figure A.1.

For any given situation (a solid under load or force), the various stress component (σ_{ij}) values depend on the coordinate system selected. Values for stress components given for a set of coordinate system (xyz) can be converted to a different set of coordinates ($x'y'z'$) through tensor transformations by knowing the angles between xx', yy', and zz':

$$\sigma'_{ij} = a_{ik}a_{jl}\sigma_{kl}, \tag{A.4}$$

where a_{ij} are direction cosines, for example, $a_{xy} = \cos(\theta_{xy})$, where θ_{xy} is the angle between x' and y coordinates. It is important to note that trace of the stress tensor (sum of diagonal terms) remains independent of the coordinate system. That means

$$\sigma'_{ii} = \sigma_{ii} \quad \text{or} \quad \sigma'_{xx} + \sigma'_{yy} + \sigma'_{zz} = \sigma_{xx} + \sigma_{yy} + \sigma_{zz}. \tag{A.5}$$

A.1 Plane Stress

In special cases such as thin plates/sheets where stresses along the thickness direction are zero and thin-walled cylinders (with outer diameter $>10 \times$ thickness) where stresses along tube wall thickness are negligible, they are called "plane stresses" meaning that stresses in one of the three directions are negligible so that the stress tensor is two dimensional with three different components (σ_{xx}, σ_{yy}, and $\sigma_{xy} = \sigma_{yx}$).

A.2
Principal Stresses

Principal planes are the planes on which maximum normal stresses act with no shear stresses and these stresses are known as "principal stresses." These are designated as σ_1, σ_2, and σ_3 implying no shear stresses or

$$\sigma_{ij} = \sigma_{ij}\delta_{ij} \quad \text{with} \quad \delta_{ij} = 0 \quad \text{for} \quad i \neq j. \tag{A.6}$$

Such principal stresses can be found for the 2D case using Mohr's circle; however, for the general 3D case, one can determine them as the three solutions to the determinant of the stress tensor:

$$\begin{Vmatrix} \sigma_{11} - \sigma & \sigma_{12} & \sigma_{13} \\ \sigma_{21} & \sigma_{22} - \sigma & \sigma_{23} \\ \sigma_{31} & \sigma_{32} & \sigma_{33} - \sigma \end{Vmatrix} = 0, \tag{A.7}$$

where the indices 1, 2, and 3 are used in place of x, y, and z. The three solutions for the determinant are the principal stresses and it is common practice to designate them to be $\sigma_1 > \sigma_2 > \sigma_3$ taking into account the sign as well.

Expansion of the determinant gives

$$\begin{aligned} 0 = \sigma^3 &- (\sigma_{11} + \sigma_{22} + \sigma_{33})\sigma^2 + (\sigma_{11}\sigma_{22} + \sigma_{22}\sigma_{33} + \sigma_{33}\sigma_{11} - \sigma_{12}^2 - \sigma_{23}^2 - \sigma_{31}^2)\sigma \\ &- (\sigma_{11}\sigma_{22}\sigma_{33} + 2\sigma_{12}\sigma_{23}\sigma_{31} - \sigma_{11}\sigma_{23}^2 - \sigma_{22}\sigma_{31}^2 - \sigma_{33}\sigma_{12}^2) \end{aligned} \tag{A.7a}$$

or

$$\sigma^3 - I_1\sigma^2 - I_2\sigma - I_3 = 0, \tag{A.7b}$$

where I's are *invariants* of the stress tensor:

$$\begin{aligned} I_1 &= (\sigma_{11} + \sigma_{22} + \sigma_{33}), \\ I_2 &= -(\sigma_{11}\sigma_{22} + \sigma_{22}\sigma_{33} + \sigma_{33}\sigma_{11} - \sigma_{12}^2 - \sigma_{23}^2 - \sigma_{31}^2), \\ I_3 &= \sigma_{11}\sigma_{22}\sigma_{33} + 2\sigma_{12}\sigma_{23}\sigma_{31} - \sigma_{11}\sigma_{23}^2 - \sigma_{22}\sigma_{31}^2 - \sigma_{33}\sigma_{12}^2. \end{aligned}$$

Note that I_1 is the *trace* of the determinant (sum of diagonal terms).

A.3
Hydrostatic and Deviatoric Stresses

Total stress tensor can be divided into two components: *hydrostatic* or *mean* stress tensor (σ_{m}) involving only pure tension or compression and *deviatoric* stress tensor

(σ'_{ij}) representing pure shear with no normal components:

$$\sigma_m = \frac{\sigma_{kk}}{3} = \frac{\sigma_1 + \sigma_2 + \sigma_3}{3}, \tag{A.8a}$$

$$\sigma_{ij} = \sigma'_{ij} + \frac{1}{3}\delta_{ij}\sigma_{kk}. \tag{A.8b}$$

It is to be noted that pure hydrostatic stress does not lead to plastic deformation and finite deviatoric stresses are needed for any plastic deformation to take place.

A.4
Normal and Shear Stresses on a Given Plane (Cut-Surface Method)

Given σ_{ij} in reference system 1 2 3, $\hat{n}$ is the *unit vector* normal to the plane $= n_1 n_2 n_3$, $\hat{m}$ is the *unit vector* in the plane $= m_1 m_2 m_3$, σ_N is the normal stress along $\bar{n}$, and τ is the shear stress along $\bar{m}$ (see Figure A.2).

Note: $\hat{n} \cdot \hat{m} = 0$, $n_1^2 + n_2^2 + n_3^2 = 1$, and $m_1^2 + m_2^2 + m_3^2 = 1$. If $\bar{n} = 1, 2, 5 \Rightarrow \hat{n} = 1/\sqrt{30}, 2/\sqrt{30}, 5/\sqrt{30} \Leftrightarrow \hat{n}$ is a *unit* vector, where $\sqrt{1^2 + 2^2 + 5^2} = \sqrt{30}$ so that $n_1^2 + n_2^2 + n_3^2 = 1$.

First, we find the stress *vector* ($\bar{S}$) {the *stress vector* is the vector force per unit area acting on the *cut*}:

$$\bar{S} = \sigma\hat{n} \Rightarrow \begin{pmatrix} S_1 \\ S_2 \\ S_3 \end{pmatrix} = \begin{pmatrix} \sigma_{11} & \sigma_{12} & \sigma_{13} \\ \sigma_{21} & \sigma_{22} & \sigma_{23} \\ \sigma_{31} & \sigma_{32} & \sigma_{33} \end{pmatrix} \begin{pmatrix} n_1 \\ n_2 \\ n_3 \end{pmatrix} \Rightarrow S_i = \sum_{k=1}^{3} \sigma_{ik} n_k;$$

that is, $S_1 = \sigma_{11}n_1 + \sigma_{12}n_2 + \sigma_{13}n_3$, and so on. σ_N and τ are given as follows: $\sigma_N = \bar{S} \cdot \hat{n} = S_1 n_1 + S_2 n_2 + S_3 n_3$ and

$$\tau = \bar{S} \cdot \hat{m} = S_1 m_1 + S_2 m_2 + S_3 m_3, \tag{A.9a}$$

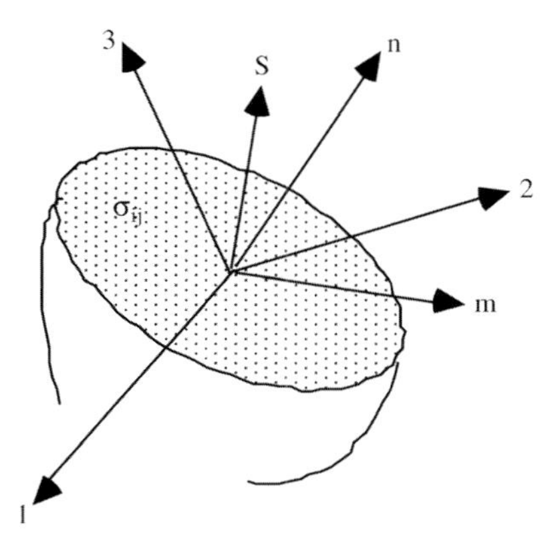

Figure A.2 Designations for normal and shear stress calculations.

and $\tau_{\max}$ occurs when n, S, and m are in the same plane:

$$|\bar{S}|^2 = \sigma_N^2 + \tau_{\max}^2. \tag{A.9b}$$

A.5 Strain Tensor

A strain tensor (ε_{ij}) is a symmetric second-rank tensor similar to stress tensor and is given by

$$\varepsilon_{ij} = \begin{pmatrix} \varepsilon_{xx} & \varepsilon_{xy} & \varepsilon_{zx} \\ \varepsilon_{xy} & \varepsilon_{yy} & \varepsilon_{yz} \\ \varepsilon_{zx} & \varepsilon_{yz} & \varepsilon_{zz} \end{pmatrix} = \begin{pmatrix} \dfrac{\partial u}{\partial x} & \dfrac{1}{2}\left(\dfrac{\partial u}{\partial y} + \dfrac{\partial v}{\partial x}\right) & \dfrac{1}{2}\left(\dfrac{\partial u}{\partial z} + \dfrac{\partial w}{\partial x}\right) \\ \dfrac{1}{2}\left(\dfrac{\partial u}{\partial y} + \dfrac{\partial v}{\partial x}\right) & \dfrac{\partial v}{\partial y} & \dfrac{1}{2}\left(\dfrac{\partial v}{\partial z} + \dfrac{\partial w}{\partial y}\right) \\ \dfrac{1}{2}\left(\dfrac{\partial u}{\partial z} + \dfrac{\partial w}{\partial x}\right) & \dfrac{1}{2}\left(\dfrac{\partial v}{\partial z} + \dfrac{\partial w}{\partial y}\right) & \dfrac{\partial w}{\partial z} \end{pmatrix}, \tag{A.10}$$

where u, v, and w are displacements along x, y, and z directions, respectively. It is to be noted that these are *true* strains and the true shear strain (ε_{ij}) is related to the engineering shear strain γ_{ij}:

$$\gamma_{ij} = \varepsilon_{ij} + \varepsilon_{ji} = 2\varepsilon_{ij}. \tag{A.11}$$

Thus,

$$\tau_{ij} = G\gamma_{ij} = G2\varepsilon_{ij}. \tag{A.12}$$

When all six components of the stress state are given, one can find the normal and shear elastic strains resulting from them realizing that the normal strain along the x-axis results not only from σ_{xx} ($=\sigma_{xx}/E$) but also from Poisson contractions due to σ_{yy} and σ_{zz} ($-\nu\sigma_{yy}/E$ and $-\nu\sigma_{zz}/E$, respectively). Thus,

$$\varepsilon_{xx} = \sigma_{xx}/E - \nu\sigma_{yy}/E - \nu\sigma_{zz}/E \tag{A.13a}$$

or

$$\varepsilon_{xx} = \frac{1}{E}\{\sigma_{xx} - \upsilon(\sigma_{yy} + \sigma_{zz})\}, \quad \varepsilon_{yy} = \frac{1}{E}\{\sigma_{yy} - \upsilon(\sigma_{xx} + \sigma_{zz})\}, \quad \text{and} \quad \varepsilon_{zz} = \frac{1}{E}\{\sigma_{zz} - \upsilon(\sigma_{xx} + \sigma_{yy})\}. \tag{A.13b}$$

The shear strains are given by equations similar to Eq. (A.12):

$$\varepsilon_{xy} = \tau_{xy}/2G, \quad \varepsilon_{yz} = \tau_{yz}/2G, \quad \varepsilon_{zx} = \tau_{zx}/2G.$$

Thus, given the stress tensor, we need only two elastic constants E and ν, since G is related to E and ν:

$$G = E/2(1+\nu). \tag{A.14}$$

We now note that the volume change or dilatation is given by

$$\Delta = (1+\varepsilon_1)(1+\varepsilon_2)(1+\varepsilon_3) - 1 = \varepsilon_1 + \varepsilon_2 + \varepsilon_3$$

since ε's $<< 1$. Note that Δ is the first invariant of the strain tensor and mean strain $\varepsilon_{\mathrm{m}} = \Delta/3$.
Thus, bulk modulus

$$K = \frac{\sigma_{\mathrm{m}}}{\Delta} = \frac{-p}{\Delta} \quad \text{and} \quad K = \frac{E}{3(1-2\nu)}. \tag{A.15}$$

We therefore note from above that given the stress state, one can evaluate the strains (in terms of short notation):

$$\varepsilon_{ij} = \frac{1+\nu}{E}\sigma_{ij} - \frac{\nu}{E}\sigma_{kk}\delta_{ij}$$

and stresses in terms of strains are given by

$$\sigma_{ij} = \frac{E}{1+\upsilon}\varepsilon_{ij} + \lambda\varepsilon_{kk}\delta_{ij},$$

where λ is Lame's constant:

$$\lambda = \frac{\upsilon E}{(1+\upsilon)(1-2\upsilon)}.$$

A.6
Strain Energy

Elastic strain energy, U_0, energy spent by the external forces in deforming an elastic body ($= (1/2)P\delta$ – area under the load–displacement curve), is given by

$$U_0 = \frac{1}{2}\sigma_x\varepsilon_x = \frac{\sigma_x^2}{2E} = \frac{E\varepsilon_x^2}{2} \tag{A.16}$$

for simple uniaxial loading. For generalized stresses and strains, $U_0 = (1/2)\sigma_{ij}\varepsilon_{ij}$ or, in expanded form,

$$U_0 = \frac{1}{2E}(\sigma_x^2 + \sigma_y^2 + \sigma_z^2) - \frac{\nu}{E}(\sigma_x\sigma_y + \sigma_y\sigma_z + \sigma_z\sigma_x) + \frac{1}{2G}(\tau_{xy}^2 + \tau_{yz}^2 + \tau_{zx}^2). \tag{A.17}$$

Note that the derivative of U_0 with respect to any strain component equals the corresponding stress component:

$$\frac{\partial U_0}{\partial \varepsilon_x} = \lambda\Delta + 2G\varepsilon_x = \sigma_x$$

and similarly

$$\frac{\partial U_0}{\partial \sigma_x} = \varepsilon_x.$$

A.7
Generalized Hooke's Law

$$\varepsilon_{ij} = S_{ijkl}\sigma_{kl}. \tag{A.18}$$

Here S_{ijkl} is the elastic compliance tensor (fourth rank) and

$$\sigma_{ij} = C_{ijkl}\varepsilon_{kl}, \tag{A.19}$$

where C_{ijkl} is elastic stiffness (or elastic constants).

Crystal symmetry reduces the number of independent terms: cubic – 3, hexagonal – 5, and so on, and these are related to E and G.

Details on these compliance and stiffness coefficients are beyond the scope of this book and may be found elsewhere.

Appendix B

B.1
SI Units

Quantity	Unit	Symbol
Length	Meter	m
Mass	Kilogram	kg
Time	Second	s
Electric current	Ampere	A
Temperature	Kelvin	K
Amount of substance	Mole	mol
Luminous intensity	Candela	cd
Plane angle	Radian	rad
Solid angle	Steradian	sr

B.2
Some Derived Units

			Equivalence in	
Quantity	Special name	Symbol	Other derived units	Base units
Force, load, weight	Newton	N	—	$kg\,m\,s^{-2}$
Stress, strength, pressure	Pascal	Pa	$N\,m^{-2}$	$kg\,m^{-1}\,s^{-2}$
Frequency	Hertz	Hz	—	s^{-1}
Energy, work, heat	Joule	J	N m	$kg\,m^2\,s^{-2}$
Power	Watt	W	$J\,s^{-1}$	$kg\,m^2\,s^{-3}$
Electric charge	Coulomb	C	—	A s
Electric potential/voltage	Volt	V	$W\,A^{-1}$	$kg\,m^2\,s^{-3}\,A^{-1}$

(continued)

An Introduction to Nuclear Materials: Fundamentals and Applications, First Edition.
K. Linga Murty and Indrajit Charit.
© 2013 Wiley-VCH Verlag GmbH & Co. KGaA. Published 2013 by Wiley-VCH Verlag GmbH & Co. KGaA.

Quantity	Special name	Symbol	Equivalence in: Other derived units	Base units
Resistance	Ohm	Ω	$V\,A^{-1}$	$kg\,m^2\,s^{-3}\,A^{-2}$
Capacitance	Farad	F	$C\,V^{-1}$	$kg^{-1}\,m^{-2}\,s^4\,A^2$
Magnetic flux	Weber	Wb	$V\,s$	$kg\,m^2\,s^{-2}\,A^{-1}$
Magnetic flux density	Tesla	T	$Wb\,m^{-2}$	$kg\,s^{-2}\,A^{-1}$
Inductance	Henry	H	$Wb\,A^{-1}$	$kg\,m^2\,s^{-2}\,A^{-2}$

B.3 Standard Unit Prefixes and Their Multiples and Submultiples

Name	Multiplication factor	Symbol
Atto	10^{-18}	a
Femto	10^{-15}	f
Pico	10^{-12}	p
Nano	10^{-9}	n
Micro	10^{-6}	μ
Milli	10^{-3}	m
Kilo	10^{3}	k
Mega	10^{6}	M
Giga	10^{9}	G
Tera	10^{12}	T

B.4 Some Unit Conversion Factors

B.4.1 Length

$1\ \text{inch} = 25.4\ \text{mm}$
$1\ \text{nm} = 10^{-9}\ \text{m}$
$1\ \mu\text{m} = 10^{-6}\ \text{m}$
$1\ \text{Å} = 10^{-10}\ \text{m} = 0.1\ \text{nm}$

B.4.2 Temperature

$T\,(\text{K}) = T\,(°\text{C}) + 273.15$
$T\,(°\text{C}) = [T\,(°\text{F}) - 32]/1.8$

B.4.3 Mass

$1\ \text{Mg} = 10^3\ \text{kg}$
$1\ \text{kg} = 10^{-3}\ \text{Mg}$
$1\ \text{kg} = 10^3\ \text{g}$
$1\ \text{kg} = 2.205\ \text{lb}_\text{m}$
$1\ \text{g} = 10^{-3}\ \text{kg}$

B.4.4 Force

$1\ \text{kgf} = 9.81\ \text{N}$
$1\ \text{lb} = 4.448\ \text{N}$
$1\ \text{dyne} = 10^{-5}\ \text{N}$

B.4.5 Stress

$1\ \text{ksi (i.e., } 10^3\ \text{psi)} = 6.89\ \text{MPa}$
$1\ \text{MPa} = 1\ \text{N mm}^{-2} = 145\ \text{psi}$
$1\ \text{Pa} = 10\ \text{dyn cm}^{-2}$
$1\ \text{ton in.}^{-2} = 15.46\ \text{MPa}$
$1\ \text{atm} = 0.101325\ \text{MPa}$
$1\ \text{bar} = 0.1\ \text{MPa}$
$1\ \text{Torr (mmHg)} = 133.3\ \text{MPa}$

B.4.6 Energy, Work, and Heat

$1\ \text{eV atom}^{-1} = 96.49\ \text{kJ mol}^{-1}$
$1\ \text{cal} = 4.184\ \text{J}$
$1\ \text{Btu} = 252.0\ \text{cal}$
$1\ \text{erg} = 10^{-7}\ \text{J}$

B.4.7 Miscellaneous

$$1^\circ = \frac{1}{57.3}\ \text{rad}$$
$1\ \text{g cm}^{-3} = 1000\ \text{kg m}^{-3}$
$1\ \text{poise} = 0.1\ \text{Pa s}$
$1\ \text{ksi in.}^{1/2} = 1.10\ \text{MN m}^{-3/2} = 1.10\ \text{MPa m}^{1/2}$

B.5
Selected Physical Properties of Metals (Including Metalloids)

Symbol	Atomic number	Atomic weight	Density at 20 °C ($g\,cm^{-3}$)	Melting point (°C)
Al	13	26.98	2.7	660
Sb	51	121.76	6.68	630.5
As	33	74.91	5.727	814
Ba	56	137.36	3.5	710
Be	4	9.013	1.845	1284
Bi	83	209.0	9.8	271.2
B	5	10.82	2.34	2300
Cd	48	112.41	8.65	320.9
Ca	20	40.08	1.54	851
Ce	58	140.13	6.66	795
Cs	55	132.91	1.873	28.5
Cr	24	52.01	7.19	1875
Co	27	58.94	8.90	1493
Nb	41	92.91	9.57	2468
Cu	29	63.54	8.94	1083
Dy	66	162.51	8.536	1407
Er	68	167.21	9.051	1497
Eu	63	152.0	5.259	826
Gd	64	157.26	7.895	1312
Ga	31	69.72	5.907	29.75
Ge	32	72.60	5.32	936
Au	79	197.2	19.32	1063
Hf	72	178.50	13.29	2150
Ho	67	164.94	8.803	1461
In	49	114.82	7.31	156.6
Ir	77	192.2	22.42	2410
Fe	26	55.85	7.87	1535
La	57	138.92	6.174	920
Pb	82	207.21	11.34	327.4
Li	3	6.940	0.534	179
Lu	71	174.99	9.842	1652
Mg	12	24.32	1.74	651
Mn	25	54.94	7.44	1244
Hg	80	200.61	13.55	−38.87
Mo	42	95.95	10.22	2610
Nd	60	144.27	7.004	1024
Ni	28	58.69	8.9	1452
Os	76	190.2	22.5	3000
Pd	46	106.4	12.02	1552

(continued)

Symbol	Atomic number	Atomic weight	Density at 20 °C ($g\,cm^{-3}$)	Melting point (°C)
Pt	78	195.09	21.40	1769
Pu	94	239.11	19.84	639.5
K	19	39.10	0.87	63.7
Pr	59	140.92	6.782	935
Pm	61	145	7.264	1035
Re	75	186.22	21.02	3180
Rh	45	102.91	12.44	1960
Ru	44	101.1	12.4	2250
Sm	62	150.35	7.536	1072
Sc	21	44.96	2.99	1539
Se	34	78.96	4.79	217
Si	14	28.09	2.33	1410
Ag	47	107.873	10.49	960.5
Na	11	22.991	0.97	97.9
Sr	38	87.63	2.6	770
Ta	73	180.95	16.6	2996
Te	52	127.61	6.25	449.5
Tb	65	158.93	8.272	1356
Tl	81	204.39	11.85	303
Th	90	232.05	11.66	1750
Tm	69	168.94	9.332	1545
Sn	50	118.7	7.3	232
Ti	22	47.90	4.54	1668
W	74	183.92	19.3	3410
U	92	238.07	19.07	1132
Yb	70	173.04	6.977	824
Y	39	88.92	4.472	1509
Zn	30	65.38	7.133	419.5
Zr	40	91.22	6.45	1852

Adapted from Ref. [1].

B.6
Thermal Neutron (0.025 eV) Absorption Cross Sections of Some Elements

Element	Absorption cross section (b)	Element	Absorption cross section (b)	Element	Absorption cross section (b)
C	0.0035	Ni	4.49	Ho	64.7
Be	0.0076	Sb	4.91	Lu	74
Bi	0.034	V	5.08	Am	75.3
Mg	0.063	Ti	6.09	Re	89.7
Si	0.171	Pd	6.90	Au	98.7
Pb	0.171	Th^{232}	7.56	Tm	100
Zr	0.185	U	7.57	Hf	104.1
Al	0.231	La	8.97	Rh	144.6
H	0.332	Pt	10.3	Np	175.9
Sn	0.626	Pr	11.5	Er	159
Ce	0.630	Se	11.7	In	193.8
Zn	0.110	Mn	13.3	Pu^{240}	289.6
Nb	1.15	Os	16.0	Ir	425
Y	1.28	W	18.3	B	767
Ge	2.20	Ta	20.6	Dy	994
Fe	2.56	Tb	23.4	Pu^{239}	1017.3
Ru	2.56	Sc	27.5	Pu^{241}	1400 (*)
Mo	2.48	Co	37.2	Cd	2520
Cr	3.05	Yb	34.8	Eu	4530
Tl	3.43	Nd	50.5	Sm	5922
Cu	3.78	Ag	63.3	Gd	49 700
Te	4.7				

[Source: Special feature section of neutron scattering lengths and cross sections of the elements and isotopes, Neutron News, 3 (1992) 29–37]

B.7
Mechanical Properties of Some Important Metals and Alloys

Alloy	Young's modulus (GPa)	Poisson's ratio	UTS (MPa)	YS (MPa)	Elongation to fracture (%)	Fracture toughness (MPa $m^{1/2}$)
Al 2024 T851	72.4	0.33	455	400	5	26.4
Al 7075 T651	72	0.33	570	505	11	24.2
Al 7178 T651	73	0.33	605	540	10	23.1
Ti-6Al-4V (grade 5)	113.8	0.342	1860	148	14	55
Alpha annealed Ti-3Al-2.5V	100	0.3	620	500	15	100
702 Zirconium	99.3	0.35	379	207	16	—
Stainless steel 4340	205	—	745	470	22	60.4
Stainless steel 304	193	0.29	505	215	70	—
Tool steel H11 (hot worked)	210	—	1990	1650	9	—
Maraging steel	200	—	1864	1737	17.4	—
Superalloy CoCrWNi	—	—	860	310	10	—
Superalloy H-X nickel	—	—	690	276	40	—

B.8
Mechanical Properties of Some Important Ceramics

Ceramic	Young's modulus (GPa)	Poisson's ratio	Hardness (HV)	Tensile strength (MPa)	Compressive strength (MPa)	Flexural strength (MPa)	Fracture toughness (MPa $m^{1/2}$)
Silicon nitride	320	0.28	1800	350–415	2100–3500	930	6
Silicon carbide	450	0.17	2300	390–450	1035–1725	634	4.3
Tungsten carbide	627	0.21	1600	344	1400–2100	1930	—
MgO-stabilized ZrO_2	200	0.3	1200	352	1750	620	11
Boron carbide	450	0.27	2700	—	470	450	3.0
Titanium diboride	556	0.11	2700	—	470	277	6.9

Note: The above values are indicative only.
Adapted from Ref. [2].

References

1 Hampel, C.A. (1961) *Rare Metals Handbook*, Chapman & Hall, London.

2 Meyers, M.A. and Chawla, K.K. (2009) *Mechanical Behavior of Materials*, Cambridge University Press.

Index

An Introduction to Nuclear Materials: Fundamentals and Applications, First Edition.
K. Linga Murty and Indrajit Charit.
© 2013 Wiley-VCH Verlag GmbH & Co. KGaA. Published 2013 by Wiley-VCH Verlag GmbH & Co. KGaA.